Synthesis Lectures on Learning, Networks, and Algorithms

Series Editor

Lei Ying, ECE, University of Michigan, Ann Arbor, MI, USA

The series publishes short books on the design, analysis, and management of complex networked systems using tools from control, communications, learning, optimization, and stochastic analysis. Each Lecture is a self-contained presentation of one topic by a leading expert. The topics include learning, networks, and algorithms, and cover a broad spectrum of applications to networked systems including communication networks, data-center networks, social, and transportation networks.

Hang Wang · Sen Lin · Junshan Zhang

Continual and Reinforcement Learning for Edge AI

Framework, Foundation, and Algorithm Design

Springer

Hang Wang
University of California
Davis, CA, USA

Sen Lin
University of Houston
Katy, TX, USA

Junshan Zhang
University of California
Davis, CA, USA

ISSN 2690-4306 ISSN 2690-4314 (electronic)
Synthesis Lectures on Learning, Networks, and Algorithms
ISBN 978-3-031-84365-5 ISBN 978-3-031-84363-1 (eBook)
https://doi.org/10.1007/978-3-031-84363-1

This Springer imprint is published by the registered company Springer Nature Switzerland AG
The registered company address is: Gewerbestrasse 11, 6330 Cham, Switzerland

Preface

With the advances in artificial intelligence (AI) and edge computing, edge-AI has recently emerged as the marriage of these two groundbreaking technologies and attracted significant interest from both industry and academia. Particularly, edge-AI seeks to push the AI frontiers to the resource-limited network edge that is closer to the data-generating sources, e.g., mobile devices and other Internet-of-Things (IoT) devices. This new inter-discipline has great potentials in enabling a wide range of on-device AI applications, spanning from video surveillance to personal assistant to autonomous driving. Nevertheless, local data is usually collected online and the underlying data distributions can continuously shift due to the environment changes, which requires the edge AI models to be adapted accordingly in a lifelong manner and efficiently considering the limited computing resources at the network edge. This online and non-stationary nature calls for a new formulation of edge AI from the perspective of continual learning, however, for which the research is still in an infancy stage and a dedicated venue for exchanging the recent advances is lacking.

To fill this void, we present in this book a survey of recent research progress, with 'bias' towards our own research efforts in edge AI, from supervised learning to reinforcement learning. More specifically, we first introduce the background and the motivation of continual learning in edge AI, followed by the potential frameworks and design considerations therein. Next, we identify and provide a detailed overview of key machine learning technologies to enable continual learning and reinforcement learning for edge AI. To better demonstrate the research directions and problems in edge continual AI, we also showcase four of our own research projects and summarize the recent progress in this field. We discuss the promising applications and future research opportunities at the end. We hope that this book will bring up more attentions, spark fruitful discussions and motivate further research ideas on continual and reinforcement learning for edge AI.

This work is supported in part by NSF Grants CNS-2203239, CNS-2148253, CCSS-2413529 and ARO Grant W911NF-2410046.

Davis, USA Hang Wang
Katy, USA Sen Lin
Davis, USA Junshan Zhang
September 2024

Contents

About the Authors

Hang Wang is currently a Ph.D. candidate in the Department of Electrical and Computer Engineering at University of California, Davis. Previously, he received the B.E. from University of Science and Technology of China (USTC) in 2018. His research aims to establish a fundamental understanding of reinforcement learning, multi-agent systems, and human-AI interaction, as well as practical applications in the domains like Autonomous Driving and Edge Computing. His contributions have been published in NeurIPS, ICML, AAMAS and his recent work on Warm-start Reinforcement Learning also garnered attention and acclaim via an oral presentation at ICML.

Sen Lin is an Assistant Professor in the Department of Computer Science at University of Houston. Previously, he was a Postdoc in the NSF AI-EDGE Institute at The Ohio State University. He received his Ph.D. degree from Arizona State University, M.S. from HKUST and B.E. from Zhejiang University. His research interests broadly fall in the intersection of machine learning and wireless networking. Currently, his research focuses on developing algorithms and theories in continual learning, meta-learning, reinforcement learning, adversarial machine learning and bilevel optimization, with applications in multiple domains, e.g., edge computing, security, network control. His research results have been published in top conferences and journals in machine learning and networking. The recognition his papers have received includes the WiOpt'18 Best Student Paper Award and Spotlight presentations in ICLR and ICML.

Junshan Zhang has been a professor in the ECE Department at University of California Davis since 2021. He received his Ph.D. degree from the School of ECE at Purdue University in August 2000, and was on the faculty of the School of ECEE at Arizona State University from 2000 to 2021. His research interests fall in the general field of information networks and data science, including edge AI, reinforcement learning, continual learning, network optimization and control, game theory. He is a Fellow of the IEEE, and a recipient of the ONR Young Investigator Award in 2005 and the NSF CAREER award in 2003.

His papers have won a few awards, including the Best Student paper atWiOPT 2018, the Kenneth C. Sevcik Outstanding Student Paper Award of ACM SIGMETRICS/IFIP Performance 2016, the Best Paper Runner up Award of IEEE INFOCOM 2009 and IEEE INFOCOM 2014, and the Best Paper Award at IEEE ICC 2008 and ICC 2017. He is currently serving as Editor-in-Chief for IEEE/ACM Transactions on Networking. He served as Editor-in-Chief for IEEE Transactions onWireless Communication during 2019–2022. Hc was TPC co-chair for IEEE INFOCOM 2012 and ACM MOBIHOC 2015.

Introduction to Continual and Reinforcement Learning for Edge AI

1

1.1 Introduction

Since its birth in 1956, artificial intelligence (AI) has gone through multiple summers and winters. Driven by the significant advancements of neural network architectures, computing power and big data, the past decade has witnessed an unparalleled growth in machine learning (ML) and AI, leading to phenomenal breakthroughs in a wide spectrum of applications, e.g., speech recognition [191], image classification [141, 276], object detection [219, 323], etc. In fact, GPU throughput and memory have increased 10× in the last four years. By leveraging the parallelism of the GPU hardware and more training data, the transformer architecture can now train much more expressive models than ever, giving rise to a new era of foundation models. It is widely recognized that these intelligent applications will significantly enrich people's lifestyle and improve human productivity.

In general, the machine learning problems can be mainly divided into three categories:

- *Supervised Learning.* Supervised learning is the process of learning a function that maps an input to an output (label) given a training dataset of inputs and their corresponding labels, such that the labels for unseen inputs can be accurately inferred based on the function. Regression and classification are examples of supervised learning, which have led to unparalleled successes in computer vision.
- *Unsupervised Learning.* In unsupervised learning, there are no labels given for the training data, and the objective is to learn the underlying structure in the data. The most common unsupervised learning tasks are clustering, e.g., finding groups in data, and density estimation, e.g., summarizing the distribution of data.
- *Self-Supervised Learning and Reinforcement Learning.* Self-supervised learning aims to build models that automatically find patterns in data and reveal these patterns explicitly with a representation. In particular, reinforcement learning (RL), a popular approach for

H. Wang et al., *Continual and Reinforcement Learning for Edge AI*,
Synthesis Lectures on Learning, Networks, and Algorithms,
https://doi.org/10.1007/978-3-031-84363-1_1

self-supervised learning, is used to train an agent that can not only perceive and understand their surroundings through vision and other sensing modalities but also can navigate and interact with their physical environments to achieve goals, engage in planning, and perform reasoning. In essence, reinforcement learning is a learning paradigm that learns how to use past data, e.g., offline data or data collected through online interaction, to enhance the future manipulation of a dynamical system, which can be cast as an optimal control problem when the system dynamics are unknown [32]. The fruitful applications, e.g., games, robotics, and substantive interactions with other disciplines, e.g., operation research, control theory and game theory, make reinforcement learning a very popular and critical research direction nowadays.

One of the key driving forces behind AI is the development of deep learning and deep neural networks (DNNs) since 2010s, which have achieved astonishing successes in solving ML problems and demonstrated great superiority over classical ML approaches, e.g., decision tree and Baysian networks. Notably, consisting of a series of layers, artificial neural networks (ANNs) can extract the underlying features from data in a hierarchical manner and provide a universal function approximator for ML problems. Multiplayer Perceptrons (MLPs) are the most basic ANNs with fully connected neurons and non-linear activation functions. To capture the spatial correlation in the input data, especially for images, Convolution Neural Networks (CNNs) [179] replace the basic linear operations in MLPs with convolution operations, making them very popular for computer vision tasks. Recurrent Neural Networks (RNNs) [386] are another type of ANNs which specialize in handling sequential data and hence are widely used for natural language processing (NLP) tasks, e.g., machine translation and question answering. Unlike feedforward neural networks such as MLPs and CNNs which process data in a single pass, RNNs are able to process data across multiple time steps, with an internal memory to remember the knowledge of previous inputs and use that for learning at current time steps. Another special kind of ANN architectures is the generative model, which aims to solve generative tasks, e.g., image generation. Generative adversarial networks (GANs) [117] are in this category, which consist of two separate networks, namely generator network and discriminator network. The generator seeks to generate new data to mimic real data in the training dataset, whereas the discriminator seeks to distinguish the fake data generated by the generator from the real data. Recently, another powerful type of generative models named as diffusion models [69] has attracted much attention, which gradually add random noise to the input data and then learn to reconstruct desired data samples from the noise by reversing the diffusion process. In 2017, a new ANN architecture, namely Transformer [335], was proposed to address the limitation of RNN-based encoder-decoder architecture in solving sequence-to-sequence tasks, by leveraging the attention mechanism to capture the long-range dependencies across data inputs in a highly parallel manner. Due to its superior performance and computational efficiency, the Transformer now becomes the mainstream architecture and is widely used in pretrained foundation models such as large language models. Based on the scaling law, the performance of ANNs often improves with larger network size and more training data samples, which can significantly increase the need of computational resources and memory size.

1.1.1 Edge AI

Recently, with the rapid proliferation of mobile computing and Internet of Things (IoT), big data is going through a radical shift of data source from the mega-scale cloud datacenters to the network edge, e.g., mobile devices and IoT devices, as illustrated in Fig. 1.1. It is anticipated that the data generated by connected devices would reach 175 zettabytes by 2025 [16], far greater than that the cloud datacenters could handle. As the personal data collected by IoT devices is sensitive in nature, there is a growing consensus that much of the personal data should be used for training the models locally and would never go beyond the network edge. Besides, many intelligent applications, such as autonomous driving and augmented reality, need to accomplish decision making with low latency, in order to meet the requirements for safety, accuracy, performance and user experience. Clearly, the conventional wisdom of transporting the data bulks from the IoT devices to the cloud datacenters for analytics would not work well, due to the extremely stringent requirements in cost, privacy and performance. As a result, it is anticipated that a high percentage of IoT data will be stored and processed locally, giving rise to a new research area, namely 'edge AI' as the marriage of edge computing and AI.

More specifically, edge AI refers to the AI model training or inference at the resource-limited network edge by leveraging available data and computational resources across edge devices and cloud datacenters. Recently, a lot of research studies have emerged to explore various aspects of edge AI:

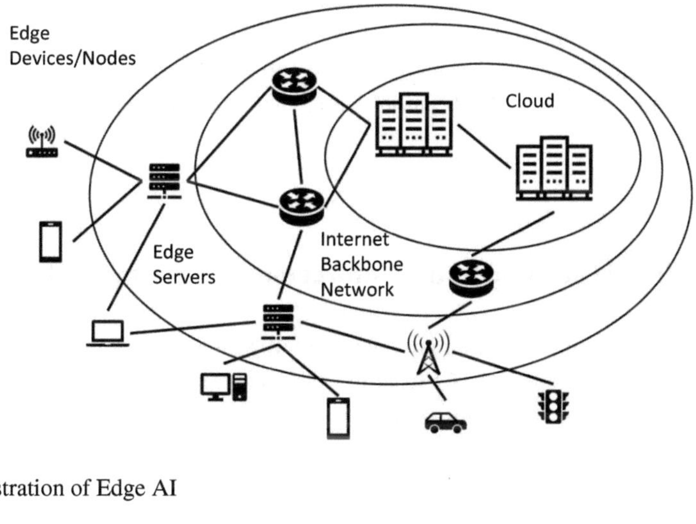

Fig. 1.1 Illustration of Edge AI

- *Model training in edge AI.* Various ML techniques have been used to enable edge AI. For example, federated learning (FL) trains a global model in a distributed manner based on the local data across multiple devices, which is widely used for model training at the network edge, e.g., [3, 169, 226, 284, 288, 358]. Split learning splits a global model into multiple sections and each edge device trains one section by using its local data, e.g., [61, 106, 255, 285]. A pre-trained model in the cloud can be transferred to an edge device based on transfer learning for further fine-tuning with local data, e.g., [46, 62, 370, 372]. Various model compression techniques are also frequently used in edge AI to improve the efficiency of memory, computation, and communication for model training, including knowledge distillation (e.g., [10, 36, 361, 404]), quantization (e.g., [60, 184, 205, 251]), and model pruning (e.g., [130, 156, 222]).
- *Model inference in edge AI.* To make full use of the available data and computation resources in edge networks, the model inference in edge AI can be implemented in different frameworks. The trained model can be split into two parts, with one in the edge devices and the other in the cloud. The intermediate results calculated in the device based on local data will be sent to the remaining part of model in the cloud to complete the inference (e.g., [161, 172, 180]). To reduce the latency and communication cost, collaborations between edge serves and edge devices have been introduced for model inference. For example, [204] studied how to accelerate inference for edge-device collaboration by jointly optimizing the model partitioning and right-sizing. Moreover, the inference can be done based on the collaborations between edge devices, which can be very useful in scenarios with high mobility or harsh environments (e.g., [84, 405]). Recent studies have also investigated efficient on-device model inference. For instance, OnceNAS [397] was proposed in to jointly optimize the number of parameters and inference latency through neural architecture search, which has significantly reduced the model size and increased the inference speed.
- *Applications of edge AI.* Due to the advantages of privacy protection, low latency, low cost, and high flexibility in personalization, edge AI can find applications in a lot of different domains, including smart healthcare [53, 123, 160, 275], smart cities [9, 66, 148, 324], smart agriculture [90, 269, 308], autonomous vehicles [54, 75, 369], recommendation systems [101], etc.

1.1.2 Continual Learning and Reinforcement Learning for Edge AI

Despite the great potential to create many novel intelligent applications and fuel the continuous prosperity of AI, pushing the AI frontier to the network edge for achieving edge AI is highly nontrivial. Specifically, running AI applications directly on edge devices to process the IoT data locally, if not designed intelligently, would suffer from poor performance and energy inefficiency, simply because many AI applications typically require high computational power that greatly outweighs the capacity of resource- and energy-constrained IoT devices. Tackling these unique challenges in edge networks successfully is vital to sustain the rapid progress of this field. To this end, an appropriate and efficient edge learning framework should possess the following properties:

- *Guarantee performance under small datasets.* While tremendous data will be generated by the ensemble of IoT devices, the amount of personal data at every edge node is limited. Therefore, to facilitate various edge-AI applications, especially on-device AI applications, the edge learning algorithms should be able to effectively extract useful knowledge from small local datasets.
- *Work efficiently with simple learning algorithms.* In contrast to more capable cloud datacenters, the edge node only has limited computational resources, which significantly restrict its capability to run complicated learning algorithms on complex models. Consequently, it is more desirable to just deploy simple algorithms, e.g., gradient descent, on a more compact model.
- *Learn quickly with high communication efficiency.* The strict requirement on performance and latency for many AI applications clearly precludes the approaches that need extensive training time, calling for a fast learning algorithm that can quickly train ML models at each edge node with performance guarantee. In the meanwhile, the communication efficiency should be taken into consideration for edge learning frameworks based on collaborations among the cloud and edge nodes.
- *Adapt continuously.* Distributions of locally generated data can shift in a lifelong manner due to the environment change. For instance, autonomous vehicles can experience different terrain and weather conditions when collecting data. The online and non-stationary nature of local data at edge nodes requires model training in edge AI to adapt continuously, such that the knowledge learned previously will not only be maintained but also be leveraged to facilitate better learning of new data.

While a lot of existing studies on edge AI have proposed various edge learning frameworks that meet the first three properties, much less attention has been paid to frameworks that can enable continuous adaptation of AI models at the network edge [350, 406]. Clearly, the standard ML paradigms that focus on learning with a fixed dataset and stationary data distributions would not work well. On one hand, directly applying the model learned for previous data distributions can fail in solving new tasks due to the data distribution shift. On the other hand, simply adapting the learned model with the new data can result in the forgetting of the knowledge of old data and sometimes may even hinder the new task learning. Note that continual learning (CL) [215] is a learning paradigm in ML which continuously adapts a single model to learn a sequence of tasks without forgetting the learned knowledge of old tasks. In particular, by leveraging the knowledge transfer across different tasks, continuoual learning (CL) can potentially solve new tasks more easily and also improve the model performance on old tasks. Therefore, a new formulation of edge learning from the perspective of CL is needed towards building edge AI that possess the four properties above.

Many edge AI applications involve in-time decision making to meet the requirements for safety and performance [80]. The goal is to find optimal control policies to maximize long-term returns. For example, autonomous vehicles (AV) should learn expert control policies to drive safely and efficiently; a virtual health assistant seeks to create a personalized wellness plan based on a patient's unique genetic profile and health history, in order to maximize his/her health. RL [32, 321] provides a promising solution to these problems by learning through interactions with the environment. However, one of the critical challenges to prevent RL from being directly used for edge AI is that RL algorithms typically require extensive online interactions with the environment for policy learning, which can be costly for resource-limited edge nodes and dangerous for safety-critical applications such as AV [170, 388] and healthcare [67, 383]. Another challenge is that most RL algorithms focus on the policy learning for stationary environments, whereas the system dynamics for edge AI applications can continuously change due to various reasons such as the physical environment changes in self-driving [253]. To address these challenges and put the great promise of RL on the ground for edge AI, continual RL recently emerged by introducing the idea of CL into RL, which can be carried out from two perspectives:

- *Warm-start RL.* To eliminate the need of online interactions, offline RL has recently attracted extensive attention by learning from offline datasets previously collected via some behavior policy [7, 200]. It is usually guaranteed that the learned policy is a better one than the behavior policy. However, the performance of pure offline RL highly depends on the quality of the offline dataset, and the learned policy from a low-quality dataset can not be directly used in real applications, even with improved performance over the behavior policy. To address this, warm-start RL combines offline RL and online RL, which continuously finetunes the learned offline policy using a few online interactions with the environment. From a different perspective, warm-start RL also enables fast online policy learning by leveraging the offline policy learned from a fixed dataset, significantly improving the usability of RL in edge AI.
- *Continual RL.* Warm-start RL typically assumes that the underlying MDP stays the same for both offline and online learning, whereas the online environment can be nonstationary in edge AI. The control problem in a stationary environment can be treated as an MDP and the MDP formulation changes when the system dynamics change. To handle the distribution shift across different MDPs, continual RL [1, 164] leverages the idea of CL to continuously finetune the online policy, such that new MDPs can be quickly solved without forgetting the policies learned for previous MDPs.

As alluded to earlier on, the research in CL and RL for edge AI is still in its infancy stage, and there is an urgent need to build a dedicated venue for exchanging the recent advances. In the rest of this chapter, we will discuss the frameworks for building efficient edge AI with CL and RL and also the important design considerations.

1.2 Framework

To fully leverage the available data and resources across the edge networks, there are various frameworks to enable CL for edge AI depending on different application scenarios.

- *Edge-only.* The learning problem locally at a single edge node can be formulated as an online continual learning problem, because of the online and nonstationary nature of local data. In this case, due to the limited data samples and constrained computation resources at the edge node, the on-edge CL algorithms need to be able to extract useful knowledge from the local data in an effective and efficient manner. Particularly, the CL algorithms should continuously adapt the local ML model to solve the new task, corresponding to the newly collected data, quickly with only a few data samples, whereas the knowledge of the learned tasks should be preserved as well in a computationally efficient way. Towards this end, a key idea is to enhance the knowledge transfer between old tasks and new tasks across the time. More specifically, appropriately leveraging the accumulated knowledge of old tasks can significantly reduce the number of data samples required for learning similar new tasks, following the same spirit as in transfer learning and meta-learning. On the other hand, by facilitating the positive knowledge transfer for the new task to similar old tasks, the model performance on these old tasks can also be potentially improved. This edge-only framework is particular important for on-device intelligent applications, such as autonomous vehicles, personalized healthcare, and audio/video surveillance.
- *Cloud-edge collaboration.* While sending a large amount of local data to the cloud for processing is not feasible for edge AI, the abundant history data and computation resources in the cloud are clearly beneficial for edge learning if leveraged through appropriate cloud-edge collaborations. In particular, there are two commonly used frameworks for cloud-edge collaboration: (1) *Cloud as a priori.* The model generalization capability closely hinges upon the number of available training data samples. With the huge amount of history data and computing resources in the cloud, a powerful ML model can be pre-trained with the capability of grasping general knowledge and features from data, such as the foundation models. These pretrained models in the cloud provide a very helpful priori for local edge learning. In particular, finetuning from (or learning an adapter with) the pretrained model using local data can not only solve edge tasks more easily by building on the learned general knowledge in the pretrained model, but also solve edge tasks better by leveraging the excellent knowledge extracting capability of the pretrained model. Based on this pretrained-finetuning paradigm, the CL at the edge can be substantially improved

by starting from the pretrained models in the cloud. As the pretrained models are typically very large in size to guarantee the generalization capability, how to efficiently finetune these model at the computation-constrained edge nodes is one key challenge herein. (2) *Cloud as a platform.* On the other hand, the cloud can serve as a platform to exchange and aggregate the information among different edge nodes, such that the CL on one edge node can potentially leverage the useful knowledge from other edge nodes with similar learning tasks. More specifically, each participating edge node has a local CL problem, and seeks to solve the new tasks based on the knowledge extracted from its past tasks and also the information from other edge nodes shared through the cloud. Following the same idea as in federated learning, the information from edge nodes should be shared in such a way that the privacy of edge nodes is protected and the communication cost is minimized. However, if not designed carefully, aggregating information from other nodes may not only hinder the new task learning but also exacerbate the forgetting of old tasks at one edge node, due to the potential interference between different edge nodes.

- *Edge-edge collaboration.* Communicating with a central cloud platform may not be always feasible, or trustworthy even if feasible. In this case, the information sharing between different edge nodes in an edge network can be done in a decentralized manner, where each edge node can only communicate with its neighbors in a certain range, e.g., self-driving cars in a vehicle-to-vehicle network. This leads to a framework of decentralized CL for which each node has a local CL problem. Similar to cloud-edge collaboration where the cloud serves as a platform, how to appropriately share information across different edge nodes is a central problem to guarantee the performance of this learning framework for edge-edge collaboration. Moreover, the communication protocol in this case is particularly important because the edge nodes share limited communication bandwidth and the information propagation delay across the edge network can significantly affect the learning performance.

1.3 Performance Measures and Efficiency

To provide guidance for algorithm design and performance evaluation of the designed algorithms for edge AI with CL and RL, it is essential to define various metrics for evaluating the learning performance and also the computational costs. In this monograph, we introduce the following metrics, which can be used to evaluate how the edge AI system performs in specific aspects:

- *Task learning performance.* Needless to say, different types of tasks may use different performance metrics. For example, test accuracy is usually used for prediction tasks such as image classification, whereas the Frechet-Inception Distance score is widely adopted for evaluating the performance of GAN models in generative tasks.

- **New task learning performance**. The metric measures how well the ML model performs after adapting with the new coming data at the edge. In general, the learning performance of a good edge CL algorithm is expected to be better than learning from scratch with the new data only, because of the forward knowledge transfer from similar tasks that the edge node has experienced in the past or useful knowledge shared by the cloud and other edge nodes.
- **Task average learning performance**. In many real applications, not only the new task learning performance matters, but also the average learning performance over all seen tasks. For example, an autonomous vehicle should be able to drive safely in all environments it has experienced. A good task average learning performance indicates that the edge AI system demonstrates competitive performance overall, which is particularly attractive when all tasks are equally important.
- **Task average forgetting**. The forgetting for a particular task evaluates the performance change between the ML model after learning this task and the ML model after learning the current task. A positive value of forgetting means that the knowledge of old tasks has been forgotten after learning new tasks, whereas a negative value implies that the knowledge gained from new tasks indeed improves the model performance on old tasks.
- **Task sample complexity**. Task sample complexity typically evaluates the number of data samples required to train a good ML model for solving a particular task. Since the amount of available data samples at a single edge node is often small, a practical edge learning algorithm should be capable of effectively extracting knowledge from limited data, leading to a low sample complexity. This metric is particularly important to evaluate the learning efficiency of RL algorithms, in terms of the number of interactions with the environment for policy learning.

• *Computational efficiency*. Unlike most standard ML algorithms, the computational efficiency is a very critical factor, which could be even more critical than the learning performance, in performance evaluation for edge learning algorithms, considering the constrained computational resources at the network edge.

- **Model training time**. Model training time for a particular task is the most direct metric to evaluate the computational efficiency of an edge learning algorithm, which evaluates the time taken from the start of the training to when a good quality ML model is obtained.
- **Number of training steps**. Computational capability of different edge nodes can be heterogeneous, which highly affects the model training time. In contrast, the number of training steps is a metric that does not depend on the computing power at the network edge but directly evaluate the convergence speed of an edge learning algorithm.

  - **Memory usage**. The memory usage of a learning algorithm typically depends on various factors, including the ML model size, intermediate training results required by the algorithm, additional data points, etc. Considering that edge nodes often have a small memory size, edge learning algorithms should leverage the limited memory in a highly efficient manner.

- *Communication efficiency.* Communication efficiency is also an important factor in characterizing the performance of edge learning algorithms, especially for cloud-edge and edge-edge collaboration frameworks. Typically, edge nodes communicate with each other and the cloud through wireless communication channels with limited spectrum. Both communication cost and communication frequency will affect the communication efficiency of the edge AI system.

  - **Communication cost**. The communication cost is usually captured by the communication bandwidth and the transmission time needed for each communication round. It highly depends on what information is shared between the cloud and the edge nodes, e.g., data, gradient, ML model. To ensure the success rate of the communication and also the overall efficiency of the edge AI system, it is important to reduce the communication cost for edge learning algorithms.
  - **Communication frequency**. The communication frequency can also be evaluated as the number of communication rounds during the entire model training process. Reducing the communication frequency in the edge learning algorithms will not only minimize the impact of unreliable communication channels but also mitigate the potential interference among different communication links in the wireless edge networks, therefore improving the communication efficiency.

In this monograph, we provide an overview of recent advances in continual learning and reinforcement learning in edge AI, with 'bias' towards our own research efforts in this area, and offer our subjective perspective on recent trends and potential cross-disciplinary developments.

Part I
Algorithmic and Theoretical Foundations

In Part I, we provide an overview of key machine learning techniques in edge AI, for enabling continuous model training, including continual learning, reinforcement learning, and meta-learning, and present a few recent research results of ours in these areas.

Continual Learning for Edge AI

2

2.1 Introduction

Continual learning (CL) [258] is a learning paradigm where an agent needs to continuously learn a sequence of tasks. To emulate the remarkable lifelong learning ability of humans, the agent is expected to leverage accumulated knowledge from previous tasks to more easily learn new ones, and further improve the learning performance of old tasks by leveraging the knowledge of new tasks. The former is referred to as forward knowledge transfer and the latter as backward knowledge transfer. One major challenge herein is the so-called *catastrophic forgetting* [231], i.e., the agent easily forgets the knowledge of old tasks when learning new tasks.

General setup In CL, different tasks arrive in a sequential manner, and each task has its own training dataset and task identity, where each data sample consists of the input feature and the corresponding label. The objective here is to train a model sequentially for learning each new task with no or limited access to old task data, such that the model can perform well on the test datasets for all seen tasks. Depending on the characterization of task datasets and availability of task identities, there are several typical CL setups:

- *Domain incremental learning*: Tasks share the same label space but have different input feature distributions.
- *Task incremental learning*: Tasks may have different input distributions and label spaces, and task identities need to be provided during both training and testing.
- *Class incremental learning*: Tasks may have different input distributions and label spaces, and task identities are known during training but not testing.
- *Online CL*: The task training data arrive as an online data stream and can only be used to update the model once.

H. Wang et al., *Continual and Reinforcement Learning for Edge AI*,
Synthesis Lectures on Learning, Networks, and Algorithms,
https://doi.org/10.1007/978-3-031-84363-1_2

Evaluation metrics Let $A_{j,i}$ is the accuracy of the model on i-th task after learning the j-th task sequentially where $j \geq i$. To evaluate the performance of the learned model in CL, there are three metrics widely used in the literature:

- Overall accuracy AA_j: measures the average test accuracy of the model learned after task j on all seen tasks

$$AA_j = \frac{1}{j}\sum_{i=1}^{j} A_{j,i}.$$

- Backward knowledge transfer BWT_j: measures the average forgetting (if negative) or the performance improvement (if positive) of old tasks after learning task $j \geq 2$

$$BWT_j = \frac{1}{j-1}\sum_{i=1}^{j-1} A_{j,i} - A_{i,i}.$$

- Forward knowledge transfer FWT_j: measures the impact of all old tasks on the learning performance of the CL model in the current task i

$$FWT_j = \frac{1}{j-1}\sum_{i=2}^{j} A_{i,i} - A_{i,*}.$$

  Here $A_{i,*}$ is the performance of some reference model, e.g., a randomly-initialized model trained using the data from task i, in task i.

2.2 Theoretical Studies on Continual Learning

The theoretical understanding of CL is still in the early stage, where only a few attempts have emerged recently. Specifically, Bennani et al. [31] and Doan et al. [87] analyzed generalization error and forgetting for the orthogonal gradient descent (OGD) approach [100] based on NTK models, and further proposed variants of OGD to address forgetting. Yin et al. [377] proposed a unified framework for the performance analysis of regularization-based CL methods, by formulating them as a second-order Taylor approximation of the loss function for each task. Asanuma et al. [22] and Lee et al. [195] studied CL in the teacher-student setup to characterize the impact of task similarity on forgetting performance. Cao et al. [47] and Li et al. [210] investigated continual representation learning with dynamically expanding feature spaces, and developed provably efficient CL methods with a characterization of the sample complexity. Chen et al. [59] characterized the lower bound of memory in CL using the PAC framework. By investigating the information flow between neural network layers,

Andle and Yasaei Sekeh [15] analyzed the selection of frozen filters based on layer sensitivity to maximize the performance of CL. Goldfarb and Hand [115] investigated the impact of overparameterization for linear models in a two-task setup. Evron et al. [94] studied CL in overparameterized linear models by analyzing the forgetting based on the training data. Li et al. [206] explored the theory in the case of applying the mixture-of-experts in CL. Nevertheless, none of these existing works show an explicit form of forgetting and generalization error, that only depends on fundamental system parameters/setups (e.g., number of tasks/samples/parameters, noise level, task similarity/order). In this chapter, we provide the first-known explicit theoretical result in a more general CL setup with an arbitrary number of tasks, which enables us to comprehensively understand which factors are relevant and how they (precisely) affect forgetting and generalization error of CL.

2.2.1 CL in Linear Models

Consider the standard CL setup where a sequence of tasks $\mathbb{T} = \{1, \ldots, T\}$ arrives sequentially in time.

Ground truth. We consider a linear ground truth [26, 94] for each task. Specifically, for task t, the output $y \in \mathbb{R}$ is given by

$$y_t = \hat{\boldsymbol{x}}_t^\top \hat{\boldsymbol{w}}_t^* + z_t, \tag{2.1}$$

where $\hat{\boldsymbol{x}}_t \in \mathbb{R}^{s_t}$ denotes the feature vector, $\hat{\boldsymbol{w}}_t^* \in \mathbb{R}^{s_t}$ denotes the model parameters, and z_t is the random noise. Here s_t denotes the number of features of ground truth (i.e., the number of true features). In practice, true features are unknown in advance. Therefore, when choosing a model to learn a certain task, people usually choose more features than enough such that all possible features are included. We write this formally into the following assumption.[1]

Assumption 2.1 We index all possible features by $1, 2, \cdots$. Let $\mathcal{W}$ denote the set of indices of all the chosen features in the model to be trained, with cardinality $|\mathcal{W}| = p$. Let $\mathcal{S}_t$ denote the set of indices of t-th task's true features, with cardinality $|\mathcal{S}_t| = s_t$. We assume that $\bigcup_{t\in\mathbb{T}} \mathcal{S}_t \subseteq \mathcal{W}$.

We next define an expanded ground-truth vector $\boldsymbol{w}_t^* \in \mathbb{R}^p$ that expands the original ground-truth vector $\hat{\boldsymbol{w}}_t^*$ from dimension s_t to dimension p by filling zeros in the positions $\mathcal{W} \setminus \mathcal{S}_t$. Let $\boldsymbol{x}_t$ be the corresponding features for $\boldsymbol{w}_t^*$. Therefore, the ground truth Eq. (2.1) can be rewritten as

$$y_t = \boldsymbol{x}_t^\top \boldsymbol{w}_t^* + z_t. \tag{2.2}$$

[1] When Assumption 2.1 does not hold, the derivation techniques for Theorem 2.1 in the next section still hold with a minor modification that treats the missing features as noise.

Data. For each task $t \in \mathbb{T}$, the training dataset is denoted as $\mathcal{D}_t = \{(\boldsymbol{x}_{t,j}, y_{t,j}) \in \mathbb{R}^p \times \mathbb{R}\}_{j\in[n_t]}$ with sample size n_t. By stacking the training data as $\boldsymbol{X}_t := [\boldsymbol{x}_{t,1}\ \boldsymbol{x}_{t,2}\ \cdots\ \boldsymbol{x}_{t,n_t}] \in \mathbb{R}^{p\times n_t}$ and $\boldsymbol{y}_t := [y_{t,1}\ y_{t,j}\ \cdots\ y_{t,n_t}]^\top \in \mathbb{R}^{n_t\times 1}$, Eq. (2.2) can be written as $\boldsymbol{y}_t = \boldsymbol{X}_t^\top \boldsymbol{w}_t^* + \boldsymbol{z}_t$.

To simplify our analysis, we consider *i.i.d.* Gaussian features and noise, which is stated in the following assumption.

Assumption 2.2 Each element of $\boldsymbol{X}_t$ for all $t \in \mathbb{T}$ follows standard Gaussian distribution $\mathcal{N}(0, 1)$ and is independent of each other. The noise $z_t \sim \mathcal{N}(\boldsymbol{0}, \sigma_t^2 \boldsymbol{I}_p)$ and is independent of each other for all $t \in \mathbb{T}$, where $\sigma_t \geq 0$ denotes the noise level.

Learning procedure. We train the model parameters $\boldsymbol{w}$ for each task sequentially. Let $\boldsymbol{w}_t$ denote the result after training for task t, which is also the initial point in the model training for task $t+1$. Let $\boldsymbol{w}_0 = \boldsymbol{0}$, i.e., task 1 starts training from zero. For each task t, the training loss is defined by mean-squared-error (MSE) with respect to (w.r.t.) $(\boldsymbol{X}_t, \boldsymbol{y}_t)$:

$$\mathcal{L}_t^{tr}(\boldsymbol{w}, \mathcal{D}_t) = \frac{1}{n_t}\|(\boldsymbol{X}_t)^\top \boldsymbol{w} - \boldsymbol{y}_t\|_2^2. \tag{2.3}$$

When underparameterized (i.e., $n_t \leq p$), minimizing Eq. (2.3) has a unique solution (with probability 1). When overparameterized (i.e., $p > n_t$), minimizing Eq. (2.3) has an infinite number of solutions that make Eq. (2.3) zero. Among all overfitted solutions, we are particularly interested in the one corresponding to the convergent point of stochastic gradient descent (SGD) for minimizing Eq. (2.3). In fact, it can be shown that such an overfitted solution has the smallest ℓ_2-norm of the change of parameters [121]. In other words, $\boldsymbol{w}_t$ corresponds to the solution to the following optimization problem:

$$\min_{\boldsymbol{w}} \ \|\boldsymbol{w} - \boldsymbol{w}_{t-1}\|_2, \quad s.t.\ (\boldsymbol{X}_t)^\top \boldsymbol{w} = \boldsymbol{y}_t. \tag{2.4}$$

The constraint in Eq. (2.4) implies that the training loss is exactly zero (i.e., overfitted).

Performance evaluation. For the described linear system, we use $\mathcal{L}_t(\boldsymbol{w})$ to denote the model error[2] for task t:

$$\mathcal{L}_t(\boldsymbol{w}) = \|\boldsymbol{w} - \boldsymbol{w}_t^*\|^2, \tag{2.5}$$

which characterizes the generalization performance of $\boldsymbol{w}$ on task t. As is standard in the empirical studies of CL, e.g., [49, 218], we evaluate the performance of CL on two key metrics, forgetting and overall generalization error, defined as below:

(1) *Forgetting*: It measures how much 'knowledge' of old tasks has been forgotten after learning the current task. Specifically, after learning task $t \in [2, T]$, the average forgetting over all old tasks $i \in [1, t-1]$ is defined as:

[2] It can be proved that the model error we defined here is equivalent to the mean-squared-error on noise-free test data.

$$F_t = \frac{1}{t-1}\sum_{i=1}^{t-1}(\mathcal{L}_i(\boldsymbol{w}_t) - \mathcal{L}_i(\boldsymbol{w}_i)). \tag{2.6}$$

In Eq. (2.6), $\mathcal{L}_i(\boldsymbol{w}_t) - \mathcal{L}_i(\boldsymbol{w}_i)$ denotes the performance difference between $\boldsymbol{w}_i$ (the result after training task i) and $\boldsymbol{w}_t$ (the result after training task t) on test data of task i.

(2) *Overall generalization error*: We evaluate the model generalization performance of the final task model $\boldsymbol{w}_T$ in terms of the average model error over all tasks:

$$G_T = \frac{1}{T}\sum_{i=1}^{T}\mathcal{L}_i(\boldsymbol{w}_T). \tag{2.7}$$

It is worth noting that the forgetting defined in Evron et al. [94] is based on the training loss, which consequently ignores the generalization performance of the learned models for old tasks. Such a definition is not only inconsistent with the evaluation metric in empirical studies, but also insufficient to capture the backward knowledge transfer because the value of forgetting therein can not be negative.

We further simplify the current setup by assuming that each task has the same number of training samples as well as the same noise level σ, stated as follows.

Assumption 2.3 $n_t = n$ and $\sigma_t = \sigma$ for all $t \in \mathbb{T}$.

Note that Assumption 2.3 is adopted only to make our results (which will be shown in the next section) easy to interpret. In fact, our analysis can be easily generalized to the situation when Assumption 2.3 does not hold.

2.2.2 Main Results and Interpretations

Although we use linear models, in order to provide hints on understanding DNNs that are usually heavily overparameterized, we are particularly interested in the performance of CL in the overparameterized region ($p > n$), where we define the overparameterized ratio as $r := 1 - \frac{n}{p}$. For ease of exposition, we define the following coefficients that will appear in our main theorem:

$$c_{i,j} := (1-r)\left(r^{T-i} - r^{j-i} + r^{T-j}\right), \tag{2.8}$$

where $1 \le i < j \le T$ are the indices of tasks. Now we are ready to state our main theorem that characterizes the expected value of forgetting and overall generalization error:

Theorem 2.1 *When $p \ge n+2$, we must have*

$$\mathbb{E}[F_T] = \frac{1}{T-1}\sum_{i=1}^{T-1}\Big[\underbrace{(r^T - r^i)\|\boldsymbol{w}_i^*\|^2}_{\text{Term F1}} + \underbrace{\sum_{j>i}^{T} c_{i,j}\|\boldsymbol{w}_i^* - \boldsymbol{w}_j^*\|^2}_{\text{Term F2}} + \underbrace{\frac{p\sigma^2}{p-n-1}(r^i - r^T)}_{\text{Term F3}}\Big] \tag{2.9}$$

$$\mathbb{E}[G_T] = \underbrace{\frac{r^T}{T}\sum_{i=1}^{T}\|w_i^*\|^2}_{Term\ G1} + \underbrace{\frac{1}{T}\sum_{i=1}^{T}\frac{nr^{T-i}}{p}\sum_{k=1}^{T}\|w_k^* - w_i^*\|^2}_{Term\ G2} + \underbrace{\frac{p\sigma^2}{p-n-1}\left(1-r^T\right)}_{Term\ G3}. \quad (2.10)$$

To the best of our knowledge, Theorem 2.1 is the first result that establishes the closed forms of forgetting and overall generalization error of CL in overparameterized linear models. In the rest of the chapter, we will see that Theorem 2.1 not only describes how CL performs on the linear system but also provides guidance on applying CL in practice that DNNs and real-world datasets. The proof of Theorem 2.1 is in Sect. 2.2.4.4. We also verify the correctness of Theorem 2.1 in Fig. 2.1 where discrete points indicated by markers in Fig. 2.1 (drawn by simulations) are very close to the curves (drawn by Theorems 2.1 and 2.2).

We can further simply Eqs. (2.9) and (2.10) by only considering two tasks, so as to better understand Theorem 2.1. The result is shown in the following corollary, which clearly characterizes the dependence on task similarity and different system parameters.

Corollary 2.1 *When $T = 2$ and $p \geq n + 2$, we must have*

$$\mathbb{E}[F_2] =(r^2 - r)\|w_1^*\|^2 + \frac{n}{p}\|w_2^* - w_1^*\|^2 + \frac{nr\sigma^2}{p-n-1}, \quad (2.11)$$

$$\mathbb{E}[G_2] =\frac{r^2}{2}\left(\|w_1^*\|^2 + \|w_2^*\|^2\right) + \frac{1-r^2}{2}\|w_1^* - w_2^*\|^2 + \frac{p\sigma^2(1-r^2)}{p-n-1}. \quad (2.12)$$

Based on Theorem 2.1, we will provide insights on the following three aspects.

(1) *Overparameterization (Sect.* 2.2.2.1*).* In order to understand the generalization power of overfitted machine learning models, much attention has focused (e.g., [26, 136, 157]) on studying the impact of overparameterization on single-task learning, whereas how overpa-

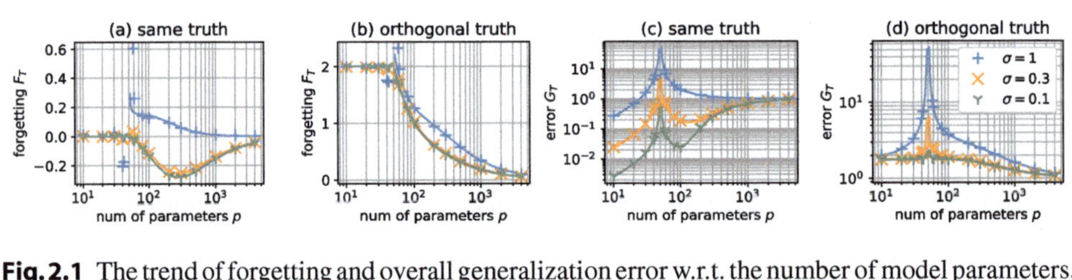

Fig. 2.1 The trend of forgetting and overall generalization error w.r.t. the number of model parameters, where $T = 8$, $n = 50$, $\hat{w}_t^* \in \mathbb{R}^{10}$ and $\|\hat{w}_t^*\|^2 = 1$ for all $t \in \mathbb{T}$. The ground truths are the same for all tasks in Subfigures (**a**) and (**c**), but are orthogonal in Subfigures (**b**) and (**d**) where $\hat{w}_t^*$ equals to t-th standard basis for all $t \in \mathbb{T}$. Here the orthogonal truth means that for two tasks i and j, they do not have common features, i.e., $\mathcal{S}_i \cap \mathcal{S}_j = \emptyset$. In this case, the inner product of their ground truth model is equal to zero. The discrete points indicated by markers are calculated by simulation and are the average of 300 random simulation runs. The curves are drawn by the theoretical expressions in Theorems 2.1 and 2.2

rameterization affects the performance of CL still remains unclear. Fortunately, the exact forms in Theorem 2.1 provide a way to directly evaluate the impact of overparameterization and the random noise on both forgetting and generalization error in CL.
(2) *Task similarity (Sect.* 2.2.2.2*).* Both forgetting and generalization error depend on the optimal model gap between any two tasks , i.e., $\|\boldsymbol{w}_k^* - \boldsymbol{w}_i^*\|^2$ for any task k and i, which defines the task similarity in this work (smaller gap means higher similarity). This model gap can be treated as the characterization of the task similarity, which depends only on the underlying true model of each task and is independent of the learnt models. Understanding the impact of task similarity is helpful to not only explain empirical observations but also guide better designs of CL in practice.
(3) *Task ordering (Sect.* 2.2.2.3*).* Given a fixed set of tasks in CL, the learning order of the task sequence clearly plays an important role in affecting both $\mathbb{E}[F_T]$ and $\mathbb{E}[G_T]$, through the task order-dependent coefficients, e.g., c_{ij} in Eq. (2.9) and r^{T-i} in Eq. (2.10). For example, suppose $\|\boldsymbol{w}_i^*\|^2$ is the same for all $i \in \mathbb{T}$, the optimal task ordering to minimize the generalization error is to learn the tasks in a decreasing order of $\sum_{k=1}^T \|\boldsymbol{w}_k^* - \boldsymbol{w}_i^*\|^2$, i.e., $i < j$ if $\sum_{k=1}^T \|\boldsymbol{w}_k^* - \boldsymbol{w}_i^*\|^2 > \sum_{k=1}^T \|\boldsymbol{w}_k^* - \boldsymbol{w}_j^*\|^2$. Intuitively, the most dissimilar task should be learnt first in this case. Investigating the impact of task ordering is particularly valuable when the agent can control the task order in CL, in the same spirit of curriculum learning [30].

In what follows, we will delve into the impact of those three crucial factors in order to provide a comprehensive understanding of CL in the linear models.

2.2.2.1 The Impact of Overparameterization

In this subsection, we show some insights about the impact of overparameterization. Specially, we will discuss what happens when p changes under a fixed n.
(1) More parameters can lead to zero forgetting and alleviate the negative impact of task dissimilarity on generalization error. As shown in Theorem 2.1, when $p \to \infty$, we can have that $\mathbb{E}[F_T] \to 0$ and Term G2 also approaches zero. In some special cases, we can further show that Term G2 is monotonically decreasing w.r.t. p. A more detailed discussion can be found in Sect. 2.2.4.3.
(2) Benign overfitting exists and is easier to observe with large noise and/or low task similarity. As we introduced in related work, benign overfitting has recently been discovered and studied in linear models as a first step towards understanding why DNNs can still generalize well even when heavily overparameterized. The concept of "benign overfitting" and "double-descent" is initially proposed for only a single task. We now show that such a phenomenon also exists in CL where there exists a sequence of tasks.

Notice that Theorem 2.1 is for the overparameterized region. For a precise comparison between the performance of overfitting and underfitting, we present the theoretical result of the underparameterized region in the following theorem.

Theorem 2.2 *When $n \geq p+2$, we must have*

$$\mathbb{E}[F_T] = \frac{1}{T-1}\sum_{i=1}^{T-1} \|\boldsymbol{w}_T^* - \boldsymbol{w}_i^*\|^2,$$

$$\mathbb{E}[G_T] = \left(\frac{1}{T}\sum_{i=1}^{T-1} \|\boldsymbol{w}_T^* - \boldsymbol{w}_i^*\|^2\right) + \frac{p\sigma^2}{n-p-1}.$$

We provide an intuitive explanation and rigorous proof of Theorem 2.2 in Sect.2.2.4.4. As shown in Theorem 2.2, $\mathbb{E}[G_T]$ becomes larger when the noise level σ is larger, and both $\mathbb{E}[F_T]$ and $\mathbb{E}[G_T]$ become larger when tasks are less similar (i.e., when $\sum_{i=1}^{T-1} \|\boldsymbol{w}_T^* - \boldsymbol{w}_i^*\|^2$ is larger). In contrast, in the overfitted situation, Term F2 and Term G2 in Theorem 2.1 (corresponding to task similarity), Term F3 and Term G3 (corresponding to noise) will go to zero when $p \to \infty$. This indicates that when the noise level is high and/or task similarity is low, the performance of CL in the overparameterized situation is more likely to be better than that in the underparameterized situation, i.e., benign overfitting exists and is easier to observe. This can be observed from Fig. 2.1. For example, the blue curve with markers "+" corresponds to the largest noise (compared with other curves in Fig. 2.1d) and the lowest task similarity (compared with Fig. 2.1c), and it has the deepest descent curve in the overparameterized region ($p > 50 = n$). This observation indicates that benign overfitting is easier to observe with larger noise and lower task similarity.

(3) A descent floor sometimes exists on forgetting and generalization error, especially when tasks are similar and noise is low. In Eq. (2.11), the term $(r^2 - r)\|\boldsymbol{w}_1^*\|^2$ first decreases and then increases as p increases from n to ∞ (i.e., r increases from 0 to 1), while the remaining two terms decrease as p increases. Thus, when $\|\boldsymbol{w}_2^* - \boldsymbol{w}_1^*\|^2$ (task similarity) and σ^2 (noise level) are relatively small, the trend of F_2 w.r.t. p will be dominated by the first term, where a descent floor of forgetting exists. In the right-hand-side of Eq. (2.12), the first term increases as p increases, while the rest two terms decrease as p increases. Taking the derivative of Eq. (2.12) on p, we have

$$\frac{\partial \mathbb{E}[G_2]}{\partial p} = \frac{2nr\boldsymbol{w}_1^{*\top}\boldsymbol{w}_2^*}{p^2} - \sigma^2\left(\frac{(n+1)(1-r^2)}{(p-n-1)^2} + \frac{2nr}{(p-n-1)p}\right).$$

Here, since $\frac{1}{p-n-1}$ is very large when p is close to n, while decreasing to zero when $p \to \infty$, we can tell that when σ^2 is relatively small w.r.t. $\boldsymbol{w}_1^{*\top}\boldsymbol{w}_2^*$, $\frac{\partial \mathbb{E}[G_2]}{\partial p}$ will be positive and then negative as p increases from $n+2$ to ∞. In other words, if these two tasks have a positive correlation (i.e., $\boldsymbol{w}_1^{*\top}\boldsymbol{w}_2^* > 0$) and noise is small, there exists a descent floor w.r.t. p on $\mathbb{E}[G_2]$. Such a phenomenon can exist in other setups besides the special case of $T = 2$. For example, in Fig. 2.1a, c where the ground truth for each task is exactly the same, we can observe a descent floor for the small noise cases $\sigma = 0.3$ and 0.1 (i.e., orange and green curves with markers "×" and "Y", respectively).

2.2.2.2 The Impact of Task Similarity

Generalization error monotonically decreases with task similarity whereas forgetting may not. Based on Theorem 2.1, it can be seen that the generalization error $G_T(\boldsymbol{w}_T)$ decreases when $\|\boldsymbol{w}_k^* - \boldsymbol{w}_i^*\|^2$ for any two different tasks k and i decreases, because of the positive coefficients in Term G2 in Eq. (2.10). Intuitively, the generalization error of CL will be smaller if the tasks are more similar with each other. In contrast, the forgetting F_T may not change monotonically with $\|\boldsymbol{w}_k^* - \boldsymbol{w}_i^*\|^2$, because the coefficients c_{ij} in Term F2 in Eq. (2.9) can be negative. To verify this result, we consider two different scenarios.

(1) Consider the case where $T = 2$. In Eq. (2.11), $\|\boldsymbol{w}_2^* - \boldsymbol{w}_1^*\|^2$ captures the task similarity between tasks 1 and 2 in terms of the optimal task models. It is clear that forgetting increases with $\|\boldsymbol{w}_2^* - \boldsymbol{w}_1^*\|^2$, i.e., less forgetting when the two tasks are more similar.

(2) Consider the case where $T = 4$. We first assume that $\|\boldsymbol{w}_i^*\|^2 = w$ for any task $i \in [1, 4]$ considering the overparameterized models [94]. Suppose that task 1 and task 2 share the same set of true features, which is orthogonal to the feature set of both task 3 and task 4, i.e., $\mathcal{S}_1 = \mathcal{S}_2$ and $\mathcal{S}_1 \cap (\mathcal{S}_3 \cup \mathcal{S}_4) = \emptyset$. Note that

$$\|\boldsymbol{w}_i^* - \boldsymbol{w}_j^*\|^2 = \|\boldsymbol{w}_i^*\|^2 + \|\boldsymbol{w}_j^*\|^2 - 2\langle \boldsymbol{w}_i^*, \boldsymbol{w}_j^* \rangle$$

where $\langle \boldsymbol{w}_i^*, \boldsymbol{w}_j^* \rangle = 0$ if $\mathcal{S}_i \cap \mathcal{S}_j = \emptyset$. Therefore, we can control the value of $\|\boldsymbol{w}_1^* - \boldsymbol{w}_2^*\|^2$ by changing $\boldsymbol{w}_2^*$, without affecting the value of $\|\boldsymbol{w}_i^* - \boldsymbol{w}_j^*\|^2$ for any pair of $\{i, j\} \neq \{1, 2\}$. Based on Theorem 2.1, it can be shown that $c_{1,2} < 0$, such that increasing $\|\boldsymbol{w}_1^* - \boldsymbol{w}_2^*\|^2$, i.e., the tasks become less similar, will surprisingly decrease forgetting.

2.2.2.3 The Impact of Task Ordering

In order to investigate the impact of task ordering on the performance of CL, we assume that $\|\boldsymbol{w}_t^*\|^2 = w$ for every task $t \in \mathbb{T}$. By ignoring the task order-independent terms in Eqs. (2.9) and (2.10), we focus on the task order-dependent terms, i.e., Term F2 and Term G2.

(1) Optimal task ordering of minimizing forgetting tends to arrange dissimilar tasks adjacently in the early stage of the sequence. As shown in Term F2, the optimal task order to minimize forgetting closely hinges upon the value of $c_{i,j}$. Based on Eq. (2.8), $c_{i,j}$ is smaller when (1) i and j are smaller and (2) they are closer. Intuitively, this implies that tasks with larger $\|\boldsymbol{w}_i^* - \boldsymbol{w}_j^*\|^2$ should be learnt adjacently with higher priority in CL, in order to minimize the impact of the task dissimilarity on the value of $\tilde{F}_T(\boldsymbol{w}_T)$. However, finding the optimal task order for the general case is highly nontrivial due to the complex coupling across $\|\boldsymbol{w}_i^* - \boldsymbol{w}_j^*\|^2$ for different tasks. To verify the implication above and better understand the structure of the optimal task order, we study several special cases of the task setups.

(1) *[Special case I: One vs Many]* There are two different categories of tasks, where tasks in the same category have the same optimal model; among the entire task set, one special task belongs to Category I while the other tasks belong to Category II. In this case, the optimal

task order is captured by the optimal learning order of the special task in Category I. We have the following result to characterize the optimal task order for Special case I.

Proposition 2.1 *Let $i^* \in [1, T]$ denote the optimal order of the special task in Category I to minimize forgetting. Suppose $p \geq n + 2$. Then (1) i^* can take any integer value between 2 and $\frac{T}{2}$, depending on the value of $\frac{n}{p}$; (2) i^* is non-decreasing with $\frac{n}{p}$.*

As indicated by Proposition 2.1, the special task will be learnt in the first half of the sequence, such that the task diversity in the first half is always larger than in the second half. Besides, with the model capacity increasing ($\frac{n}{p} \to 0$), the order of the special task will move towards the beginning of the sequence, because (1) the model is less concerned about the special task since it is powerful enough to learn different features and (2) the model focuses on the performance of the majority and seeks to learn more tasks from Category II at the end of the sequence for better performance.

(2) *[Special case II: Equal Occurrence]* There are two different categories (C_1 and C_2) of tasks, where tasks in the same category have the same optimal model; particularly, two categories contain the same number of tasks. If task $1 \in C_1$ and task $2 \in C_2$, we will denote the task order as (C_1, C_2). The following proposition characterizes the optimal task order in this case:

Proposition 2.2 *Suppose $p \geq n + 2$. For $T = 4$ and $T = 6$, the optimal task order to minimize forgetting is the perfectly alternating order, i.e., (C_i, C_j, C_i, C_j) and $(C_i, C_j, C_i, C_j, C_i, C_j)$, where $i, j \in \{1, 2\}$ and $i \neq j$.*

Proposition 2.2 clearly shows that adjacent tasks always belong to different categories in the optimal task order, which leads to a more diverse task learning sequence. Intuitively, the alternating order maximizes the memorization of each category by keeping practicing on different tasks. It can be further proved that the perfectly alternating order is also optimal for $T = 6$ with three different categories (Sect. 2.2.4.3). Based on these results, we expect that such an alternating order may minimize forgetting for more general scenarios where the tasks contain multiple categories with equal cross-category task model distance.

The findings on the optimal task order indeed share similar insights with the surprising impact of task correlation on forgetting mentioned earlier. Intuitively, learning more dissimilar tasks in the early stage facilitates the exploration of a larger feature space and expands the learnt feature space in CL, which can make the learning of similar tasks in the future much easier. In the meanwhile, the impact of task similarity among the early tasks continuously diminishes in CL with T increasing, as suggested by the coefficients $c_{i,j}$ (which can be smaller for smaller i, j) in Theorem 2.1. Therefore, the negative impact of learning more dissimilar tasks on forgetting is weaker when they are learnt in the early stage, compared to being learnt in the late stage.

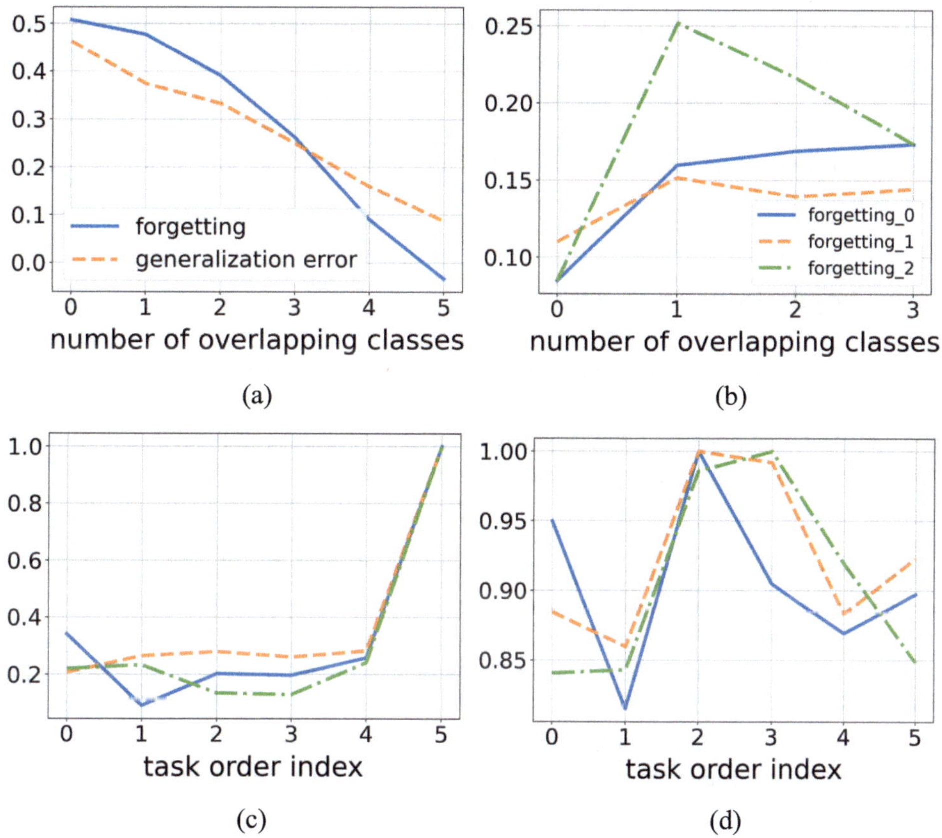

Fig. 2.2 Impact of task similarity and task order. **a** When $T = 2$, both forgetting and generalization error decrease when two tasks have more overlapping classes. **b** When $T = 4$, the forgetting surprisingly increases when the first two tasks have more overlapping classes; 'forgetting_0', 'forgetting_1' and 'forgetting_2' correspond to three cases of the task setups (also in (**c**) and (**d**)). **c** Consider one special task and five same tasks in CL with $T = 6$; the task order index shows the order of the special task, and the smallest forgetting is achieved always when the special task is learnt in the first half for each case (we normalize the forgetting w.r.t. the worst forgetting in each case). **d** Consider two categories of tasks in CL with $T = 4$; the task order indices 0 and 1 refer to the perfectly alternating orders, one of which always achieve the smallest forgetting among all possible orders. All the results are averaged over 3 random seeds

(2) The optimal task ordering for minimizing forgetting and for minimizing generalization error are not always the same. Consider Special case I and Special case II. It can be shown that the optimal task orders for minimizing forgetting and generalization error are different in Special case I but same in Special case II. This would open up an interesting direction of finding the task order with balanced impact on forgetting and generalization error. A more detailed discussion can be found in Sect. 2.2.4.3.

2.2.3 Implications on CL with DNN

So far, we have explored different aspects that affect the performance of CL in overparameterized linear models. More interestingly, we will show next that Theorem 2.1 can also shed light on CL in practice with DNNs, by reflecting on recent empirical observations and guiding improved designs therein. More experimental details are in Sect. 2.2.4.1.

2.2.3.1 Forgetting Is Not Always Monotonic with Task Similarity

To see if our understandings about the impact of task similarity on forgetting can be carried over to CL with DNN, we conduct experiments on MNIST [190] using a convolutional neural network to investigate the impact of task similarity therein. More specifically, we consider each task i as a binary classification problem which seeks to decide if an image belongs to a task-specific label subset Y_i of the classes, i.e., $Y_i \subset \{0, \ldots, 9\}$ in MNIST, and we control the task similarity through the degree of class overlapping between the task-specific subsets, e.g., task i and j are more similar if the cardinality of $Y_i \cap Y_j$ is larger.

We first consider the case with two tasks, where we fix Y_1 for task 1 as $\{0, 1, 2, 3, 4\}$ and change Y_2 for task 2 to have different numbers of overlapping classes with Y_1. As shown in Fig. 2.2a, both forgetting and generalization error decrease when the number of overlapping classes increases, i.e., the two tasks are more similar, which is indeed consistent with our analysis for the overparameterized linear models for $T = 2$. More interestingly, this result also agrees with some recent studies [94, 195, 272], which found that 'intermediate task similarity' leads to the worst forgetting in a two-task setup using various notions of task similarity (different from our definition of task similarity using the optimal model gap), through either empirical studies or analyzing the upper bound of forgetting. We can build the connection based on the closed form of forgetting F_2 in Eq. (2.11).

Note that in Eq. (2.11)

$$\|w_2^* - w_1^*\|^2 = \|w_2^*\|^2 + \|w_1^*\|^2 - 2\langle w_1^*, w_2^*\rangle$$

and we can divide the task correlation into three cases depending on the value of $\langle w_1^*, w_2^*\rangle$: (1) $\langle w_1^*, w_2^*\rangle = 0$: Two tasks are orthogonal in the sense that they share no common features, i.e., $\mathcal{S}_1 \cap \mathcal{S}_2 = \emptyset$; (2) $\langle w_1^*, w_2^*\rangle > 0$: Two tasks share some common features and are 'positively' correlated; (3) $\langle w_1^*, w_2^*\rangle < 0$: Two tasks share some common features but are 'negatively' correlated. Compared to the first case when two tasks are orthogonal, it can be easily shown that forgetting is worse when two tasks are negatively correlated even if they share some common features, which indeed corresponds to 'the intermediate task similarity' in [94, 195, 272]. The reason behind is that in this case task 2 updates the model in the opposite direction to the model update of task 1, which inevitably leads to more forgetting in CL. Note that in Fig. 2.2a, the non-overlapping case means that task 1 and 2 are negatively correlated because in this two-task case the image that is not in Y_1 must be in Y_2. On the other hand, the forgetting can even be negative when the two tasks are positively correlated.

We next consider the case with $T = 4$, where we control the task similarity by changing Y_2 while fixing Y_1, Y_3 and Y_4. Here we let $(Y_1 \cup Y_2) \cap (Y_3 \cup Y_4) = \emptyset$ as in Sect. 2.2.2.2. As shown in Fig. 2.2b, forgetting surprisingly increases when task 1 and task 2 have more overlapping classes, which is also consistent with our analysis for the linear models. Indeed, this also justifies our observation that forgetting can decrease when the adjacent tasks are more dissimilar when studying the impact of task order.

2.2.3.2 Diversify the Tasks in the Early Stage and Order Dissimilar Tasks Adjacently

We also evaluate the impact of task ordering on forgetting in CL with DNN, by constructing the tasks using a similar strategy as in Sect. 2.2.3.1. More specifically, we consider two different scenarios: (1) $T = 6$, where the task sequence includes one special task and five same tasks; (2) $T = 4$, where the task sequence includes two categories of tasks and each has two same tasks.

Figure 2.2c demonstrates forgetting in the first scenario w.r.t. the learning order of the special task, and three plots correspond to three different cases, respectively. It is clear that for all three cases, the optimal order of the special task to minimize forgetting is always in the first half of the sequence. For the second scenario, we evaluate forgetting in Fig. 2.2d for all six possible task orders, where task indices 0 and 1 refer to the perfectly alternating order. We can see that the smallest forgetting is also achieved in the perfectly alternating order. These results indicate that our findings in Sect. 2.2.2.3 for the overparameterized linear models can also be carried over to CL with DNN, i.e., the optimal task order should diversify the tasks in the early stage and learn more different tasks adjacently. Such an implication is indeed consistent with the empirical observations in recent studies [28, 210]. Note that in both Fig. 2.2c and d, we normalize forgetting w.r.t. the worst forgetting in each case.

2.2.3.3 Weight the Fresher Old Tasks More in Forward Knowledge Transfer

Recently, there has been increasing interest in CL on leveraging task correlation to facilitate knowledge transfer [162, 217, 218], which first selects the most correlated old tasks with the current task and then designs algorithms to directly increase the knowledge transfer between correlated tasks. By investigating knowledge transfer in the linear models, we show that improved algorithms can be motivated to achieve better knowledge transfer.

Table 2.1 The averaged final testing accuracy (ACC) and backward transfer (BWT: negative value of forgetting, larger is better) over all the tasks on different datasets

Method	PMNIST		Split CIFAR-100	
	ACC (%)	BWT (%)	ACC (%)	BWT (%)
TRGP	96.34	–0.8	74.46	–0.9
TRGP+	**96.75**	**–0.46**	**75.31**	**0.13**

Given a task t in CL, the forward knowledge transfer [336] in the linear model can be defined as

$$\mathbb{E}[\|\boldsymbol{w}_t - \boldsymbol{w}_t^*\|^2] - \mathbb{E}[\|\boldsymbol{w}_t^r - \boldsymbol{w}_t^*\|^2], \tag{2.13}$$

where $\boldsymbol{w}_t^r$ is the learnt model of task t by starting from a random model. Intuitively, Eq. (2.13) characterizes the gap in the testing performance between $\boldsymbol{w}_t$ learnt in CL and $\boldsymbol{w}_t^r$ learnt from scratch, for which a positive value means that the accumulated knowledge in CL benefits the learning of the current task. As the second term in Eq. (2.13) is independent with CL, it suffices to analyze $\mathbb{E}[\|\boldsymbol{w}_t - \boldsymbol{w}_t^*\|^2]$ for understanding the forward knowledge transfer. Based on Lemma 2.2, we can obtain

$$\mathbb{E}[\|\boldsymbol{w}_t - \boldsymbol{w}_t^*\|^2] = r^t\|\boldsymbol{w}_t^*\|^2 + \sum_{i=1}^{t}\frac{nr^{t-i}}{p}\|\boldsymbol{w}_i^* - \boldsymbol{w}_t^*\|^2 + \frac{p\sigma^2}{p-n-1}.$$

While it is intuitive that better forward knowledge transfer can be achieved when $\|\boldsymbol{w}_i^* - \boldsymbol{w}_t^*\|^2$ is smaller for the current task t and the old task i, the impact of different old tasks on the current task is non-uniform, in the sense that a more recent old task i (i.e., $t-i$ is smaller) has a larger effect on the forward knowledge transfer to task t. This result implies that fresher old tasks should contribute more when designing algorithms to leverage correlated old tasks to facilitate better forward knowledge transfer.

To verify this insight, we consider the TRGP algorithm proposed in Lin et al. [218]. Specifically, TRGP first selects the most correlated old tasks with the current task and reuses their knowledge through a scaled weight projection to facilitate forward knowledge transfer, where all the selected old tasks are treated equivalently. We slightly modify TRGP by assigning a larger weight to the selected old task that is more recent to the current task, named as TRGP+, and evaluate its performance on standard CL benchmarks (PMNIST [227] and Split CIFAR-100 [178]) and DNN architectures. As shown in Table 2.1, TRGP+ outperforms TRGP in both accuracy and forgetting. Assigning a larger weight to the more recent correlated old task not only improves the forward knowledge transfer, but also increases the backward knowledge transfer by forcing the learnt model of the current task to be closer to the model of those highly correlated old tasks.

2.2.4 Experimental Evaluation

2.2.4.1 Experimental Details for Sects. 2.2.3.1 and 2.2.3.2

Datasets. We consider the MNIST dataset. For each task, we randomly select 200 samples for training and 1000 samples for testing. Different tasks have different subsets of classes.

DNN architecture and training details. We use a five-layer neural network with two convolutional layers and three fully-connected layers. Relu is used for the first four layers and Sigmoid is used for the last layer. The first convolutional layer is followed by 2D max-pooling operation with stride of 2. We learn each task by using SGD with a learning rate of 0.1 for 600 epochs. The forgetting and overall generalization error are evaluated as in Eqs. (2.6) and (2.7), respectively, while here $\mathcal{L}_t(\boldsymbol{w})$ is defined as the mean-squared test error instead of Eq. (2.5).

Task setups. For Fig. 2.2a, we consider the following setup:

- task 1: $(0, 1, 2, 3, 4)$.
- task 2: $(5, 6, 7, 8, 9)$, $(4, 5, 6, 7, 8)$, $(3, 4, 5, 6, 7)$, $(2, 3, 4, 5, 6)$, $(1, 2, 3, 4, 5)$, $(0, 1, 2, 3, 4)$, which correspond to the different numbers of overlapping classes with task 1.

For Fig. 2.2b, we randomly select three different setups:

- 'forgetting_0':
  - task 1: $(0, 1, 2)$.
  - task 2: $(3, 4, 5)$, $(2, 3, 4)$, $(1, 2, 3)$, $(0, 1, 2)$, which correspond to the different numbers of overlapping classes with task 1.
  - task 3: $(7, 8, 9)$.
  - task 4: $(7, 8, 9)$.
- 'forgetting_1':
  - task 1: $(3, 4, 5)$.
  - task 2: $(0, 1, 2)$, $(1, 2, 3)$, $(2, 3, 4)$, $(3, 4, 5)$, which correspond to the different numbers of overlapping classes with task 1.
  - task 3: $(6, 7, 8)$.
  - task 4: $(7, 8, 9)$.
- 'forgetting_2':
  - task 1: $(0, 1, 2)$.

- task 2: (7, 8, 9), (2, 7, 8), (1, 2, 7), (0, 1, 2), which correspond to the different numbers of overlapping classes with task 1.
- task 3: (4, 5, 6).
- task 4: (4, 5, 6).

For Fig. 2.2c, we randomly select three different setups:

- 'forgetting_0': the special task is (4, 5, 6, 7) and the other tasks are (0, 1, 2, 3).
- 'forgetting_1': the special task is (0, 1, 2, 3) and the other tasks are (5, 6, 7, 8).
- 'forgetting_2': the special task is (3, 4, 5, 6) and the other tasks are (1, 2, 7, 8).

For Fig. 2.2d, we randomly select three different setups:

- 'forgetting_0': the two task categories are (4, 5, 6, 7) and (1, 2, 4, 5), and the task order indices are:
  - '0': (4, 5, 6, 7), (1, 2, 4, 5), (4, 5, 6, 7), (1, 2, 4, 5).
  - '1': (1, 2, 4, 5), (4, 5, 6, 7), (1, 2, 4, 5), (4, 5, 6, 7).
  - '2': (4, 5, 6, 7), (4, 5, 6, 7), (1, 2, 4, 5), (1, 2, 4, 5).
  - '3': (1, 2, 4, 5), (1, 2, 4, 5), (4, 5, 6, 7), (4, 5, 6, 7).
  - '4': (4, 5, 6, 7), (1, 2, 4, 5), (1, 2, 4, 5), (4, 5, 6, 7).
  - '5': (1, 2, 4, 5), (4, 5, 6, 7), (4, 5, 6, 7), (1, 2, 4, 5).
- 'forgetting_1': the two task categories are (4, 5, 6, 7) and (2, 3, 4, 5), and the task order indices are:
  - '0': (4, 5, 6, 7), (2, 3, 4, 5), (4, 5, 6, 7), (2, 3, 4, 5).
  - '1': (2, 3, 4, 5), (4, 5, 6, 7), (2, 3, 4, 5), (4, 5, 6, 7).
  - '2': (4, 5, 6, 7), (4, 5, 6, 7), (2, 3, 4, 5), (2, 3, 4, 5).
  - '3': (2, 3, 4, 5), (2, 3, 4, 5), (4, 5, 6, 7), (4, 5, 6, 7).
  - '4': (4, 5, 6, 7), (2, 3, 4, 5), (2, 3, 4, 5), (4, 5, 6, 7).
  - '5': (2, 3, 4, 5), (4, 5, 6, 7), (4, 5, 6, 7), (2, 3, 4, 5).
- 'forgetting_2': the two task categories are (6, 7, 8, 9) and (3, 4, 5, 6), and the task order indices are:
  - '0': (6, 7, 8, 9), (3, 4, 5, 6), (6, 7, 8, 9), (3, 4, 5, 6).
  - '1': (3, 4, 5, 6), (6, 7, 8, 9), (3, 4, 5, 6), (6, 7, 8, 9).
  - '2': (6, 7, 8, 9), (6, 7, 8, 9), (3, 4, 5, 6), (3, 4, 5, 6).
  - '3': (3, 4, 5, 6), (3, 4, 5, 6), (6, 7, 8, 9), (6, 7, 8, 9).

- '4': (6, 7, 8, 9), (3, 4, 5, 6), (3, 4, 5, 6), (6, 7, 8, 9).
- '5': (3, 4, 5, 6), (6, 7, 8, 9), (6, 7, 8, 9), (3, 4, 5, 6).

2.2.4.2 Experimental Details for Sect. 2.2.3.3

TRGP vs TRGP+. TRGP [218] seeks to solve the following optimization problem for the current task t:

$$\begin{aligned} \min_{\{w^l\}_l, \{Q^l_{j,t}\}_{l, j\in \mathcal{TR}^l_t}} \quad & \mathcal{L}(\{w^l_{eff}\}_l, \mathcal{D}_t), \\ s.t \quad & w^l_{eff} = w^l + \sum\nolimits_{j\in\mathcal{TR}^l_t} [\text{proj}^Q_{S^l_j}(w^l) - \text{proj}_{S^l_j}(w^l)], \end{aligned} \tag{2.14}$$

where w^l is the DNN weight for the layer l, and S^l_j denotes the input subspace of the layer l for the old task $j < t$, which can be constructed by using SVD on the representation matrix for that layer. Two important designs are introduced in Eq. (2.14):

- The trust region $\mathcal{TR}^l_t$: $\mathcal{TR}^l_t$ denotes the set of the most correlated old tasks selected for task t based on some correlation evaluation metric in a layer-wise manner. The purpose here is to select the most correlated old tasks and facilitate the forward knowledge transfer by reusing the learnt knowledge of the old tasks in $\mathcal{TR}^l_t$.
- The scaled weight projection $\text{proj}^Q_{S^l_j}(w^l)$: $\text{proj}^Q_{S^l_j}(w^l)$ is developed to reuse the learnt model of the selected old tasks in $\mathcal{TR}^l_t$. Specifically, for any $j \in \mathcal{TR}^l_t$,

$$\text{proj}^Q_{S^l_j}(w^l) = w^l_{t-1} B^l_j Q^l_{j,t} (B^l_j)'$$

where B^l_j is the bases matrix for the subspace S^l_j, and $Q^l_{j,t}$ is the scaling matrix to scale the weight projection onto S^l_j. In contrast, $\text{proj}_{S^l_j}(w^l) = w^l_{t-1} B^l_j (B^l_j)'$ is the standard weight projection onto S^l_j. Since the learnt knowledge for the old task j is indeed $\text{proj}_{S^l_j}(w^l)$, scaling the projection provides a way to reuse this knowledge directly for learning the task t. Intuitively, $\text{proj}^Q_{S^l_j}(w^l) - \text{proj}_{S^l_j}(w^l)$ characterizes the boosted forward knowledge transfer from old task $j \in \mathcal{TR}^l_t$ to task t.

However, as shown in Eq. (2.14), all the selected old tasks in $\mathcal{TR}^l_t$ are treated equivalently in the effective weight w^l_{eff}, which could be suboptimal. As suggested by our theoretical results, we proposed a slightly modified version of TRGP, i.e., TRGP+, by assigning non-uniform weights for the most correlated old tasks selected in $\mathcal{TR}^l_t$:

$$\begin{aligned} \min_{\{w^l\}_l, \{Q^l_{j,t}\}_{l, j\in \mathcal{TR}^l_t}} \quad & \mathcal{L}(\{w^l_{eff}\}_l, \mathcal{D}_t), \\ s.t \quad & w^l_{eff} = w^l + \sum\nolimits_{j\in\mathcal{TR}^l_t} \lambda_j [\text{proj}^Q_{S^l_j}(w^l) - \text{proj}_{S^l_j}(w^l)], \end{aligned} \tag{2.15}$$

where $\lambda_j > \lambda_{j'}$ if $t - j < t - j'$ for both $j, j' \in \mathcal{TR}^l_t$.

Datasets. We consider two standard benchmarks in CL: (1) **PMNIST**: 10 sequential tasks will be created using different permutations, where each task has 10 classes; (2) **Split CIFAR-100**: The entire dataset of CIFAR-100 will be splitted into 10 group, where each task is a 10-way multi-class classification problem for each group.

DNN architecture and training details. Following [218], we use a 3-layer fully-connected network with 2 hidden layer of 100 units for PMNIST, and train the network for 5 epochs with a batch size of 10 for each task. For Split CIFAR-100, we use a version of 5-layer AlexNet, and train the network for a maximum of 200 epochs with early stopping for each task. Two most correlated old tasks are selected for the current task for each layer, and we assign a larger weight of 1.2 to the more recent old task and 0.8 to the other one.

Evaluation metrics. The performance is evaluated based on ACC, the average final accuracy over all tasks, and Backward Transfer (BWT) which measures the forgetting of old tasks when learning new tasks. Specfically, ACC and BWT are defined as:

$$ACC = \frac{1}{T}\sum\nolimits_{i=1}^{T} A_{T,i}, BWT = \frac{1}{T-1}\sum\nolimits_{i=1}^{T-1} A_{T,i} - A_{i,i} \tag{2.16}$$

where $A_{T,i}$ is the accuracy of the model on i-th task after learning the T-th task sequentially.

2.2.4.3 Additional Results

Characterization of negative forgetting As shown in Fig. 2.2a, the forgetting can even be negative when the two tasks are positively correlated. Intuitively, because the common features play a similar role in these two tasks, task 2 updates the model in a favorable direction for task 1, which could even result in better performance of task 1 due to the backward knowledge transfer herein. A formal quantification of the condition for better performance of task 1 can be found in the following proposition:

Proposition 2.3 *Suppose* $\sigma^2 < \frac{p-n-1}{p}\|w^*_1\|^2$ *and* $p \geq n+2$. *The learning of task 2 would lead to a better model for task 1, i.e.,* $\mathbb{E}[F_2] \leq 0$, *if*

$$2\langle w^*_{1,S_1}, w^*_{2,S_2}\rangle \geq \frac{n}{p}\|w^*_1\|^2 + \|w^*_2\|^2 + \frac{(p-n)\sigma^2}{p-n-1}.$$

Evolution of forgetting We can also characterize the evolution of forgetting after learning new tasks. Based on the definition of forgetting, we have

$$\mathbb{E}[F_t] = \frac{1}{t-1}\sum_{i=1}^{t-1}\mathbb{E}\left[\|\boldsymbol{w}_t - \boldsymbol{w}_i^*\|^2 - \|\boldsymbol{w}_i - \boldsymbol{w}_i^*\|^2\right],$$

$$\mathbb{E}[F_{t+1}] = \frac{1}{t}\sum_{i=1}^{t}\mathbb{E}\left[\|\boldsymbol{w}_{t+1} - \boldsymbol{w}_i^*\|^2 - \|\boldsymbol{w}_i - \boldsymbol{w}_i^*\|^2\right].$$

Rearranging the above equations gives

$$\sum_{i=1}^{t-1}\mathbb{E}[\|\boldsymbol{w}_t - \boldsymbol{w}_i^*\|^2] = (t-1)\mathbb{E}[F_t] + \sum_{i=1}^{t-1}\mathbb{E}[\|\boldsymbol{w}_i - \boldsymbol{w}_i^*\|^2],$$

$$\sum_{i=1}^{t-1}\mathbb{E}[\|\boldsymbol{w}_{t+1} - \boldsymbol{w}_i^*\|^2] = t\mathbb{E}[F_{t+1}] + \sum_{i=1}^{t}\mathbb{E}[\|\boldsymbol{w}_i - \boldsymbol{w}_i^*\|^2] - \mathbb{E}[\|\boldsymbol{w}_{t+1} - \boldsymbol{w}_t^*\|^2].$$

Based on the relationship between $\mathbb{E}[\|\boldsymbol{w}_t - \boldsymbol{w}_i^*\|^2]$ and $\mathbb{E}[\|\boldsymbol{w}_{t+1} - \boldsymbol{w}_i^*\|^2]$ characterized in Lemma 2.2, it can be seen that

$$\begin{aligned}
&\sum_{i=1}^{t-1}\mathbb{E}[\|\boldsymbol{w}_{t+1} - \boldsymbol{w}_i^*\|^2] \\
=&t\mathbb{E}[F_{t+1}] + \sum_{i=1}^{t}\mathbb{E}[\|\boldsymbol{w}_i - \boldsymbol{w}_i^*\|^2] - \mathbb{E}[\|\boldsymbol{w}_{t+1} - \boldsymbol{w}_t^*\|^2] \\
=&\sum_{i=1}^{t-1}\left\{\left(1-\frac{n}{p}\right)\mathbb{E}[\|\boldsymbol{w}_t - \boldsymbol{w}_i^*\|^2] + \frac{n}{p}\|\boldsymbol{w}_{t+1}^* - \boldsymbol{w}_i^*\|^2 + \frac{n\sigma^2}{p-n-1}\right\} \\
=&\left(1-\frac{n}{p}\right)\sum_{i=1}^{t-1}\mathbb{E}[\|\boldsymbol{w}_t - \boldsymbol{w}_i^*\|^2] + \frac{n}{p}\sum_{i=1}^{t-1}\|\boldsymbol{w}_{t+1}^* - \boldsymbol{w}_i^*\|^2 + \frac{n\sigma^2(t-1)}{p-n-1} \\
=&\left(1-\frac{n}{p}\right)\left\{(t-1)\mathbb{E}[F_t] + \sum_{i=1}^{t-1}\mathbb{E}[\|\boldsymbol{w}_i - \boldsymbol{w}_i^*\|^2]\right\} + \frac{n}{p}\sum_{i=1}^{t-1}\|\boldsymbol{w}_{t+1}^* - \boldsymbol{w}_i^*\|^2 + \frac{n\sigma^2(t-1)}{p-n-1},
\end{aligned}$$

such that

$$\begin{aligned}
t\mathbb{E}[F_{t+1}] =&(t-1)\left(1-\frac{n}{p}\right)\mathbb{E}[F_t] + \left(1-\frac{n}{p}\right)\sum_{i=1}^{t-1}\mathbb{E}[\|\boldsymbol{w}_i - \boldsymbol{w}_i^*\|^2] + \frac{n}{p}\sum_{i=1}^{t-1}\|\boldsymbol{w}_{t+1}^* - \boldsymbol{w}_i^*\|^2 \\
&+ \frac{n\sigma^2(t-1)}{p-n-1} - \sum_{i=1}^{t}\mathbb{E}[\|\boldsymbol{w}_i - \boldsymbol{w}_i^*\|^2] + \mathbb{E}[\|\boldsymbol{w}_{t+1} - \boldsymbol{w}_t^*\|^2] \\
=&(t-1)\left(1-\frac{n}{p}\right)\mathbb{E}[F_t] - \frac{n}{p}\sum_{i=1}^{t-1}\mathbb{E}[\|\boldsymbol{w}_i - \boldsymbol{w}_i^*\|^2] - \mathbb{E}[\|\boldsymbol{w}_t - \boldsymbol{w}_t^*\|^2] \\
&+ \frac{n}{p}\sum_{i=1}^{t-1}\|\boldsymbol{w}_{t+1}^* - \boldsymbol{w}_i^*\|^2 + \mathbb{E}[\|\boldsymbol{w}_{t+1} - \boldsymbol{w}_t^*\|^2] + \frac{n\sigma^2(t-1)}{p-n-1}.
\end{aligned} \tag{2.17}$$

Let $i = t$ in Lemma 2.2. We can show that

$$\begin{aligned}&\mathbb{E}[\|\boldsymbol{w}_{t+1} - \boldsymbol{w}_t^*\|^2 - \|\boldsymbol{w}_t - \boldsymbol{w}_t^*\|^2]\\ =&\frac{n}{p}\|\boldsymbol{w}_{t+1}^* - \boldsymbol{w}_t^*\|^2 - \frac{n}{p}\mathbb{E}[\|\boldsymbol{w}_t - \boldsymbol{w}_t^*\|^2] + \frac{n\sigma^2}{p-n-1}.\end{aligned} \tag{2.18}$$

By substituting Eq. (2.18) back to Eq. (2.17), we can have

$$\begin{aligned}\mathbb{E}[F_{t+1}] =&\frac{t-1}{t}\left(1-\frac{n}{p}\right)\mathbb{E}[F_t] + \frac{n}{tp}\sum_{i=1}^{t-1}\{\|\boldsymbol{w}_{t+1}^* - \boldsymbol{w}_i^*\|^2 - \mathbb{E}[\|\boldsymbol{w}_i - \boldsymbol{w}_i^*\|^2]\}\\ &+ \frac{n}{tp}\{\|\boldsymbol{w}_{t+1}^* - \boldsymbol{w}_t^*\|^2 - \mathbb{E}[\|\boldsymbol{w}_t - \boldsymbol{w}_t^*\|^2]\} + \frac{n\sigma^2}{p-n-1}\\ =&\frac{t-1}{t}\left(1-\frac{n}{p}\right)\mathbb{E}[F_t] + \frac{n}{tp}\sum_{i=1}^{t}\{\|\boldsymbol{w}_{t+1}^* - \boldsymbol{w}_i^*\|^2 - \mathbb{E}[\|\boldsymbol{w}_i - \boldsymbol{w}_i^*\|^2]\} + \frac{n\sigma^2}{p-n-1}.\end{aligned} \tag{2.19}$$

Impact of overparameterization **(1) Forgetting approaches zero with more parameters.** In Eq. (2.9), when $p \to \infty$, we have $r \to 1$, which implies that $(r^T - r^i) \to 0$ and $c_{i,j} \to 0$. Therefore, we can conclude that $\mathbb{E}[F_T] \to 0$ when $p \to \infty$. An intuitive explanation is that with more parameters, the model has a larger "memory" such that it can remember all knowledge of previous tasks, i.e., zero forgetting.

(2) More parameters can alleviate the negative impact of task dissimilarity on generalization error. Term G2 in Eq. (2.10) describes the effect of task dissimilarity on G_T. When $p \to \infty$, Term G2 approaches zero, which indicates that the negative impact of task dissimilarity on generalization error diminishes. In some special cases, we can further show that Term G2 is monotonically decreasing with respect to p, e.g., $T = 2$ shown in Eq. (2.12). A more general[3] case is when $\sum_{k=1}^{T}\|\boldsymbol{w}_k^* - \boldsymbol{w}_i^*\|^2 = C$ for all task i, we have Term G2 $= \frac{1-r^T}{T}C$ which is also monotonically decreasing w.r.t. p.

Impact of task order

(1) **[Special cases]** There are three categories (C_1, C_2 and C_3) of tasks: each category contains the same number of tasks; the tasks are same in the same category but different across categories. Without loss of generality, we assume that for any task i and j

$$\|\boldsymbol{w}_i^* - \boldsymbol{w}_j^*\|^2 = \begin{cases}0, & \text{if } i, j \in C_m \text{ for } m \in \{1,2,3\};\\ 1, & \text{else.}\end{cases}$$

Based on Theorem 2.1, we can show that the optimal task order for Special case III follows a similar structure of that for Special case II, as characterized in the following proposition:

[3] For general T, this requirement holds if the ground truth of each task has the same power and is orthogonal to each other, i.e., $\|\boldsymbol{w}_i^*\|^2 = \|\boldsymbol{w}_j^*\|^2$ and $(\boldsymbol{w}_i^*)^T\boldsymbol{w}_j^* = 0$ for all $i \neq j$.

Proposition 2.4 *Suppose $p \geq n + 2$. For $T = 6$, the optimal task order to minimize forgetting is the perfectly alternating order, i.e., $(C_i, C_j, C_k, C_i, C_j, C_k)$, where $i, j, k \in \{1, 2, 3\}$, $i \neq j$, $i \neq k$ and $j \neq k$.*

(2) **[The optimal task order can be different for minimizing forgetting and generalization error]**

[Special case I] As shown in Proposition 2.1, the optimal task order to minimize forgetting is to learn the special task between the $2nd$ place and the $\frac{T}{2}th$ place. In stark contrast, this special task, which has the largest value of $\sum_{k=1}^{T} \|\boldsymbol{w}_k^* - \boldsymbol{w}_i^*\|^2$, should be always learnt in the very first place in order to minimize the generalization error, i.e., $i = 1$. The underlying rationale is that the generalization error characterizes the average testing performance of the final model on all tasks, which is maximized when the final model works the best for the majority. Therefore, in this case the optimal order for minimizing forgetting is different from that for minimizing generalization error.

[Special case II] As shown in Proposition 2.2, the optimal task order to minimize forgetting is the perfectly alternating order. In contrast, the task order indeed does not affect the generalization performance, because $\sum_{k=1}^{T} \|\boldsymbol{w}_k^* - \boldsymbol{w}_i^*\|^2$ is same for every task $i \in \mathbb{T}$. In this case, the optimal task order for minimizing forgetting is also 'optimal' for minimizing generalization error. That is to say, we can find an optimal task order to minimize forgetting and generalization error simultaneously.

2.2.4.4 Proofs

The following lemma characterizes the solution to the optimization problem Eq. (2.4) for task t:

Lemma 2.1 *The solution to the optimization problem Eq.* (2.4), *i.e., the learnt model for task t, is given by*

$$\boldsymbol{w}_t = \boldsymbol{w}_{t-1} + \boldsymbol{X}_t(\boldsymbol{X}_t^\top \boldsymbol{X}_t)^{-1} \left(\boldsymbol{y}_t - \boldsymbol{X}_t^\top \boldsymbol{w}_{t-1}\right). \tag{2.20}$$

In the overparameterized case, multiple $\boldsymbol{w}_t$ exist to perfectly fit $(\boldsymbol{X}_t)^\top \boldsymbol{w} = \boldsymbol{y}_t$, and solving Eq. (2.4) picks the one that has minimum l^2 distance to $\boldsymbol{w}_{t-1}$. Therefore, the solution in Eq. (2.20) not only incorporates the information of current task t through $\mathcal{D}_t$ but also depends on the previous model evolution trajectory in CL.

By leveraging the recent advance in [27], we can have the following lemma about the evolution of $\mathbb{E}[\|\boldsymbol{w}_t - \boldsymbol{w}_i^*\|^2]$:

Lemma 2.2 *Suppose $p \geq n+2$. For any task $t \in [1, T-1]$ and any old task $i \in [1, t]$, the following equation holds:*

$$\begin{aligned}&\mathbb{E}[\|\boldsymbol{w}_{t+1} - \boldsymbol{w}_i^*\|^2] \\ &= \left(1 - \frac{n}{p}\right)\mathbb{E}[\|\boldsymbol{w}_t - \boldsymbol{w}_i^*\|^2] + \frac{n}{p}\|\boldsymbol{w}_{t+1}^* - \boldsymbol{w}_i^*\|^2 + \frac{n\sigma^2}{p-n-1}.\end{aligned}$$

***Proof of Lemma* 2.1** Let $\hat{\boldsymbol{w}} = \boldsymbol{w} - \boldsymbol{w}_{t-1}$. It is clear that Eq. (2.4) can be reformulated as

$$\begin{aligned}&\min \ \|\hat{\boldsymbol{w}}\|_2, \\ &s.t. \ \ \boldsymbol{X}_t^\top \hat{\boldsymbol{w}} = \boldsymbol{y}_t - \boldsymbol{X}_t^\top \boldsymbol{w}_{t-1}.\end{aligned} \tag{2.21}$$

For the overparameterized case, $\boldsymbol{X}_t^\top \boldsymbol{X}_t$ is invertible. Using the Lagrange multipliers, we can get

$$\min_{\hat{\boldsymbol{w}}, \lambda} \ \frac{\hat{\boldsymbol{w}}^\top \hat{\boldsymbol{w}}}{2} + \lambda^T [\boldsymbol{X}_t^\top \hat{\boldsymbol{w}} - (\boldsymbol{y}_t - \boldsymbol{X}_t^\top \boldsymbol{w}_{t-1})].$$

By setting the derivative w.r.t. $\hat{\boldsymbol{w}}$ to 0, it follows that

$$\hat{\boldsymbol{w}}^* = -\boldsymbol{X}_t \lambda \tag{2.22}$$

such that

$$\boldsymbol{X}_t^\top \hat{\boldsymbol{w}}^* = -\boldsymbol{X}_t^\top \boldsymbol{X}_t \lambda = \boldsymbol{y}_t - \boldsymbol{X}_t^\top \boldsymbol{w}_{t-1}.$$

Therefore,

$$\lambda = -(\boldsymbol{X}_t^\top \boldsymbol{X}_t)^{-1}(\boldsymbol{y}_t - \boldsymbol{X}_t^\top \boldsymbol{w}_{t-1}). \tag{2.23}$$

By substituting Eq. (2.23) into Eq. (2.22), we can have

$$\hat{\boldsymbol{w}}^* = \boldsymbol{X}_t(\boldsymbol{X}_t^\top \boldsymbol{X}_t)^{-1}(\boldsymbol{y}_t - \boldsymbol{X}_t^\top \boldsymbol{w}_{t-1})$$

such that

$$\boldsymbol{w}_t = \boldsymbol{w}_{t-1} + \boldsymbol{X}_t(\boldsymbol{X}_t^\top \boldsymbol{X}_t)^{-1}(\boldsymbol{y}_t - \boldsymbol{X}_t^\top \boldsymbol{w}_{t-1}).$$

***Proof of Lemma* 2.2** Let $\boldsymbol{P}_t := \boldsymbol{X}_t(\boldsymbol{X}_t^\top \boldsymbol{X}_t)^{-1}\boldsymbol{X}_t^\top$ and $\boldsymbol{X}_t^\dagger := \boldsymbol{X}_t(\boldsymbol{X}_t^\top \boldsymbol{X}_t)^{-1}$ for any $t \in \mathbb{T}$, where $\boldsymbol{P}_t$ characterizes the projection onto the row space of $\boldsymbol{X}_t^\top$. Based on Lemma 2.1, we have

$$\boldsymbol{w}_{t+1} = (\boldsymbol{I} - \boldsymbol{P}_{t+1})\boldsymbol{w}_t + \boldsymbol{P}_{t+1}\boldsymbol{w}_{t+1}^* + \boldsymbol{X}_{t+1}^\dagger z_{t+1}. \tag{2.24}$$

Intuitively, the learnt model $\boldsymbol{w}_{t+1}$ for task $t+1$ is an 'interpolation' between the learnt model $\boldsymbol{w}_t$ for task t and the optimal task model $\boldsymbol{w}^*_{t+1}$ for task $t+1$, while being perturbed by the random noise z_{t+1}.

Let $H = (\boldsymbol{I} - \boldsymbol{P}_{t+1})(\boldsymbol{w}_t - \boldsymbol{w}^*_i) + \boldsymbol{P}_{t+1}(\boldsymbol{w}^*_{t+1} - \boldsymbol{w}^*_i)$. Based on Eq. (2.24), we can know that

$$
\begin{aligned}
&\mathbb{E}[\|\boldsymbol{w}_{t+1} - \boldsymbol{w}^*_i\|^2] \\
=&\mathbb{E}[\|(\boldsymbol{I} - \boldsymbol{P}_{t+1})\boldsymbol{w}_t + \boldsymbol{P}_{t+1}\boldsymbol{w}^*_{t+1} + \boldsymbol{X}^\dagger_{t+1}z_{t+1} - \boldsymbol{w}^*_i\|^2] \\
=&\mathbb{E}[\|(\boldsymbol{I} - \boldsymbol{P}_{t+1})(\boldsymbol{w}_t - \boldsymbol{w}^*_i) + \boldsymbol{P}_{t+1}(\boldsymbol{w}^*_{t+1} - \boldsymbol{w}^*_i) + \boldsymbol{X}^\dagger_{t+1}z_{t+1}\|^2] \\
=&\mathbb{E}[\|H + \boldsymbol{X}^\dagger_{t+1}z_{t+1}\|^2] \\
=&\underbrace{\mathbb{E}[\|H\|^2]}_{(a)} + \underbrace{2\mathbb{E}[\langle H, \boldsymbol{X}^\dagger_{t+1}z_{t+1}\rangle]}_{(b)} + \underbrace{\mathbb{E}[\|\boldsymbol{X}^\dagger_{t+1}z_{t+1}\|^2]}_{(c)}.
\end{aligned}
\tag{2.25}
$$

(1) For the term (a), we have

$$
\begin{aligned}
\mathbb{E}[\|H\|^2] =&\mathbb{E}[\|(\boldsymbol{I} - \boldsymbol{P}_{t+1})(\boldsymbol{w}_t - \boldsymbol{w}^*_i) + \boldsymbol{P}_{t+1}(\boldsymbol{w}^*_{t+1} - \boldsymbol{w}^*_i)\|^2] \\
=&\mathbb{E}[\|(\boldsymbol{I} - \boldsymbol{P}_{t+1})(\boldsymbol{w}_t - \boldsymbol{w}^*_i)\|^2] + \mathbb{E}[\|\boldsymbol{P}_{t+1}(\boldsymbol{w}^*_{t+1} - \boldsymbol{w}^*_i)\|^2] \\
&+ 2\mathbb{E}[\langle(\boldsymbol{I} - \boldsymbol{P}_{t+1})(\boldsymbol{w}_t - \boldsymbol{w}^*_i), \boldsymbol{P}_{t+1}(\boldsymbol{w}^*_{t+1} - \boldsymbol{w}^*_i)\rangle] \\
\overset{(a)}{=}&\mathbb{E}[\|(\boldsymbol{I} - \boldsymbol{P}_{t+1})(\boldsymbol{w}_t - \boldsymbol{w}^*_i)\|^2] + \mathbb{E}[\|\boldsymbol{P}_{t+1}(\boldsymbol{w}^*_{t+1} - \boldsymbol{w}^*_i)\|^2] \\
\overset{(b)}{=}&\mathbb{E}[\|\boldsymbol{w}_t - \boldsymbol{w}^*_i\|^2] - \mathbb{E}[\|\boldsymbol{P}_{t+1}(\boldsymbol{w}_t - \boldsymbol{w}^*_i)\|^2] + \mathbb{E}[\|\boldsymbol{P}_{t+1}(\boldsymbol{w}^*_{t+1} - \boldsymbol{w}^*_i)\|^2]
\end{aligned}
\tag{2.26}
$$

where (a) is because of the orthogonality between $\boldsymbol{I} - \boldsymbol{P}_{t+1}$ and $\boldsymbol{P}_{t+1}$, and (b) is due to the Pythagorean theorem.

Because $\boldsymbol{P}_{t+1}$ is the orthogonal projection matrix for the row space of $\boldsymbol{X}_{t+1}$, based on the rotational symmetry of the standard normal distribution, it follows that

$$
\mathbb{E}[\|\boldsymbol{P}_{t+1}(\boldsymbol{w}^*_{t+1} - \boldsymbol{w}^*_i)\|^2] = \frac{n}{p}\|\boldsymbol{w}^*_{t+1} - \boldsymbol{w}^*_i\|^2,
\tag{2.27}
$$

and

$$
\mathbb{E}[\|\boldsymbol{P}_{t+1}(\boldsymbol{w}_t - \boldsymbol{w}^*_i)\|^2] = \frac{n}{p}\mathbb{E}[\|\boldsymbol{w}_t - \boldsymbol{w}^*_i\|^2],
\tag{2.28}
$$

since $\boldsymbol{P}_{t+1}$ is independent with $\boldsymbol{w}_t$.

By substituting Eqs. (2.27) and (2.28) back to Eq. (2.26), we can obtain that

$$
\mathbb{E}[\|H\|^2] = \left(1 - \frac{n}{p}\right)\mathbb{E}[\|\boldsymbol{w}_t - \boldsymbol{w}^*_i\|^2] + \frac{n}{p}\|\boldsymbol{w}^*_{t+1} - \boldsymbol{w}^*_i\|^2.
\tag{2.29}
$$

(2) For the term (b), we have

$$\begin{aligned}\mathbb{E}[\langle H, \boldsymbol{X}_{t+1}^{\dagger} z_{t+1}\rangle] =&\mathbb{E}[\langle (\boldsymbol{I}-\boldsymbol{P}_{t+1})(\boldsymbol{w}_t-\boldsymbol{w}_i^*)+\boldsymbol{P}_{t+1}(\boldsymbol{w}_{t+1}^*-\boldsymbol{w}_i^*), \boldsymbol{X}_{t+1}^{\dagger} z_{t+1}\rangle]\\ =&\mathbb{E}[\langle (\boldsymbol{I}-\boldsymbol{P}_{t+1})(\boldsymbol{w}_t-\boldsymbol{w}_i^*), \boldsymbol{X}_{t+1}^{\dagger} z_{t+1}\rangle]+\mathbb{E}[\langle \boldsymbol{P}_{t+1}(\boldsymbol{w}_{t+1}^*-\boldsymbol{w}_i^*), \boldsymbol{X}_{t+1}^{\dagger} z_{t+1}\rangle].\end{aligned}$$

Because $(\boldsymbol{I}-\boldsymbol{P}_{t+1})$ is the projection onto the null space of $\boldsymbol{X}_{t+1}^{\top}$ and $\boldsymbol{X}_{t+1}^{\dagger} z_{t+1}$ is a vector in the row space of $\boldsymbol{X}_{t+1}^{\top}$, it follows that

$$\mathbb{E}[\langle (\boldsymbol{I}-\boldsymbol{P}_{t+1})(\boldsymbol{w}_t-\boldsymbol{w}_i^*), \boldsymbol{X}_{t+1}^{\dagger} z_{t+1}\rangle]=0. \tag{2.30}$$

And since

$$\mathbb{E}[\langle \boldsymbol{P}_{t+1}(\boldsymbol{w}_{t+1}^*-\boldsymbol{w}_i^*), \boldsymbol{X}_{t+1}^{\dagger} z_{t+1}\rangle]=\mathbb{E}[\langle (\boldsymbol{X}_{t+1}^{\dagger})^{\top}\boldsymbol{P}_{t+1}(\boldsymbol{w}_{t+1}^*-\boldsymbol{w}_i^*), z_{t+1}\rangle]=0.$$

we can know that

$$\mathbb{E}[\langle H, \boldsymbol{X}_{t+1}^{\dagger} z_{t+1}\rangle]=0. \tag{2.31}$$

(3) For the term (c), we apply the “trace trick" by following [27]. Specifically, it can be first seen that

$$\begin{aligned}\|\boldsymbol{X}_{t+1}^{\dagger} z_{t+1}\|^2 =&\|\boldsymbol{X}_{t+1}(\boldsymbol{X}_{t+1}^{\top}\boldsymbol{X}_{t+1})^{-1} z_{t+1}\|^2\\ =&tr((\boldsymbol{X}_{t+1}^{\top}\boldsymbol{X}_{t+1})^{-1}(\boldsymbol{X}_{t+1}^{\top}\boldsymbol{X}_{t+1})(\boldsymbol{X}_{t+1}^{\top}\boldsymbol{X}_{t+1})^{-1} z_{t+1} z_{t+1}^{\top})\\ =&tr((\boldsymbol{X}_{t+1}^{\top}\boldsymbol{X}_{t+1})^{-1} z_{t+1} z_{t+1}^{\top})\end{aligned}$$

Due to the independence between $\boldsymbol{X}_{t+1}$ and the random noise z_{t+1}, we can have that

$$\begin{aligned}\mathbb{E}[\|\boldsymbol{X}_{t+1}^{\dagger} z_{t+1}\|^2] =&\mathbb{E}[tr((\boldsymbol{X}_{t+1}^{\top}\boldsymbol{X}_{t+1})^{-1} z_{t+1} z_{t+1}^{\top}))]\\ =&tr[\mathbb{E}[(\boldsymbol{X}_{t+1}^{\top}\boldsymbol{X}_{t+1})^{-1} z_{t+1} z_{t+1}^{\top}]]\\ =&tr(\mathbb{E}[(\boldsymbol{X}_{t+1}^{\top}\boldsymbol{X}_{t+1})^{-1}]\mathbb{E}[z_{t+1} z_{t+1}^{\top}])\\ =&\sigma^2 tr(\mathbb{E}[(\boldsymbol{X}_{t+1}^{\top}\boldsymbol{X}_{t+1})^{-1}]).\end{aligned}$$

Since $(\boldsymbol{X}_{t+1}^{\top}\boldsymbol{X}_{t+1})^{-1}$ follows the inverse-Wishart distribution with identity scale matrix $\boldsymbol{I}\in\mathbb{R}^{n\times n}$ and p degrees-of-freedom, and each diagonal entry of $(\boldsymbol{X}_{t+1}^{\top}\boldsymbol{X}_{t+1})^{-1}$ has a reciprocal that follows the χ^2 distribution with $p-n+1$ degrees-of-freedom. Therefore, for $p\geq n+2$,

$$tr(\mathbb{E}[(\boldsymbol{X}_{t+1}^{\top}\boldsymbol{X}_{t+1})^{-1}])=\frac{n}{p-n+1},$$

such that

$$\mathbb{E}[\|\boldsymbol{X}_{t+1}^{\dagger} z_{t+1}\|^2]=\frac{n\sigma^2}{p-n+1}. \tag{2.32}$$

Lemma 2.2 can be proved by substituting Eqs. (2.29), (2.31) and (2.32)–(2.25).

Proof of Theorem 2.1 Based on Lemma 2.2, we can have that

$$\begin{aligned}
\mathbb{E}[\|\boldsymbol{w}_t - \boldsymbol{w}_i^*\|^2] &= \left(1 - \frac{n}{p}\right)^t \|\boldsymbol{w}_0 - \boldsymbol{w}_i^*\|^2 + \sum_{k=1}^{t}\left(1 - \frac{n}{p}\right)^{t-k} \frac{n}{p}\|\boldsymbol{w}_k^* - \boldsymbol{w}_i^*\|^2 \\
&+ \frac{n\sigma^2}{p-n-1}\sum_{k=1}^{t}\left(1 - \frac{n}{p}\right)^{t-k} \\
&= \left(1 - \frac{n}{p}\right)^t \|\boldsymbol{w}_i^*\|^2 + \sum_{k=1}^{t}\left(1 - \frac{n}{p}\right)^{t-k} \frac{n}{p}\|\boldsymbol{w}_k^* - \boldsymbol{w}_i^*\|^2 \\
&+ \frac{n\sigma^2}{p-n-1}\sum_{k=1}^{t}\left(1 - \frac{n}{p}\right)^{t-k} \quad (\text{since } \boldsymbol{w}_0 = \boldsymbol{0}). \qquad (2.33)
\end{aligned}$$

Let $t = i$ in Eq. (2.33). We have

$$\begin{aligned}
\mathbb{E}[\|\boldsymbol{w}_i - \boldsymbol{w}_i^*\|^2] &= \left(1 - \frac{n}{p}\right)^i \|\boldsymbol{w}_i^*\|^2 + \sum_{k=1}^{i}\left(1 - \frac{n}{p}\right)^{i-k} \frac{n}{p}\|\boldsymbol{w}_k^* - \boldsymbol{w}_i^*\|^2 \\
&+ \frac{n\sigma^2}{p-n-1}\sum_{k=1}^{i}\left(1 - \frac{n}{p}\right)^{i-k}. \qquad (2.34)
\end{aligned}$$

Based on Eqs. (2.33) and (2.34), we can obtain the closed form of $\mathbb{E}[F_T]$:

$$\begin{aligned}
&\mathbb{E}[F_T] \\
=&\frac{1}{T-1}\sum_{i=1}^{T-1}\mathbb{E}\left[\|\boldsymbol{w}_T - \boldsymbol{w}_i^*\|^2 - \|\boldsymbol{w}_i - \boldsymbol{w}_i^*\|^2\right] \\
=&\frac{1}{T-1}\sum_{i=1}^{T-1}\left\{\left(1-\frac{n}{p}\right)^T\|\boldsymbol{w}_i^*\|^2 + \sum_{k=1}^{T}\left(1-\frac{n}{p}\right)^{T-k}\frac{n}{p}\|\boldsymbol{w}_k^* - \boldsymbol{w}_i^*\|^2 + \frac{n\sigma^2}{p-n-1}\sum_{k=1}^{T}\left(1-\frac{n}{p}\right)^{T-k}\right. \\
&\left. - \left(1-\frac{n}{p}\right)^i\|\boldsymbol{w}_i^*\|^2 - \sum_{k=1}^{i}\left(1-\frac{n}{p}\right)^{i-k}\frac{n}{p}\|\boldsymbol{w}_k^* - \boldsymbol{w}_i^*\|^2 - \frac{n\sigma^2}{p-n-1}\sum_{k=1}^{i}\left(1-\frac{n}{p}\right)^{i-k}\right\} \\
=&\frac{1}{T-1}\sum_{i=1}^{T-1}\left\{\left[\left(1-\frac{n}{p}\right)^T - \left(1-\frac{n}{p}\right)^i\right]\|\boldsymbol{w}_i^*\|^2 + \sum_{k=1}^{i}\frac{n}{p}\left[\left(1-\frac{n}{p}\right)^{T-k} - \left(1-\frac{n}{p}\right)^{i-k}\right]\|\boldsymbol{w}_k^* - \boldsymbol{w}_i^*\|^2\right. \\
&+ \sum_{k=i+1}^{T}\frac{n}{p}\left(1-\frac{n}{p}\right)^{T-k}\|\boldsymbol{w}_k^* - \boldsymbol{w}_i^*\|^2 + \frac{n\sigma^2}{p-n-1}\sum_{k=1}^{i}\left[\left(1-\frac{n}{p}\right)^{T-k} - \left(1-\frac{n}{p}\right)^{i-k}\right] \\
&\left. + \frac{n\sigma^2}{p-n-1}\sum_{k=i+1}^{T}\left(1-\frac{n}{p}\right)^{T-k}\right\} \\
=&\frac{1}{T-1}\sum_{i=1}^{T-1}\left\{\left[\left(1-\frac{n}{p}\right)^T - \left(1-\frac{n}{p}\right)^i\right]\|\boldsymbol{w}_i^*\|^2 + \sum_{j>i}^{T} c_{i,j}\|\boldsymbol{w}_i^* - \boldsymbol{w}_j^*\|^2\right. \\
&\left. + \frac{n\sigma^2}{p-n-1}\sum_{k=1}^{i}\left[\left(1-\frac{n}{p}\right)^{T-k} - \left(1-\frac{n}{p}\right)^{i-k}\right] + \frac{n\sigma^2}{p-n-1}\sum_{k=i+1}^{T}\left(1-\frac{n}{p}\right)^{T-k}\right\}
\end{aligned}$$

$$
\begin{aligned}
&= \frac{1}{T-1}\sum_{i=1}^{T-1}\left\{\left[\left(1-\frac{n}{p}\right)^{T}-\left(1-\frac{n}{p}\right)^{i}\right]\|\boldsymbol{w}_i^*\|^2+\sum_{j>i}^{T}c_{i,j}\|\boldsymbol{w}_i^*-\boldsymbol{w}_j^*\|^2\right.\\
&\quad\left.+\frac{n\sigma^2}{p-n-1}\left[\sum_{k=1}^{T}\left(1-\frac{n}{p}\right)^{T-k}-\sum_{k=1}^{i}\left(1-\frac{n}{p}\right)^{i-k}\right]\right\}\\
&= \frac{1}{T-1}\sum_{i=1}^{T-1}\left\{\left[\left(1-\frac{n}{p}\right)^{T}-\left(1-\frac{n}{p}\right)^{i}\right]\|\boldsymbol{w}_i^*\|^2+\sum_{j>i}^{T}c_{i,j}\|\boldsymbol{w}_i^*-\boldsymbol{w}_j^*\|^2\right.\\
&\quad\left.+\frac{n\sigma^2}{p-n-1}\left[\frac{1-\left(1-\frac{n}{p}\right)^{T}}{1-\left(1-\frac{n}{p}\right)}-\frac{1-\left(1-\frac{n}{p}\right)^{i}}{1-\left(1-\frac{n}{p}\right)}\right]\right\}\\
&= \frac{1}{T-1}\sum_{i=1}^{T-1}\left\{\left[\left(1-\frac{n}{p}\right)^{T}-\left(1-\frac{n}{p}\right)^{i}\right]\|\boldsymbol{w}_i^*\|^2+\sum_{j>i}^{T}c_{i,j}\|\boldsymbol{w}_i^*-\boldsymbol{w}_j^*\|^2\right.\\
&\quad\left.+\frac{n\sigma^2}{p-n-1}\frac{p}{n}\left[\left(1-\left(1-\frac{n}{p}\right)^{T}\right)-\left(1-\left(1-\frac{n}{p}\right)^{i}\right)\right]\right\}\\
&= \frac{1}{T-1}\sum_{i=1}^{T-1}\left\{\left[\left(1-\frac{n}{p}\right)^{T}-\left(1-\frac{n}{p}\right)^{i}\right]\|\boldsymbol{w}_i^*\|^2+\sum_{j>i}^{T}c_{i,j}\|\boldsymbol{w}_i^*-\boldsymbol{w}_j^*\|^2\right.\\
&\quad\left.+\frac{p\sigma^2}{p-n-1}\left[\left(1-\frac{n}{p}\right)^{i}-\left(1-\frac{n}{p}\right)^{T}\right]\right\}\\
&= \frac{1}{T-1}\sum_{i=1}^{T-1}\left\{(r^T-r^i)\|\boldsymbol{w}_i^*\|^2+\sum_{j>i}^{T}c_{i,j}\|\boldsymbol{w}_i^*-\boldsymbol{w}_j^*\|^2+\frac{p\sigma^2}{p-n-1}\left(r^i-r^T\right)\right\}.
\end{aligned}
$$

Based on Eq. (2.33), we can also obtain the exact form of the generalization error. Specifically,

$$
\begin{aligned}
&\mathbb{E}[\|\boldsymbol{w}_T-\boldsymbol{w}_i^*\|^2]\\
&=\left(1-\frac{n}{p}\right)^{T}\|\boldsymbol{w}_i^*\|^2+\sum_{k=1}^{T}\frac{n}{p}\left(1-\frac{n}{p}\right)^{T-k}\|\boldsymbol{w}_k^*-\boldsymbol{w}_i^*\|^2+\frac{n\sigma^2}{p-n-1}\sum_{k=1}^{T}\left(1-\frac{n}{p}\right)^{T-k},
\end{aligned}
$$

such that

$$
\begin{aligned}
\mathbb{E}[G_T] &= \frac{1}{T}\sum_{i=1}^{T}\mathbb{E}[\|\boldsymbol{w}_T-\boldsymbol{w}_i^*\|^2]\\
&= \frac{1}{T}\left(1-\frac{n}{p}\right)^{T}\sum_{i=1}^{T}\|\boldsymbol{w}_i^*\|^2+\frac{1}{T}\sum_{k=1}^{T}\frac{n}{p}\left(1-\frac{n}{p}\right)^{T-k}\sum_{i=1}^{T}\|\boldsymbol{w}_k^*-\boldsymbol{w}_i^*\|^2\\
&\quad+\frac{n\sigma^2}{p-n-1}\sum_{k=1}^{T}\left(1-\frac{n}{p}\right)^{T-k}\\
&= \frac{1}{T}\left(1-\frac{n}{p}\right)^{T}\sum_{i=1}^{T}\|\boldsymbol{w}_i^*\|^2+\frac{1}{T}\sum_{k=1}^{T}\frac{n}{p}\left(1-\frac{n}{p}\right)^{T-k}\sum_{i=1}^{T}\|\boldsymbol{w}_k^*-\boldsymbol{w}_i^*\|^2
\end{aligned}
$$

$$
\begin{aligned}
&+\frac{n\sigma^2}{p-n-1}\frac{1-\left(1-\frac{n}{p}\right)^T}{1-\left(1-\frac{n}{p}\right)} \\
=&\frac{1}{T}\left(1-\frac{n}{p}\right)^T\sum_{i=1}^{T}\|\boldsymbol{w}_i^*\|^2+\frac{1}{T}\sum_{k=1}^{T}\frac{n}{p}\left(1-\frac{n}{p}\right)^{T-k}\sum_{i=1}^{T}\|\boldsymbol{w}_k^*-\boldsymbol{w}_i^*\|^2 \\
&+\frac{p\sigma^2}{p-n-1}\left[1-\left(1-\frac{n}{p}\right)^T\right] \\
=&\frac{r^T}{T}\sum_{i=1}^{T}\|\boldsymbol{w}_i^*\|^2+\frac{1}{T}\sum_{i=1}^{T}\frac{nr^{T-i}}{p}\sum_{k=1}^{T}\|\boldsymbol{w}_k^*-\boldsymbol{w}_i^*\|^2+\frac{p\sigma^2}{p-n-1}\left(1-r^T\right).
\end{aligned}
$$

Proof of Proposition 2.3 Based on Theorem 2.1, it follows that

$$
\begin{aligned}
\mathbb{E}[F_2]=&(r^2-r)\|\boldsymbol{w}_1^*\|^2+\frac{n}{p}\|\boldsymbol{w}_1^*-\boldsymbol{w}_2^*\|^2+\frac{nr\sigma^2}{p-n-1} \\
=&-\left(1-\frac{n}{p}\right)\frac{n}{p}\|\boldsymbol{w}_{1,s}^*\|^2+\frac{n}{p}\|\boldsymbol{w}_{1,s}^*\|^2+\frac{n}{p}\|\boldsymbol{w}_{2,s}^*\|^2-2\frac{n}{p}\langle\boldsymbol{w}_{1,s}^*,\boldsymbol{w}_{2,s}^*\rangle+\frac{nr\sigma^2}{p-n-1} \\
=&\left(\frac{n}{p}\right)^2\|\boldsymbol{w}_{1,s}^*\|^2+\frac{n}{p}\|\boldsymbol{w}_{2,s}^*\|^2-2\frac{n}{p}\langle\boldsymbol{w}_{1,s}^*,\boldsymbol{w}_{2,s}^*\rangle+\frac{nr\sigma^2}{p-n-1}.
\end{aligned}
$$

When $\sigma^2<\frac{p-n-1}{p}\|\boldsymbol{w}_1^*\|^2$,

$$
\frac{n}{p}\|\boldsymbol{w}_1^*\|^2+\|\boldsymbol{w}_2^*\|^2+\frac{(p-n)\sigma^2}{p-n-1}\le\|\boldsymbol{w}_1^*\|^2+\|\boldsymbol{w}_2^*\|^2,
$$

such that $\mathbb{E}[F_2]\le 0$ if

$$
2\langle\boldsymbol{w}_{1,S_1}^*,\boldsymbol{w}_{2,S_2}^*\rangle\ge\frac{n}{p}\|\boldsymbol{w}_1^*\|^2+\|\boldsymbol{w}_2^*\|^2+\frac{(p-n)\sigma^2}{p-n-1}.
$$

Proof of Proposition 2.1 Without loss of generality, we assume that $\|\boldsymbol{w}_i^*-\boldsymbol{w}_j^*\|=1$ for task i in Category I and task j in Category II. It follows that

$$
\begin{aligned}
\tilde{F}_T(\boldsymbol{w}_T)=&\sum_{i<i^*}c_{i,i^*}+\sum_{j>i^*}c_{i^*,j} \\
=&(1-r)\left(\sum_{i=1}^{i^*-1}(r^{T-i}-r^{i^*-i}+r^{T-i^*})+\sum_{j=i^*+1}^{T}(r^{T-i^*}-r^{j-i^*}+r^{T-j})\right) \\
=&(1-r)\left((T-1)\cdot r^{T-i^*}+r^{T-i^*+1}\frac{r^{i^*-1}-1}{r-1}-r\frac{r^{i^*-1}-1}{r-1}+1-r^{T-i^*}\right) \\
=&(1-r)(T-2)r^{T-i^*}+(r^{T-i^*}-1)(1-r^{i^*-1})r+(1-r).
\end{aligned}
$$

Letting $\alpha := r^{T-i^*}$. Then minimizing $\tilde{F}_T(\boldsymbol{w}_T)$ is equivalent to minimize

$$
\begin{aligned}
&(1-r)(T-2)\alpha + (\alpha - 1)(1 - \frac{r^{T-1}}{\alpha})r \\
=&((1-r)(T-2)+r)\alpha + \frac{r^T}{\alpha} - r^T - r.
\end{aligned}
$$

By setting the derivative w.r.t. α to 0, we can have that the optimal value of α is

$$
\alpha = \sqrt{\frac{r^T}{T-2-(T-1)r}} \tag{2.35}
$$

which is clearly increasing with r. Therefore, the optimal order of the special task i^* is non-increasing with r, i.e., non-decreasing with $\frac{n}{p}$.

Proof of Proposition 2.2 Without loss of generality, we assume that for any task i and j

$$
\|\boldsymbol{w}_i^* - \boldsymbol{w}_j^*\|^2 = \begin{cases} 0, & \text{if task } i \text{ and } j \text{ are in the same category;} \\ 1, & \text{if task } i \text{ and } j \text{ are in the different categories.} \end{cases}
$$

Based on the closed form of forgetting, it suffices to minimize $\sum_{i=1}^{T-1}\sum_{j>i}^{T} c_{i,j}\|\boldsymbol{w}_i^* - \boldsymbol{w}_j^*\|^2$ in order to minimize the forgetting $F_T(\boldsymbol{w}_T)$, where $c_{i,j} = (1-r)(r^{T-i} - r^{j-i} + r^{T-j})$. Besides, since whenever we change the order between the i-th task and the j-th task, the value of $r^{T-i} + r^{T-j}$ does not change. In other words, only the term r^{j-i} affects the optimal task order, which should minimize $\sum_{i=1}^{T-1}\sum_{j>i}^{T}(-r^{j-i})\|\boldsymbol{w}_i^* - \boldsymbol{w}_j^*\|^2$.

(1) For the case $T = 4$, there are three effective task orders: (1) task $1 \in C_1$, task $2 \in C_1$, task $3 \in C_2$, task $4 \in C_2$ ((C_1, C_1, C_2, C_2) for simplicity); (2) (C_1, C_2, C_1, C_2); (3) (C_1, C_2, C_2, C_1). Swapping all tasks in C_1 with all tasks in C_2 does not change the value of forgetting, e.g., (C_1, C_1, C_2, C_2) has the same forgetting with (C_2, C_2, C_1, C_1). In what follows, we compare $\sum_{i=1}^{T-1}\sum_{j>i}^{T}(-r^{j-i})\|\boldsymbol{w}_i^* - \boldsymbol{w}_j^*\|^2$ among these three orders.

(a) For (C_1, C_1, C_2, C_2),

$$
\sum_{i=1}^{T-1}\sum_{j>i}^{T}(-r^{j-i})\|\boldsymbol{w}_i^* - \boldsymbol{w}_j^*\|^2 = -(r^2 + r^3 + r + r^2).
$$

(b) For (C_1, C_2, C_1, C_2),

$$
\sum_{i=1}^{T-1}\sum_{j>i}^{T}(-r^{j-i})\|\boldsymbol{w}_i^* - \boldsymbol{w}_j^*\|^2 = -(r + r^3 + r + r).
$$

(c) For (C_1, C_2, C_2, C_1),

$$\sum_{i=1}^{T-1}\sum_{j>i}^{T}(-r^{j-i})\|\boldsymbol{w}_i^* - \boldsymbol{w}_j^*\|^2 = -(r + r^2 + r + r^2).$$

It is clear that the alternating task order, i.e., (C_1, C_2, C_1, C_2) and (C_2, C_1, C_2, C_1), is the optimal order for this special case.

(2) For the case $T = 6$, based on the closed form of forgetting in Theorem 2.1, we can use computer programming to show that besides the perfectly alternating task order, i.e., $(C_1, C_2, C_1, C_2, C_1, C_2)$ and $(C_2, C_1, C_2, C_1, C_2, C_1)$, there are 10 effective task orders as illustrated in Table 2.2. We further evaluate the difference of forgetting between each task order in Table 2.2 and the perfectly alternating task order, where a positive difference means that the corresponding task order will lead a larger forgetting than the perfectly alternating task order. It can be verified that the difference of forgetting is positive for all the task orders in Table 2.2, which indicates that the optimal task order is the perfectly alternating task order.

Proof of Proposition 2.4 Following the same strategy with Special case II, we can have Table 2.3 to show all effective task orders and their difference of forgetting with the perfectly alternating task order, i.e., $(C_1, C_2, C_3, C_1, C_2, C_3)$ and its 'equivalent' task orders (e.g., $(C_1, C_3, C_2, C_1, C_3, C_2)$). It can also be verified that the perfectly alternating task order is the optimal task order in this case.

Table 2.2 Evaluation of the difference of forgetting between each effective task order and the perfectly alternating task order $(C_1, C_2, C_1, C_2, C_1, C_2)$, where a positive difference means that the corresponding task order will lead a larger forgetting than the perfectly alternating task order

Index	Order	Difference of forgetting
1	$(C_1, C_2, C_1, C_2, C_1, C_2)$	0
2	$(C_1, C_1, C_2, C_1, C_2, C_2)$	$r\left(2 - 2r + 2r^2 - 2r^3\right)$
3	$(C_1, C_1, C_2, C_2, C_1, C_2)$	$r\left(2 - 3r + 2r^2 - r^3\right)$
4	$(C_1, C_1, C_2, C_2, C_2, C_1)$	$r\left(3 - 3r - r^3 + r^4\right)$
5	$(C_1, C_2, C_2, C_1, C_1, C_2)$	$r\left(2 - 4r + 2r^2\right)$
6	$(C_1, C_2, C_2, C_1, C_2, C_1)$	$r\left(1 - 2r + 2r^2 - 2r^3 + r^4\right)$
7	$(C_1, C_1, C_1, C_2, C_2, C_2)$	$r\left(4 - 2r - 2r^3\right)$
8	$(C_1, C_2, C_1, C_2, C_2, C_1)$	$r\left(1 - 2r + 2r^2 - 2r^3 + r^4\right)$
9	$(C_1, C_2, C_1, C_1, C_2, C_2)$	$r\left(2 - 3r + 2r^2 - r^3\right)$
10	$(C_1, C_2, C_2, C_2, C_1, C_1)$	$r\left(3 - 3r - r^3 + r^4\right)$

Table 2.3 Evaluation of the difference of forgetting between each effective task order and the perfectly alternating task order $(C_1, C_2, C_3, C_1, C_2, C_3)$, where a positive difference means that the corresponding task order will lead a larger forgetting than the perfectly alternating task order

Index	Order	Difference of forgetting
1	$(C_1, C_2, C_3, C_1, C_2, C_3)$	0
2	$(C_1, C_2, C_1, C_2, C_3, C_3)$	$r\left(1+2r-3r^2\right)$
3	$(C_1, C_2, C_2, C_3, C_3, C_1)$	$r\left(2-3r^2+r^4\right)$
4	$(C_1, C_2, C_1, C_3, C_2, C_3)$	$r^2(2-2r)$
5	$(C_1, C_2, C_3, C_2, C_1, C_3)$	$r^2\left(1-2r+r^2\right)$
6	$(C_1, C_2, C_3, C_1, C_3, C_2)$	$r^2\left(1-2r+r^2\right)$
7	$(C_1, C_1, C_2, C_3, C_2, C_3)$	$r\left(1+2r-3r^2\right)$
8	$(C_1, C_2, C_2, C_1, C_3, C_3)$	$r\left(2-2r^2\right)$
9	$(C_1, C_1, C_2, C_2, C_3, C_3)$	$r\left(3-3r^2\right)$
10	$(C_1, C_2, C_1, C_3, C_3, C_2)$	$r\left(1+r-3r^2+r^3\right)$
11	$(C_1, C_2, C_3, C_3, C_1, C_2)$	$r\left(1-3r^2+2r^3\right)$
12	$(C_1, C_2, C_3, C_3, C_2, C_1)$	$r\left(1-2r^2+r^4\right)$
13	$(C_1, C_2, C_2, C_3, C_1, C_3)$	$r\left(1+r-3r^2+r^3\right)$
14	$(C_1, C_1, C_2, C_3, C_3, C_2)$	$r\left(2-2r^2\right)$
15	$(C_1, C_2, C_3, C_2, C_3, C_1)$	$r^2\left(2-3r+r^3\right)$

Proof of Theorem 2.2 *Intuitive explanation of Theorem* 2.2*:* In the underparameterized region, minimizing the loss Eq. (2.3) for the current task t will lead to a unique solution for this task, which does not depend on the learning process and the learned model of previous tasks. That is to say, the task learning is independent among all tasks, such that (i) the learning order of the first $T-1$ tasks does not matter, and (ii) both forgetting and generalization performance depend only on the model distance between the last task and the other tasks, i.e., $\sum_{i=1}^{T-1}\|\boldsymbol{w}_T^* - \boldsymbol{w}_i^*\|^2$.

Now we formally prove Theorem 2.2.

For the underparameterized regime, the solution of minimizing the training loss is

$$
\begin{aligned}
\boldsymbol{w}_t =&(\boldsymbol{X}_t\boldsymbol{X}_t^\top)^{-1}\boldsymbol{X}_t\boldsymbol{y}_t\\
=&(\boldsymbol{X}_t\boldsymbol{X}_t^\top)^{-1}\boldsymbol{X}_t\left(\boldsymbol{X}_t^\top\boldsymbol{w}_t^* + \boldsymbol{z}_t\right)\\
=&\boldsymbol{w}_t^* + (\boldsymbol{X}_t\boldsymbol{X}_t^\top)^{-1}\boldsymbol{X}_t\boldsymbol{z}_t.
\end{aligned}
$$

It follows that

$$\boldsymbol{w}_T - \boldsymbol{w}_i^* = \boldsymbol{w}_T^* - \boldsymbol{w}_i^* + (\boldsymbol{X}_T\boldsymbol{X}_T^\top)^{-1}\boldsymbol{X}_T\boldsymbol{z}_T,$$

such that the model error for the i-th task can be represented as:

$$\|\boldsymbol{w}_T - \boldsymbol{w}_i^*\|^2 = \|\boldsymbol{w}_T^* - \boldsymbol{w}_i^*\|^2 + \|(\boldsymbol{X}_T\boldsymbol{X}_T^\top)^{-1}\boldsymbol{X}_T\boldsymbol{z}_T\|^2.$$

By taking expectation on both sides, we can have

$$\mathbb{E}\|\boldsymbol{w}_T - \boldsymbol{w}_i^*\|^2 = \|\boldsymbol{w}_T^* - \boldsymbol{w}_i^*\|^2 + \frac{p\sigma^2}{n-p-1}.$$

Therefore, it can be shown that

$$\mathbb{E}[G_T] = \mathbb{E}\frac{1}{T}\sum_{i=1}^{T}\|\boldsymbol{w}_T - \boldsymbol{w}_i^*\|^2 = \left(\frac{1}{T}\sum_{i=1}^{T}\|\boldsymbol{w}_T^* - \boldsymbol{w}_i^*\|^2\right) + \frac{p\sigma^2}{n-p-1}$$

and

$$\begin{aligned}\mathbb{E}[F_T] =& \frac{1}{T-1}\sum_{i=1}^{T-1}\mathbb{E}\left[\|\boldsymbol{w}_T - \boldsymbol{w}_i^*\|^2 - \|\boldsymbol{w}_i - \boldsymbol{w}_i^*\|^2\right] \\ =& \frac{1}{T-1}\sum_{i=1}^{T-1}\|\boldsymbol{w}_T^* - \boldsymbol{w}_i^*\|^2.\end{aligned}$$

2.3 Discussions on Experimental Studies on Continual Learning

On the other hand, there have been extensive experimental studies on CL in order to address the catastrophic forgetting problem. Generally, the proposed CL algorithms can be divided into several different categories.

Regularization-based methods Regularization-based methods protect the knowledge of old tasks by adding explicit regularization terms in the loss function to balance between remembering old tasks and learning new tasks. This is usually done by regularizing the model change on the important weights to old tasks. Notably, in the seminal work [171], a regularization method named as Elastic Weight Consolidation (EWC) is proposed to determine the weight importance by leveraging Fisher information matrix (FIM). Motivated by the complex molecular machinery from biological neural networks, Synaptic Intelligence (SI) [391] maintains an online estimate of the weight importance by tracking the difference between the past and the current weight value. Memory Aware Synapses (MAS) [12] calculates the weight importance in an unsupervised and online manner, by evaluating the model outputs sensitivity to the change of weights for a given sample. To address the issue that the FIM is generally not diagonal, violating the assumption in EWC, R-EWC [225] is proposed to approximately diagonalize the FIM through a factorized rotation of the network parameter space. XK-FAC [193] developes a more practical Hessian approximation method based on a Kronecker-factored approximate curvature, by taking inter-sample relations into consideration, for CL with batch normalization. The idea of knowledge distillation (KD) is also widely used in regularization-based CL methods, where the previously learned model is the teacher and the currently trained model is the student. For instance, LwM [83] introduces an information preserving penalty based on attention distillation loss to penalize the changes in the attention maps for new task learning.

Expansion-based methods Expansion-based methods (e.g., [147, 208, 211, 281, 282, 336, 380]) dynamically expand the network capacity to reduce the interference between the new tasks and the old ones. Progressive Neural Network (PNN) [282] expands the network architecture for new tasks and preserves the weights of old tasks. Learning Without Forgetting (LWF) [211] splits the model layers into two parts, i.e., the shared part co-used by all tasks, and the task-specific part which grows for new tasks. Dynamic-Expansion Net (DEN) [380] and Compacting-Picking-Growing (CPG) [147] combine the strategies of model compression/pruning, weight selection and model expansion. In order to find the optimal structure for each of the sequential tasks, Reinforced Continual Learning (RCL) [365] leverages reinforcement learning and [208] adapts architecture search. APD [381] adds additional task-specific parameters for each task and selectively learns the task-shared parameters. PathNet [102] dynamically selects and combines modules from a pool of modules to create task-specific pathways, allowing efficient learning and avoiding interference.

Memory-based methods Depending on if data of old tasks is utilized when learning new tasks, memory-based methods can be further divided into the following two categories. *1) Experience-replay based methods.* This class of methods replays the old tasks data along with the current task data to mitigate catastrophic forgetting. Gradient Episodic Memory (GEM) [227] and Averaged GEM (A-GEM) [49] alter the current gradient based on the gradient computed with data in the memory. A unified view of episodic memory based methods is proposed in [124], based on new approaches are developed to balance between old tasks and the new task. Tiny episodic memory is considered in [50] and meta-learning is leveraged in [278]. iCaRL [277] maintains a subset of exemplars from each class and uses these exemplars for rehearsal during training on new tasks. It also integrates a nearest-class-mean classifier for incremental learning. Generative replay [301] trains a generative model to produce synthetic data that resembles past experiences, enabling rehearsal without storing the original data. Unlike traditional replay methods that store and replay raw input data from previous tasks, Latent Replay [250] stores and replays activations from an intermediate layer of the neural network. These activations represent the internal representations learned by the network for previous tasks, with the idea being that the first part of the model is frozen but later layers are trainable. *2) Orthogonal-projection based method.* To eliminate the need of storing data of old tasks, recently a series work [100, 283, 390] updates the model in the orthogonal direction of old tasks, and has shown remarkable performance. Particularly,

Orthogonal Weight Modulation (OWM) [390] learns a projector matrix to multiply with the new gradients. Orthogonal Gradient Descent (OGD) [100] stores the gradient directions of old tasks and projects the new gradients on the directions orthogonal to the subspace spanned by the old gradients. Gradient Projection Memory (GPM) [283] stores the bases of the subspaces spanned by old task data and projects the new gradients on the directions orthogonal to these subspaces.

2.4 Conclusion

In this chapter, we introduced one of the key techniques for enabling continuous AI model training at the network edge, namely continual learning. We first discussed the theoretical development behind CL, where we presented one of our own studies in building the theoretical foundations of CL. By investigating CL in the overparameterized linear models where each task is a linear regression problem and solved by using SGD, we derived the exact forms of both forgetting and generalization error, which built the key foundations of understanding the performance of CL. In particular, we investigated the impact of overparameterization, task similarity, and task ordering on both forgetting and generalization error. Experimental results on real datasets with DNNs indicated that our findings in linear models can even be carried over to CL in practice and leveraged to develop better algorithms. In the end, we also gave a brief overview of the recent advances in experimental studies on CL.

3 Reinforcement Learning for Edge AI

In this chapter on Reinforcement Learning for Edge AI, we explore Warm-Start Reinforcement Learning (RL), which integrates offline training to boost the efficiency of online learning in real-world applications. Warm-Start RL leverages a prior policy, derived from offline training, to give the learning process a significant head start. This method holds promise for edge AI systems, where computational resources are limited and fast adaptation to dynamic environments is essential. However, despite its potential, Warm-Start RL faces challenges in achieving consistent improvements. As empirical studies have shown, while performance can improve rapidly in some cases, it may also stagnate, especially when function approximation is involved.

To address these challenges, this chapter investigates whether and when Warm-Start RL can truly accelerate online learning. We specifically focus on the widely used Actor-Critic (A-C) method and examine how errors, encompassing function approximation errors, optimization errors, and estimation errors, affect both the Actor and Critic updates. *For ease of exposition, we refer to these errors together as approximation errors, acknowledging the potential imprecision in terminology.* By casting the Warm-Start A-C algorithm as a form of Newton's method with perturbation, we analyze the influence of these errors on the learning process, deriving upper bounds that highlight the conditions under which finite-time learning performance can be achieved. Our findings reveal that reducing algorithm bias and controlling error propagation are key to optimizing the Warm-Start A-C approach, making it a valuable tool for edge AI systems that require both rapid adaptation and computational efficiency.

H. Wang et al., *Continual and Reinforcement Learning for Edge AI*,
Synthesis Lectures on Learning, Networks, and Algorithms,
https://doi.org/10.1007/978-3-031-84363-1_3

3.1 Introduction

Online reinforcement learning (RL) [158, 321] often faces the formidable challenge of high sample complexity and intensive computational cost [182, 364], which hinders its applicability in real-world tasks. Indeed, this is the case in portfolio management [63], vehicles control [298, 359] and other time-sensitive settings [111, 209]. To tackle this challenge, Warm-Start RL has recently garnered much attention [113, 238, 332], by enabling online policy adaptation from an initial policy pre-trained using offline data (e.g., via behavior cloning or offline RL). One main insight of Warm-Start RL is that online learning can be significantly accelerated, thanks to the bootstrapping by an initial policy.

Despite the encouraging empirical successes [303, 304, 332], a fundamental understanding of the learning performance of Warm-Start RL is lacking, especially in the practical settings with function approximation by neural networks. In this work, we focus on the widely used Actor-Critic (A-C) method [120, 261], which combines the merits of both policy iteration and value iteration approaches [321] and has great potential for RL applications [332]. Notably, in the framework of abstract dynamic programming (ADP) [33], the policy iteration method [322] has been studied extensively, for warm-start learning *under the assumption of accurate updates*. In such a setting, policy iteration can be regarded as a second-order method in convex optimization [118] from the perspective of ADP, and can achieve *super-linear* convergence rate [39, 264, 287]. Nevertheless, when the A-C method is implemented in practical applications, *the approximation errors are inevitable in the Actor/Critic updates* due to many implementation issues, including function approximation using neural networks, the finite sample size, and the finite number of gradient iterations. Moreover, the error propagation from iteration to iteration may exacerbate the 'slowing down' of the convergence and have intricate impact therein. Clearly, the (stochastic) accumulated errors may throttle the convergence rate significantly and degrade the learning performance dramatically [74, 99, 108, 188]. Thus, it is of great importance to characterize the learning performance of Warm-Start RL in practical scenarios; and the primary objective of this study is to take steps to build a fundamental understanding of the impact of the approximation errors on the finite-time sub-optimality gap for the Warm-Start A-C algorithm, i.e.,

Whether and when online learning can be significantly accelerated by a warm-start policy from offline RL?

To this end, we address the question in two steps:

(1) We first focus on the characterization of the approximation errors via finite time analysis, based on which we quantify its impact on the sub-optimality gap of the A-C algorithm in Warm-Start RL. In particular, we analyze the A-C algorithm in a more realistic setting where the samples are Markovian in the rollout trajectories for the Critic update (different from the widely used i.i.d. assumption). Further, we consider that the Actor update and the Critic update take place on the *single-time scale*, indicating that the time-scale decomposition is not applicable to the finite-time analysis here. We tackle these challenges using recent advances on Bernstein's Inequality for Markovian samples [98, 154]. By delving into the coupling due to the interleaved updates of the Actor and the Critic, we provide upper bounds on the approximation errors in the Critic update and the Actor update of online exploration, respectively, from which we pinpoint the root causes of the approximation errors.

(2) We analyze the impact of the approximation errors on the finite-time learning performance of Warm-Start A-C. Based on the approximation error characterization, we treat the Warm-Start A-C algorithm as Newton's method with perturbation, and study the impact of the approximation errors on the finite-time learning performance of Warm-Start A-C. We first establish the upper bound of the bias term in the perturbation. Then we derive the upper bounds on the learning performance gap for both biased and unbiased cases. Our findings reveal that it is essential to reduce the algorithm bias in online learning. When the approximation errors are biased, we derive lower bounds on the sub-optimality gap, which reveals that even with a sufficiently good warm-start, the performance gap of online policy adaptation to the optimal policy is still bounded away from zero when the biases are not negligible. We remark that the primary objective of this work is to understand the convergence behavior, which is essential before answering further questions related to the convergence rate and sampling complexity.

3.1.1 Related Work

(Warm-Start RL) The Warm-start RL considered in our work has the same setup as in Bertsekas [33] and recent successful applications including AlphaZero [303], where the offline pretrained policy is utilized as the initialization for online learning and this policy is *updated* while interacting with the MDP online. In a line of very recent works [125, 150, 168] on Warm-Start RL, the policy is initialized via behavior cloning from offline data and then is fine-tuned with online reinforcement learning. A variant of this scheme is proposed in Advanced Weighted Actor Critic [238] which enables quick learning of skills across a suite of benchmark tasks. In the same spirit, Offline-Online Ensemble [196] leverages multiple Q-functions trained pessimistically offline as the initial function approximation for online learning. However, we remark the theoretical characterization of the finite-time performance of Warm-Start RL is still lacking. Our work aims to take steps to quantify the impact of approximation error on online RL with a warm-start policy.

In particular, it is worth to mention that some works [24, 332, 364] consider a different warm-start setting from ours. For instance, Xie et al. [364] considers the case where the reference policy is used to collect samples but remains *fixed* during the online learning. Under this setting, Xie et al. [364] provides a quantitative understanding on the policy fine-tuning problem in episodic Markov Decision Processes (MDPs) and establishes the lower bound for the sample complexity, where function approximation is not used. Jump-start RL [332] utilizes a guided-policy to initialize online RL in the early phase with a separate online exploration-policy.

Meanwhile, we remark the major differences from "offline-focus" works, which aim to derive conditions on the quality of the offline part in the warm-start RL, e.g., coverage. Notably, the focus of Wagenmaker and Pacchiano [344] and Song et al. [315] is on the offline policy quality while requiring the online learning part to satisfy certain conditions (either through delicate design or assumptions), e.g., Song et al. [315] requires the Bellman error to be upper bounded and Wagenmaker and Pacchiano [344] requires the online exploration to satisfy certain conditions. In [364], the online algorithm needs to output a lower value estimate which is not available in standard online RL algorithms. On the contrary, motivated by recent empirical studies, which have demonstrated that a "good" warm-start policy does not necessary improve the online learning performance, especially when the function approximation is used [238, 332], we consider the widely used Actor-Critic (A-C) method for online learning and aim to build a deep understanding on how the approximation errors in the *online* Actor and Critic step has impact on the learning performance. Furthermore, we summarize the comparison between our work and related work in Table 3.1. The detailed comparison in terms of the assumptions on the MDP and the function approximation is available in Sect. 3.1.

(Actor-Critic as Newton's Method) The intrinsic connection between the A-C method and Newton's method can be traced back to the convergence analysis of policy iteration in MDPs with continuous action spaces [264]. The connection is further examined later in a special MDP with discretized continuous state space [287]. Recent work [34] points out that the success of Warm-Start RL, e.g., AlphaZero, can be attributed to the equivalence between policy iteration and Newton's method in the ADP framework, which leads to the superlinear convergence rate for online policy adaptation. Under the generalized differentiable assumption, it has also been proved theoretically that policy iteration is the instances of semi-smooth Newton-type methods to solve the Bellman equation [112]. While some

Table 3.1 Related work in terms of (1) Warm-start setting, (2) Actor function approximation and (3) Critic function approximation

Paper	Warm-start	Actor	Critic
Munos [237]			√
Farahmand et al. [99]		√	√
Lazaric et al. [188]		√	√
Fu et al. [107]		√	√
Xie et al. [364]	√		
Bertsekas [34]	√		√
This work	√	√	√

prior works [118] have provided theoretical investigation of the connections between policy iteration and Newton's Method, the studies are carried out in the abstract dynamic programming (ADP) framework, assuming accurate updates in iterations. Departing from the ADP framework, this work treats the A-C algorithm as Newton's method in the presence of approximation errors, and focuses on the finite-time learning performance of Warm-Start RL.

(Finite-time analysis for Actor-Critic methods) Among the existing works on the finite time analysis of A-C methods with function approximation, Yang et al. [374] establishes the global convergence under the linear quadratic regulator. Kumar et al. [183] considers the sample complexity under i.i.d. assumptions where the Actor update and Critic update can be 'decoupled'. Khodadadian et al. [165] considers the two-timescale setting with Markovian samples. Fu et al. [107] focuses on the more general single-time scale setting but constrains the policy function approximation in the energy based function class. While the analysis in approximate policy/value iteration [99, 188, 237] present the error propagation in the upper bound, it is unclear how the error from each update step behave. In this chapter, we provide the analysis on the approximation error for each learning step explicitly and based on which we establish the error propagation in both the upper bound and lower bound.

3.1.2 Comparison with Related Work

In this chapter, we consider the same warm-start RL setup as in Bertsekas [33] and recent successful applications including AlphaZero, where the offline pretrained policy is utilized as the initialization for online learning and this policy is updated while interacting with the MDP online. Policy improvement via online adaptation (finetuning) plays a critical role in addressing the notorious challenge of "distribution shift" between offline training and online learning, and this is one main motivation for our study on Warm-start RL. In stark contrast, the reference policy in Xie et al. [364] and Bagnell et al. [24] is used to collect samples but remains fixed during the online learning. It is clear that if one queries 0 samples from the reference policy, Algorithm 2 [364] would NOT reduce to the proposed warm-start learning algorithm in our setting. Meanwhile, Algorithms 1 and 2 [332] assume the episodic MDP setting, which is different from the MDP setting in our study. On the other hand, the hybrid RL setting [315, 344] mainly focuses on the usage of the offline dataset while the initial policy is not initialized by any warm-start policy (e.g., Algorithm 1 [315], Sect. 6 [344]).

Moreover, the finite time analysis with function approximation errors in both Actor and Critic updates has not been studied before under this warm-start RL setting. From a theoretic perspective, our work has contributed to developing a fundamental understanding of the impact of the function approximation errors in the general MDP settings (Ref. Table 3.2), beyond the references listed.

Table 3.2 Detailed Comparison with related work in terms of MDP and function approximation settings

Reference	Previous Work	Our Work
Xie et al. [364]	Episodic MDP setting and no function approximation error during the policy finetuning	We consider general MDP with the linear value function (Critic) approximation and general Actor function approximation
Bagnell et al. [24] and Uchendu et al. [332]	Bagnell et al. [24] and Uchendu et al. [332] gives the results with either the approximation error from policy update (Theorem 1 [24]) or value function update (Sect. 4.2 [24], Assumption A.6 [332]) through term ε	Our work first characterizes ε explicitly (which is not available in Bagnell et al. [24] and Uchendu et al. [332]) and also studies how the approximation errors from "both Actor update and Critic update" affect the learning performance at the same time (through bias term $\mathcal{B}(t)$)
Song et al. [315]	Song et al. [315] considers Q-function approximation and assume the greedy policy can be obtained exactly (line 3, Algorithm 1)	We consider Actor-Critic and consider the approximation error in the Actor and Critic, respectively
Wagenmaker and Pacchiano [344]	Wagenmaker and Pacchiano [344] requires the underlying MDP structure to be linear and only considers linear Softmax Policy (Actor)	We consider general MDP and general Actor approximation

3.2 Background

Markov Decision Processes. We consider a MDP defined by a tuple $(\mathcal{S}, \mathcal{A}, P, r, \gamma)$, where $\mathcal{S} = \{1, 2, \ldots, n\}$, $n < \infty$ and $\mathcal{A} = \{1, 2, \ldots, A\}$, $A < \infty$ represent the finite state space and finite action space, respectively. $P(s'|s, a) : \mathcal{S} \times \mathcal{A} \times \mathcal{S} \to [0, 1]$ is the probability of the transition from state s to state s' by applying action a and $r(s, a) : \mathcal{S} \times \mathcal{A} \to \mathbb{R}$ is the corresponding reward. $\gamma \in (0, 1)$ is the discount factor. At each step t, an agent moves from the current state s_t to next state s_{t+1} by taking an action a_t following the policy $\pi \in \Pi : \mathcal{S} \to \mathcal{A}$ and receives the reward r_t. In the Warm-Start RL, we assume that the initial policy π_0 is given, e.g., in the form of a neural network [209], and obtained by offline training. For brevity, we use bold symbols $\boldsymbol{r}_\pi \in \mathbb{R}^n : [r_\pi]_s = r(s, \pi(s))$ and $\boldsymbol{P}_\pi \in \mathbb{R}^{n\times n} : [P^\pi]_{s,s'} \triangleq P(s'|s, \pi(s))$ to denote the reward vector and the transition matrix induced by policy π. We further denote by $d^\pi : \mathcal{S} \to [0, 1]$ and $\rho^\pi : \mathcal{S} \times \mathcal{A} \to [0, 1]$ the stationary state distribution and state-action transition distribution induced by policy π. We use ρ_0 to represent the initial state distribution. We use $\|\cdot\|$ or $\|\cdot\|_2$ to represent the Euclidean norm.

Value Functions. For any policy π, define the value function $v^{\pi}(s) : \mathcal{S} \rightarrow \mathbb{R}$ as $v^{\pi}(s) = \mathbf{E}_{a_t \sim \pi(\cdot|s_t), s_{t+1} \sim P(\cdot|s_t, a_t)} \left[\sum_{t=0}^{\infty} \gamma^t r_t | s_0 = s \right]$ to measure the average accumulative reward staring from state s by following policy π. We define Q-function $Q^{\pi}(s, a) : \mathcal{S} \times \mathcal{A} \rightarrow \mathbb{R}$ as $Q^{\pi}(s, a) = \mathbf{E}[\sum_{t=0}^{\infty} \gamma^t r_t | s_0 = s, a_0 = a]$ to represent the expected return when the action a is chosen at the state s. By using the transition matrix and reward vector defined above, we have the compact form of the value function $\boldsymbol{v}^{\pi} = (\boldsymbol{I} - \gamma \boldsymbol{P}_{\pi})^{-1} \boldsymbol{r}_{\pi}$, where $\boldsymbol{I} \in \mathbb{R}^{n \times n}$ is the identity matrix and $\boldsymbol{v}^{\pi} \in \mathbb{R}^n$ is the value vector with the component-wise values $[v^{\pi}]_s \triangleq v^{\pi}(s)$, with

$$v^{\pi}(s) \triangleq \mathbf{E}_{a \sim \pi(\cdot|s)}[Q^{\pi}(s, a)].$$

The main objective is to find an optimal policy π^* such that the value function is maximized, i.e.,

$$\max_{\pi} \mathbf{E}_{s \sim \rho_0}[v^{\pi}(s)] \triangleq \max_{\pi} \mathbf{E}_{s \sim \rho_0, a \sim \pi(\cdot|s)}[Q^{\pi}(s, a)]. \tag{3.1}$$

In what follows, we use both Q-function and value function $v(s)$ for convenience, and the relation between the two is given in Eq. 3.1.

Bellman Operator. For $\boldsymbol{v} \in \mathbb{R}^n$, define the Bellman evaluation operator $T^{\pi} : \mathbb{R}^n \rightarrow \mathbb{R}^n$ and the Bellman operator $T : \mathbb{R}^n \rightarrow \mathbb{R}^n$ as

$$\begin{aligned} T^{\pi}(\boldsymbol{v}) &= \boldsymbol{r}_{\pi} + \gamma \boldsymbol{P}_{\pi} \boldsymbol{v}, \\ T(\boldsymbol{v}) &= \max_{\pi} \{\boldsymbol{r}_{\pi} + \gamma \boldsymbol{P}_{\pi} \boldsymbol{v}\} = \max_{\pi} T^{\pi}(\boldsymbol{v}). \end{aligned}$$

It is well known that the Bellman operator T is a contraction mapping and has order-preserving property. Note that the Bellman operator T may not be differentiable everywhere due to the max operator, and the value $\boldsymbol{v}^*$ of the optimal policy π^* is the only fixed point of the Bellman operator T [263]. From the definition of the Bellman Evaluation Operator T^{π}, we have $\boldsymbol{v}^{\pi}$ to be the fixed point of T^{π}, i.e., $\boldsymbol{v}^{\pi} = T^{\pi}(\boldsymbol{v}^{\pi})$.

3.2.1 Policy Iteration as Newton's Method in Abstract Dynamic Programming

Policy iteration carries out policy learning by alternating between two steps: policy improvement and policy evaluation. At time t, the policy evaluation step seeks to learn the value function $\boldsymbol{v}^{\pi_t}$ for the current policy π_t by solving the fixed point equation of the Bellman evaluation operator:

$$\boldsymbol{v} = T^{\pi_t}(\boldsymbol{v}).$$

Denote $\boldsymbol{v}_t = \boldsymbol{v}^{\pi_t}$ for simplicity. Then in the policy improvement step, a new policy π_{t+1} is obtained by maximizing the learnt value function $\boldsymbol{v}_t$ in the policy evaluation step, in a greedy manner, i.e.,

$$\pi_{t+1} = \arg\max_{\pi} T^{\pi}(\boldsymbol{v}_t). \tag{3.2}$$

To introduce the connection between policy iteration and Newton's Method, we first define operator $F : \boldsymbol{v} \to \boldsymbol{v} - T(\boldsymbol{v})$ for convenience. As in [118, 263], F can be treated as the "gradient" of an unknown function. Under the assumption that $F(\boldsymbol{v})$ is differentiable at $\boldsymbol{v}$, the Jacobian $\boldsymbol{J}_{\boldsymbol{v}}$ of F at $\boldsymbol{v}$ can be obtained as $\boldsymbol{J}_{\boldsymbol{v}} = I - \gamma \boldsymbol{P}_{\pi(\boldsymbol{v})}$, where $\pi(\boldsymbol{v}) \triangleq \arg\max_{\pi} T^{\pi}(\boldsymbol{v})$. Note that $\boldsymbol{J}_{\boldsymbol{v}}^{-1} = \sum_{i=1}^{\infty}(\gamma \boldsymbol{P}_{\pi(\boldsymbol{v})})^i$ is invertible [263]. Since it can be shown that $\boldsymbol{v}^{\pi_{t+1}} = (\boldsymbol{I} - \gamma \boldsymbol{P}_{\pi_{t+1}})^{-1} \boldsymbol{r}_{\pi_{t+1}} = \boldsymbol{J}_{\boldsymbol{v}^{\pi_t}}^{-1} \boldsymbol{r}_{\pi_{t+1}}$ for the policy evaluation of π_{t+1}, we have that,

$$\boldsymbol{v}^{\pi_{t+1}} = \boldsymbol{v}^{\pi_t} - \boldsymbol{J}_{\boldsymbol{v}^{\pi_t}}^{-1} F(\boldsymbol{v}^{\pi_t}), \tag{3.3}$$

which indicates that the analytic representation of policy iteration in the abstract dynamic programming framework reduces to Newton's Method. It is worth mentioning that the convergence behavior of policy iteration near the optimal value $\boldsymbol{v}^*$ cannot be directly obtained by using the results from convex optimization [39] since the Bellman operator T may not be differentiable at any given value vector $\boldsymbol{v}$.

Proof In waht follows, we provide the proof details on the illustrative example in Sect. 3.2.1. We use the notation defined in Fig. 3.1.

Policy Iteration as Newton's Method. Based on [118, 264], we first build the relation between policy iteration and Newton's Method in the abstract dynamic programming (ADP) framework, assuming accurate updates.

From the definition of the value function $\boldsymbol{v}$, we have that for any policy π,

$$\boldsymbol{v}^{\pi} = \boldsymbol{r}_{\pi} + \gamma \boldsymbol{P}_{\pi} \boldsymbol{v}^{\pi}.$$

Recall the definition of Bellman evaluation operator $T^{\pi}(\cdot)$ and the Bellman operator $T(\cdot)$,

$$T^{\pi}(\boldsymbol{v}) = \boldsymbol{r}_{\pi} + \gamma \boldsymbol{P}_{\pi} \boldsymbol{v}, \; T(\boldsymbol{v}) = \max_{\pi}\{\boldsymbol{r}_{\pi} + \gamma \boldsymbol{P}_{\pi} \boldsymbol{v}\} = \max_{\pi} T^{\pi}(\boldsymbol{v}).$$

It follows that

$$\begin{aligned}
\boldsymbol{v}^{\pi_{t+1}} &= \boldsymbol{J}_{\boldsymbol{v}^{\pi_t}}^{-1} \boldsymbol{r}_{\pi_{t+1}} \\
&= \boldsymbol{v}^{\pi_t} - \boldsymbol{v}^{\pi_t} + \boldsymbol{J}_{\boldsymbol{v}^{\pi_t}}^{-1} \boldsymbol{r}_{\pi_{t+1}} \\
&= \boldsymbol{v}^{\pi_t} - \boldsymbol{J}_{\boldsymbol{v}^{\pi_t}}^{-1} \boldsymbol{J}_{\boldsymbol{v}^{\pi_t}} \boldsymbol{v}^{\pi_t} + \boldsymbol{J}_{\boldsymbol{v}^{\pi_t}}^{-1} \boldsymbol{r}_{\pi_{t+1}} \\
&= \boldsymbol{v}^{\pi_t} - \boldsymbol{J}_{\boldsymbol{v}^{\pi_t}}^{-1} \left(-\boldsymbol{r}_{\pi_{t+1}} + \boldsymbol{J}_{\boldsymbol{v}^{\pi_t}} \boldsymbol{v}^{\pi_t}\right) \\
&= \boldsymbol{v}^{\pi_t} - \boldsymbol{J}_{\boldsymbol{v}^{\pi_t}}^{-1} \left(-\boldsymbol{r}_{\pi_{t+1}} + \left(\boldsymbol{I} - \gamma \boldsymbol{P}_{\pi_{t+1}}\right) \boldsymbol{v}^{\pi_t}\right) \\
&= \boldsymbol{v}^{\pi_t} - \boldsymbol{J}_{\boldsymbol{v}^{\pi_t}}^{-1} \left(\boldsymbol{v}^{\pi_t} - \boldsymbol{r}_{\pi_{t+1}} - \gamma \boldsymbol{P}_{\pi_{t+1}} \boldsymbol{v}^{\pi_t}\right) \\
&= \boldsymbol{v}^{\pi_t} - \boldsymbol{J}_{\boldsymbol{v}^{\pi_t}}^{-1} \left(\boldsymbol{v}^{\pi_t} - T\left(\boldsymbol{v}^{\pi_t}\right)\right),
\end{aligned} \tag{3.4}$$

where $\boldsymbol{J}_{\boldsymbol{v}} = \boldsymbol{I} - \gamma \boldsymbol{P}_{\pi(\boldsymbol{v})}$ and $\pi(\boldsymbol{v})$ attains the max in $T(\boldsymbol{v})$. Equation (3.4) establishes a connection between policy iteration under ADP and Newton's Method. Specifically, if we assume function $F : \boldsymbol{v} \rightarrow \boldsymbol{v} - T(\boldsymbol{v})$ is differentiable at any vector $\boldsymbol{v}$ visited by policy iteration, then we have $\boldsymbol{v}_{t+1} = \boldsymbol{v}_t + \boldsymbol{J}_{\boldsymbol{v}_t}^{-1} F(\boldsymbol{v}_t)$, which is exactly the update of the Newton's Method in convex optimization [39]. Due to the fact that $F(\cdot)$ may not be differentiable at all $\boldsymbol{v}$ in policy iteration, the assumptions on the Lipschitzness of $\boldsymbol{v} \rightarrow \boldsymbol{J}_{\boldsymbol{v}}$ is commonly used to prove the convergence of the policy iteration (see Assumption 3.5). Following the same line, next we show the case when function approximation is used in the A-C algorithm.

A-C Updates with Function Approximation. Next we outline the main differences between the A-C update with function approximation and the policy iteration in the ADP framework, and cast A-C based policy iteration with function approximation as Newton's Method with perturbation. Specifically,

$$\begin{aligned}
\boldsymbol{v}^{\widetilde{\pi}_{t+1}} &= \boldsymbol{J}_{\boldsymbol{v}^{\widetilde{\pi}_t}}^{-1} \boldsymbol{r}_{\widetilde{\pi}_{t+1}} \\
&= \boldsymbol{v}^{\pi_t} - \boldsymbol{v}^{\pi_t} + \boldsymbol{J}_{\boldsymbol{v}^{\widetilde{\pi}_t}}^{-1} \boldsymbol{r}_{\widetilde{\pi}_{t+1}} \\
&= \boldsymbol{v}^{\pi_t} - \boldsymbol{J}_{\boldsymbol{v}^{\widetilde{\pi}_t}}^{-1} \boldsymbol{J}_{\boldsymbol{v}^{\widetilde{\pi}_t}} \boldsymbol{v}^{\pi_t} + \boldsymbol{J}_{\boldsymbol{v}^{\widetilde{\pi}_t}}^{-1} \boldsymbol{r}_{\widetilde{\pi}_{t+1}} \\
&= \boldsymbol{v}^{\pi_t} - \boldsymbol{J}_{\boldsymbol{v}^{\widetilde{\pi}_t}}^{-1} \left(-\boldsymbol{r}_{\widetilde{\pi}_{t+1}} + \boldsymbol{J}_{\boldsymbol{v}^{\widetilde{\pi}_t}} \boldsymbol{v}^{\pi_t}\right) \\
&= \boldsymbol{v}^{\pi_t} - \boldsymbol{J}_{\boldsymbol{v}^{\widetilde{\pi}_t}}^{-1} \left(-\boldsymbol{r}_{\widetilde{\pi}_{t+1}} + \left(\boldsymbol{I} - \gamma \boldsymbol{P}_{\widetilde{\pi}_{t+1}}\right) \boldsymbol{v}^{\pi_t}\right) \\
&= \boldsymbol{v}^{\pi_t} - \boldsymbol{J}_{\boldsymbol{v}^{\widetilde{\pi}_t}}^{-1} \left(\boldsymbol{v}^{\pi_t} - (\boldsymbol{r}_{\widetilde{\pi}_{t+1}} + \gamma \boldsymbol{P}_{\widetilde{\pi}_{t+1}} \boldsymbol{v}^{\pi_t})\right) \\
&\triangleq \boldsymbol{v}^{\pi_t} - \boldsymbol{J}_{\boldsymbol{v}^{\widetilde{\pi}_t}}^{-1} \left(\boldsymbol{v}^{\pi_t} - \widetilde{T}\left(\boldsymbol{v}^{\pi_t}\right)\right),
\end{aligned}$$

where $\boldsymbol{J}_{\boldsymbol{v}^{\widetilde{\pi}_t}} = \boldsymbol{I} - \gamma \boldsymbol{P}_{\pi(\boldsymbol{v}^{\widetilde{\pi}_t})}$ and $\pi(\boldsymbol{v})$ attains the max in $T(\boldsymbol{v})$ (not $\widetilde{T}(v)$), with the following two operators defined as

$$\begin{aligned}
T(\boldsymbol{v}_t) &\triangleq \boldsymbol{r}_{\pi_{t+1}} + \gamma \boldsymbol{P}_{\pi_{t+1}} \boldsymbol{v}_t, \\
\widetilde{T}(\boldsymbol{v}_t) &\triangleq \boldsymbol{r}_{\widetilde{\pi}_{t+1}} + \gamma \boldsymbol{P}_{\widetilde{\pi}_{t+1}} \boldsymbol{v}_t.
\end{aligned}$$

For convenience, let $\mathcal{E}_{T,t}$ and $\mathcal{E}_{J,t}$ denote the approximation errors in the Bellman operator T and the Jacobian $\boldsymbol{J}_{\boldsymbol{v}}$, i.e.,

$$\begin{aligned}\widetilde{T}(\boldsymbol{v}_t) - T(\boldsymbol{v}_t) =& (\boldsymbol{r}_{\widetilde{\pi}_{t+1}} + \gamma \boldsymbol{P}_{\widetilde{\pi}_{t+1}} \boldsymbol{v}_t) - (\boldsymbol{r}_{\pi_{t+1}} + \gamma \boldsymbol{P}_{\pi_{t+1}} \boldsymbol{v}_t) \triangleq \mathcal{E}_{T,t},\\ \boldsymbol{J}_{\widetilde{\boldsymbol{v}}_t}^{-1} - \boldsymbol{J}_{\boldsymbol{v}_t}^{-1} =& (\boldsymbol{I} - \gamma \boldsymbol{P}_{\widetilde{\pi}_{t+1}})^{-1} - (\boldsymbol{I} - \gamma \boldsymbol{P}_{\pi_{t+1}})^{-1} \triangleq \mathcal{E}_{J,t},\end{aligned}$$

and define

$$\boldsymbol{v}^{\hat{\pi}_{t+1}} \triangleq \boldsymbol{v}^{\widetilde{\pi}_{t+1}} + \mathcal{E}_{a,t},$$

where $\mathcal{E}_{a,t}$ capture the error induced by inaccurate policy improvement (the greedy step, e.g., Eq. (3.6)) in the Actor update. Then we have that

$$\begin{aligned}\boldsymbol{v}^{\widetilde{\pi}_{t+1}} &= \boldsymbol{v}^{\pi_t} - \boldsymbol{J}_{\boldsymbol{v}^{\widetilde{\pi}_t}}^{-1} \left(\boldsymbol{v}^{\pi_t} - \widetilde{T}\left(\boldsymbol{v}^{\pi_t}\right)\right)\\ &= \boldsymbol{v}_t - (\boldsymbol{J}_{\boldsymbol{v}_t}^{-1} + \mathcal{E}_{J,t})\left(\boldsymbol{v}_t - T(\boldsymbol{v}_t) - \mathcal{E}_{T,t}\right)\\ &= \underbrace{\boldsymbol{v}_t - \boldsymbol{J}_{\boldsymbol{v}_t}^{-1}(\boldsymbol{v}_t - T(\boldsymbol{v}_t))}_{\text{Exact Newton Step}} \underbrace{-\mathcal{E}_{J,t}(\boldsymbol{v}_t - T(\boldsymbol{v}_t)) + (\boldsymbol{J}_{\boldsymbol{v}_t}^{-1} + \mathcal{E}_{J,t})\mathcal{E}_{T,t}}_{\text{Perturbation}}\\ &\triangleq \underbrace{\boldsymbol{v}_t - \boldsymbol{J}_{\boldsymbol{v}_t}^{-1}(\boldsymbol{v}_t - T(\boldsymbol{v}_t))}_{\text{Exact Newton Step}} + \mathcal{E}_t\\ &= \boldsymbol{v}^{\pi_{t+1}} + \mathcal{E}_t.\end{aligned}$$

In a nutshell, we have that

$$\boldsymbol{v}^{\hat{\pi}_{t+1}} = \boldsymbol{v}^{\pi_{t+1}} + \mathcal{E}_{c,t} + \mathcal{E}_{a,t},$$

where

$$\mathcal{E}_{c,t} \triangleq -\mathcal{E}_{J,t}(\boldsymbol{v}_t - T(\boldsymbol{v}_t)) + (\boldsymbol{J}_{\boldsymbol{v}_t}^{-1} + \mathcal{E}_{J,t})\mathcal{E}_{T,t}.$$

□

Off-policy A-C Algorithem as Newton's Method with Perturbation We note that the actor and critic updates in Eqs. (9) and (8) are a general template that admits both off- and on-policy method. More specifically, denote the target policy by π_{tar} and the behavior policy by π_{bhv}. When the off-policy method is used, then the updates in Eqs. (9) and (8) are given by

$$\omega_{t+1} \leftarrow \arg\min_{\omega} \mathbf{E}_{(s,a)\sim\rho^{\pi_{\text{bhv}}}} \left[Q_{\omega,\pi_{\text{tar}t+1}}(s,a) - \omega^{\top}\phi(s,a)\right]^2,$$

$$\pi_{t+1} \leftarrow \arg\max_{\pi} \mathbf{E}_{(s,a)\sim\rho^{\pi_{\text{bhv}}}} \left[Q_{\omega_{t+1},\pi_{\text{tar},t}}(s,a)\right].$$

This is in contrast to the updates given below when the on-policy method is used:

$$\omega_{t+1} \leftarrow \arg\min_{\omega} \mathbf{E}_{(s,a)\sim\rho^{\pi_{\text{tar}}}} \left[Q_{\omega,\pi_{\text{tar}t+1}}(s,a) - \omega^\top \phi(s,a) \right]^2,$$

$$\pi_{t+1} \leftarrow \arg\max_{\pi} \mathbf{E}_{(s,a)\sim\rho^{\pi_{\text{tar}}}} \left[Q_{\omega_{t+1},\pi_{\text{tar},t}}(s,a) \right].$$

- One major challenge of the off-policy analysis lies in the fact that the behavior policy can be arbitrary [321, 322] and hence it is impossible to develop a unifying framework. For example, the behavior policy can be obtained by human demonstration (a similar idea is used in an early version of AlphaGo), deriving from the target policy as in Q-learning/DQN or from a previous behavior policy. Meanwhile, the key drawback of off-policy method is that it does not stably interact with the function approximation and is generally of greater variance and slower convergence rate [321]. In this regard, modern off-policy deep RL requires techniques such as growing batch learning, importance sampling or ensemble method to stabilize the algorithm. Thus, for ease of exposition, we only include the on-policy analysis in our work.
- Our framework and theoretical results are able to be applied to off-policy setting with the extra assumption on the behavior policy. In particular, we assume the behavior policy is in the neighborhood of the target policy, i.e., in each Actor and Critic update step,

$$\|\mathcal{E}_{\text{bhv-tar},t}\| := \|\pi_{\text{tar}},t - \pi_{\text{bhv},t}\| \leq C_{bt},$$

where $C_{tb} \geq 0$ is a constant. In this way, we can write the A-C update in the off-policy setting as a Newton Method with perturbation, i.e.,

$$\boldsymbol{v}_{\pi_{\text{tar}},t+1} = \boldsymbol{v}_{\pi_{\text{tar}},t} - (\boldsymbol{J}_{\boldsymbol{v}_{\pi_{\text{tar}},t}}^{-1}(\boldsymbol{v}_{\pi_{\text{tar}},t} - T(\boldsymbol{v}_{\pi_{\text{tar}},t})) - \mathcal{E}_t),$$

where $\mathcal{E}_t$ is the perturbation which captures the approximation error from Actor update, Critic update and the behavior policy. Explicitly, we have the perturbation with the following form,

$$\mathcal{E}_t = \mathcal{E}_{v,t} + \mathcal{E}_{\hat{J},t}(\boldsymbol{v}^{\hat{\pi}_{t+1}} - (\boldsymbol{r}_{\tilde{\pi}_{t+1}} + \gamma \boldsymbol{P}_{\tilde{\pi}_{t+1}} \boldsymbol{v}^{\hat{\pi}_{t+1}})) - \boldsymbol{J}_{\hat{\boldsymbol{v}}_t}^{-1}(\mathcal{E}_{r,t} + \mathcal{E}_{bhv-tar,t} \\ + \gamma(\mathcal{E}_{P,t} + \mathcal{E}_{bhv-tar,t})\boldsymbol{v}^{\hat{\pi}_t}).$$

Thus, the off-policy analysis is similar to the on-policy case but with the 'error' induced by the behavior policy.

3.2.2 An Illustrative Example of the Error Propagation in Actor-Critic Updates

The A-C method can be viewed as a generalization of policy iteration in ADP, where the Critic update corresponds to the policy evaluation of the current policy and the Actor update performs the policy improvement. In practice, function approximation (e.g., via neural networks) is often used to approximate both the Critic and the Actor, which inevitably incurs approximation errors for the policy update and evaluation. Moreover, the approximation errors could propagate along with the iterative updates in the A-C method. We have the illustrative example to get a more concrete sense of the impact of the approximation errors on the policy update.

As illustrated in Fig. 3.1, for a given policy π_t with the underlying true policy value $\boldsymbol{v}^{\pi_t}$, we denote $\widetilde{\boldsymbol{v}}^{\pi_t}$ as the learnt value estimation of $\boldsymbol{v}^{\pi_t}$ in the Critic step. We further denote π_{t+1} and $\widetilde{\pi}_{t+1}$ as the greedy policy obtained in the Actor update Eq. (3.2) by using $\boldsymbol{v}^{\pi_t}$ and $\widetilde{\boldsymbol{v}}^{\pi_t}$, respectively. Let $\hat{\pi}_{t+1}$ be the policy estimation of $\widetilde{\pi}_{t+1}$ with function approximation in the Actor step. Intuitively, π_{t+1} is the underlying true policy update from π_t using one step policy iteration without any error, $\widetilde{\pi}_{t+1}$ is the policy update from π_t with approximation errors in the Critic update, and $\hat{\pi}_{t+1}$ is the policy update from π_t with approximation errors in both the Critic step and the Actor step. To characterize the impact of the approximation errors on the policy update, i.e., the difference between $\boldsymbol{v}^{\pi_{t+1}}$ and $\boldsymbol{v}^{\hat{\pi}_{t+1}}$, we evaluate the Critic error, i.e., the difference between $\boldsymbol{v}^{\pi_{t+1}}$ and $\boldsymbol{v}^{\widetilde{\pi}_{t+1}}$, and the Actor error, i.e., the difference between

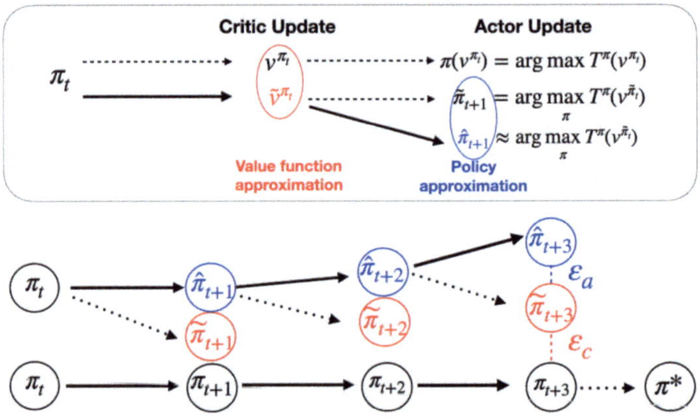

Fig. 3.1 Illustration of error propagation effect in the A-C method: The approximation errors from Critic update ($\mathcal{E}_c$) and Actor update ($\mathcal{E}_a$) are carried forward and may get amplified due to accumulation. (To distinguish the approximation errors between Critic update and Actor update, we use tilde symbol ($\tilde{}$) above variables, such as policy $\widetilde{\pi}$ and value vector $\widetilde{\boldsymbol{v}}$, to represent the policy and the value vector obtained in the presence of Critic update error. We use hat symbol ($\hat{}$) above the variables to represent the results with approximation error in Actor update)

$\boldsymbol{v}^{\widetilde{\pi}_{t+1}}$ and $\boldsymbol{v}^{\hat{\pi}_{t+1}}$, in a separate manner. More specifically, to quantify the Critic error, we can first have the following update based on the same reasoning with Eq. (3.3):

$$\begin{aligned}\boldsymbol{v}^{\widetilde{\pi}_{t+1}} &= \boldsymbol{v}_t - \boldsymbol{J}_{\widetilde{\boldsymbol{v}}_t}^{-1}\left(\boldsymbol{v}_t - (\boldsymbol{r}_{\widetilde{\pi}_{t+1}} + \gamma \boldsymbol{P}_{\widetilde{\pi}_{t+1}} \boldsymbol{v}_t)\right) \\ &\triangleq \boldsymbol{v}_t - \boldsymbol{J}_{\widetilde{\boldsymbol{v}}_t}^{-1}\left(\boldsymbol{v}_t - \widetilde{T}(\boldsymbol{v}_t)\right),\end{aligned}$$

where $\widetilde{T}(\boldsymbol{v}_t) = \boldsymbol{r}_{\widetilde{\pi}_{t+1}} + \gamma \boldsymbol{P}_{\widetilde{\pi}_{t+1}} \boldsymbol{v}_t$ and $\boldsymbol{J}_{\widetilde{\boldsymbol{v}}_t} = \boldsymbol{I} - \gamma \boldsymbol{P}_{\widetilde{\pi}_{t+1}}$. Denote the approximation error (random variable) in the Bellman operator and the Jacobian by $\mathcal{E}_{T,t}$ and $\mathcal{E}_{J,t}$, i.e.,

$$\widetilde{T}(\boldsymbol{v}_t) - T(\boldsymbol{v}_t) \triangleq \mathcal{E}_{T,t}, \quad \boldsymbol{J}_{\widetilde{\boldsymbol{v}}_t}^{-1} - \boldsymbol{J}_{\boldsymbol{v}_t}^{-1} \triangleq \mathcal{E}_{J,t},$$

where it is clear that both error terms stem from the function approximation errors in the Critic update. To quantify the Actor error, we assume that

$$\boldsymbol{v}^{\hat{\pi}_{t+1}} = \boldsymbol{v}^{\widetilde{\pi}_{t+1}} + \mathcal{E}_{a,t},$$

where $\mathcal{E}_{a,t}$ is the error term. Therefore, by casting the A-C method as Newton's method with perturbation, we can characterize the approximation errors on the policy update:

$$\boldsymbol{v}^{\hat{\pi}_{t+1}} = \boldsymbol{v}^{\pi_{t+1}} + \mathcal{E}_{c,t} + \mathcal{E}_{a,t},$$

where $\mathcal{E}_{c,t} \triangleq -\mathcal{E}_{J,t}(\boldsymbol{v}_t - T(\boldsymbol{v}_t)) + (\boldsymbol{J}_{\boldsymbol{v}_t}^{-1} + \mathcal{E}_{J,t})\mathcal{E}_{T,t}$ and $\mathcal{E}_{a,t}$ capture the impact of the approximation error from Critic update step and Actor update step, respectively. Intuitively, as illustrated in Fig. 3.1, both errors from the previous update in the A-C method may propagate to the next update and thus affect the convergence behavior of the algorithm substantially, in contrast to idealized policy iteration without approximation errors. This phenomenon has also been observed in the empirical results [108, 328]. In this work, we strive to systematically analyze the impact of the approximation errors, through (1) a detailed characterization of the approximation errors in the Critic update and the Actor update in Sect. 3.3 and (2) a thorough analysis of the error propagation effect and biases in Sect. 3.4. We also provide the illustration on our theoretical results in Fig. 3.2.

3.3 Characterization of Approximation Errors

Actor-Critic Methods with Function Approximation. In what follows, we consider that the policy is parameterized by $\theta \in \Theta$, which in general corresponds to a non-linear function class. Following [175, 182, 183, 261, 287], the Q-function is parameterized by a linear function class with feature vector $\phi(s,a) : \mathcal{S} \times \mathcal{A} \to \mathbb{R}^d$ and parameter $\omega \in \Omega \subset \mathbb{R}^d$, i.e., $Q_\omega(s,a) = \omega^\top \phi(s,a)$. We note that the modeling of the Q-function via linear value function is often used to extract insight in the A-C method. Similar to the policy iteration, the update in the A-C method alternates between the following two steps.

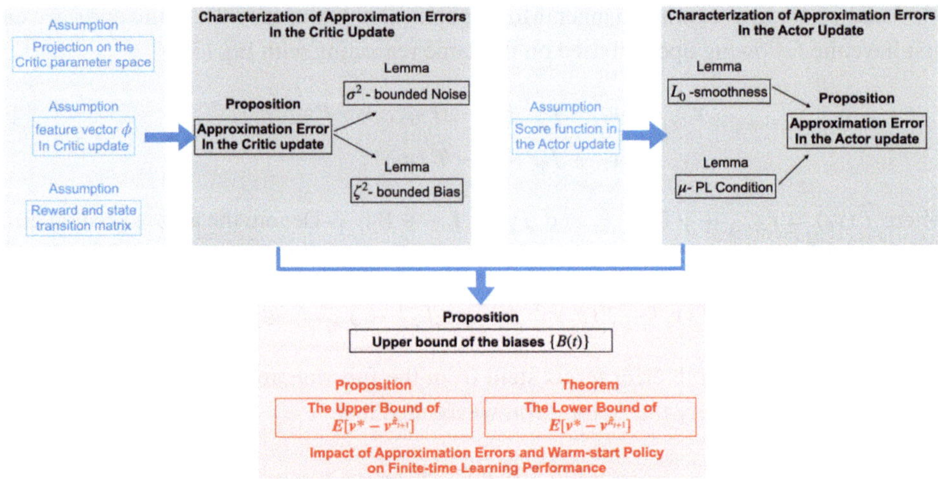

Fig. 3.2 Illustration of the theoretical analysis

Critic update: The Critic updates its parameter ω to evaluate the current policy π_t, e.g., through m-step ($m \geq 1$) Bellman evaluation operator T^{π} to the current Q-function estimator (namely, m-step return), which leads to the following update rule at time step t,

$$\begin{aligned} Q_{t+1}(s,a) &\leftarrow \mathbf{E}_{\pi_t}[(1-\gamma)\cdot\sum_{i=0}^{m-1}\gamma^i r(s_i,a_i) \\ &\quad + \gamma^m \cdot Q_{\omega_t}(s_m,a_m) \mid s_0 = s, a_0 = a], \\ \omega_{t+1} &\leftarrow \arg\min_{\omega} \mathbf{E}_{(s,a)\sim\rho^{\pi_t}}\left[Q_{t+1} - \omega^{\top}\phi\right]^2(s,a). \end{aligned} \tag{3.5}$$

Actor update: The Actor is updated through a greedy step to maximize Q-function $Q_{\omega_{t+1}}$, i.e.,

$$\pi_{t+1} \leftarrow \arg\max_{\pi} \mathbf{E}_{(s,a)\sim\rho^{\pi}}\left[Q_{\omega_{t+1}}(s,a)\right]. \tag{3.6}$$

Next, we first show that the noise in Actor update is bounded and the bias in Critic update is also bounded.

Proof of Bounded Noise in the Actor Update Based on Proposition 3.1, we have the following two lemmas for the upper bounds on the bias term $b = \mathbf{E}[C_{k,t}]$ and the error term $\beta = f_{k,t} + C_{k,t} - \mathbf{E}[C_{k,t}] - \mathbf{E}[f_{k,t}]$ in the stochastic gradient update Eq. (3.16), respectively.

Lemma 3.1 (σ^2-bounded noise) *Suppose Assumptions 3.1, 3.2, 3.3 hold. Then with probability at least* $1 - p$, $\mathbf{E}[\|\beta\|^2] \leq \|\nabla_\theta h(\omega, \theta) + b\|^2 + \sigma^2,\ \forall\theta$, *where* $\sigma^2 \geq 0$ *is a constant and depends on* p.

Recall $\beta = f_{k,t} + C_{k,t} - \mathbf{E}[C_{k,t}] - \mathbf{E}[f_{k,t}]$. We also have the following definitions:

$$\begin{aligned}
C_{k,t,1} \triangleq& 1/l \sum_{i=1}^{l} (Q_{\omega_{t+1}}(s_i, a_i)\nabla_\theta \pi_{\theta_k}(a_i|s_i) - Q_{\widetilde{\omega}_{t+1}}(s_i, a_i)\nabla_\theta \pi_{\theta_k}(a_i|s_i)),\\
C_{k,t,2} \triangleq& 1/l \sum_{i=1}^{l} (Q_{\widetilde{\omega}_{t+1}}(s_i, a_i)\nabla_\theta \pi_{\theta_k}(a_i|s_i) - Q^{\pi_{\theta_t}}(s_i, a_i)\nabla_\theta \pi_{\theta_k}(a_i|s_i)),\\
f_{k,t} \triangleq& 1/l \sum_{i=1}^{l} Q^{\pi_{\theta_t}}(s_i, a_i)\nabla_\theta \pi_{\theta_k}(a_i|s_i),\\
C_{k,t} \triangleq& C_{k,t,1} + C_{k,t,2}.
\end{aligned}$$

Next we evaluate $\mathbf{E}[\|f_{k,t} + C_{k,t} - \mathbf{E}[C_{k,t}] - \mathbf{E}[f_{k,t}]\|^2]$ as follows:

$$\begin{aligned}
&\|f_{k,t} + C_{k,t} - \mathbf{E}[C_{k,t}] - \mathbf{E}[f_{k,t}]\|^2\\
=&(f_{k,t} + C_{k,t})(f_{k,t} + C_{k,t})^\top + (\mathbf{E}[C_{k,t}] + \mathbf{E}[f_{k,t}])(\mathbf{E}[C_{k,t}] + \mathbf{E}[f_{k,t}])^\top\\
&- 2(f_{k,t} + C_{k,t})(\mathbf{E}[C_{k,t}] + \mathbf{E}[f_{k,t}])^\top\\
\leq&(f_{k,t} + C_{k,t})(f_{k,t} + C_{k,t})^\top + (\mathbf{E}[C_{k,t}] + \mathbf{E}[f_{k,t}])(\mathbf{E}[C_{k,t}] + \mathbf{E}[f_{k,t}])^\top. \qquad (3.7)
\end{aligned}$$

Note that $C_{k,t}$ and $f_{k,t}$ are both bounded above since Q-function is bounded and $\nabla_\theta \pi_\theta(a|s)$ is bounded (see Assumption 3.4), i.e.,

$$\begin{aligned}
\|\nabla\pi(a|s)\| \leq& C_\psi,\\
\|Q(s, a)\| \leq& \sum_{t=1}^{\infty} \gamma^t r_{\max} = \frac{r_{\max}}{1-\gamma}.
\end{aligned}$$

Then we have the following bounds for $C_{k,t}$ and $f_{k,t}$:

$$\begin{aligned}
\|C_{k,t}\| \leq& 2C_\psi \frac{r_{\max}}{1-\gamma},\\
\|f_{k,t}\| \leq& C_\psi \frac{r_{\max}}{1-\gamma}.
\end{aligned}$$

Then we have

$$\begin{aligned}
(f_{k,t} + C_{k,t})(f_{k,t} + C_{k,t})^\top \leq& \|f_{k,t}\|^2 + \|C_{k,t}\|^2 + 2\|f_{k,t}\|\|C_{k,t}\|\\
\leq& 9C_\psi^2 (\frac{r_{\max}}{1-\gamma})^2
\end{aligned}$$

Taking expectation over both sides of the inequality (3.7), we have that

$$\mathbf{E}[\|\beta\|^2] \leq 1 \cdot \|\mathbf{E}[C_{k,t}] + \mathbf{E}[f_{k,t}]\|^2 + \mathbf{E}[(f_{k,t} + C_{k,t})(f_{k,t} + C_{k,t})^\top].$$

Let $M_n = 1$ and $\sigma^2 = 9C_\psi^2(\frac{r_{\max}}{1-\gamma})^2$. Then we have that

$$\mathbf{E}[\|\beta\|^2] \leq M_n \cdot \|\nabla_\theta h(\omega, \theta) + \mathbf{E}[C_{k,t}]\| + \sigma^2.$$

Proof of Bounded Bias in the Actor Update.

Lemma 3.2 (ζ^2-bounded bias) *Suppose Assumptions 3.1, 3.2, 3.3 hold. Then with probability at least* $1 - p$, $\|b\|^2 \leq \zeta^2, \forall\theta$, *where* $\zeta^2 \geq 0$ *is a constant and depends on* p.

Recall that $b = \mathbf{E}[C_{k,t}]$ and

$$\begin{aligned}C_{k,t} :=&C_{k,t,1} + C_{k,t,2}\\ =&1/l \sum_{i=1}^{l} \Big(Q_{\omega_{t+1}}(s_i, a_i)\nabla_\theta \pi_{\theta_k}(a_i|s_i) - Q_{\tilde{\omega}_{t+1}}(s_i, a_i)\nabla_\theta \pi_{\theta_k}(a_i|s_i) +\\ &(Q_{\tilde{\omega}_{t+1}}(s_i, a_i)\nabla_\theta \pi_{\theta_k}(a_i|s_i) - Q^{\pi_{\theta_t}}(s_i, a_i)\nabla_\theta \pi_{\theta_k}(a_i|s_i)\Big).\end{aligned}$$

Next, we evaluate $\|b\|^2$. Observe that

$$\|Q_{\tilde{\omega}_{t+1}}(s_i, a_i)\nabla_\theta \pi_{\theta_k}(a_i|s_i) - Q^{\pi_{\theta_t}}(s_i, a_i)\nabla_\theta \pi_{\theta_k}(a_i|s_i)\| \leq 2C_\psi \frac{r_{\max}}{1-\gamma}.$$

Meanwhile, recall the results from Proposition 3.1 Eq. (3.14), we have that, for any $(s, a) \in \mathcal{S} \times \mathcal{A}$,

$$\|Q_\omega - Q_{\tilde{\omega}}\| \leq \varepsilon_Q.$$

Then we have,

$$\|Q_{\omega_{t+1}}(s_i, a_i)\nabla_\theta \pi_{\theta_k}(a_i|s_i) - Q_{\tilde{\omega}_{t+1}}(s_i, a_i)\nabla_\theta \pi_{\theta_k}(a_i|s_i)\| \leq c_\psi \varepsilon_Q,$$

where ε_Q depends on p.

Let $\zeta = c_\psi \varepsilon_Q + 2C_\psi \frac{r_{\max}}{1-\gamma}$. Then we have

$$\|b\|^2 = \|\mathbf{E}[C_{k,t}]\|^2 \leq \mathbf{E}[\|C_{k,t}\|^2] \leq \zeta^2.$$

3.3.1 Approximation Error in the Critic Update

Solving the minimization problem in Eq. (3.5) involves the expectation over the stationary state-action distribution ρ^{π_t} induced by the current policy π_t, which can be approximated by sample average in practice. Therefore, we consider the Critic update below based on two groups of samples, $\{(s_l, a_l)\}_{l=1}^N$ and $\{\tau_l\}_{l=1}^N$ where $\tau_l = \{s_{l,t}, a_{l,t}, r_{l,t}\}_{t=0}^m$, which are collected by following π_t:

$$\omega_{t+1} = \Gamma_R\left\{ \left(\sum_{l=1}^{N} \phi(s_l, a_l)\, \phi(s_l, a_l)^\top \right)^{-1} \cdot \sum_{l=1}^{N} \left((1-\gamma) \sum_{i=0}^{m-1} \gamma^i r_{l,i} + \gamma^m Q_{\omega_t} \right) \phi(s_{l,m}, a_{l,m}) \right\}, \tag{3.8}$$

where Γ is the projection operator onto the Critic parameter space Ω with radius R in $\mathbb{R}^d$. Since the samples in each trajectory τ_l are obtained via rolllouts, in general the samples in each trajectory follow a Markovian process [73, 183]. We assume the samples are from the stationary distribution induced by the current policy.

In what follows, we use ω and $\tilde{\omega}$ to distinguish the difference between the sample-based update and the solution from Eq. (3.5), such that the approximation error in the Critic update can be quantified as $|Q_{\tilde{\omega}_t} - Q_{\omega_t}|$. We first impose the following standard assumptions on the Bellman evaluation operator T^π, the feature vector $\phi(s, a)$ and the MDP.

Assumption 3.1 For given Critic parameter ω and policy parameter θ, the following condition holds:

$$\inf_{\tilde{\omega}\in\Omega} \mathbf{E}_{\rho^{\pi_\theta}}\left[\left((T^{\pi_\theta})^m Q_\omega - \tilde{\omega}^\top \phi\right)(s, a)\right] = 0,$$

where ρ^{π_θ} is the stationary state-action transition probability induced by policy π_θ.

Assumption 3.1 [107] indicates that the solution of the Critic update given in Eq. 3.5 lies in the Critic parameter space Ω. We note that this assumption is used for ease of exposition, and our results can be modified by incorporating an additional constant term when this assumption does not hold.

Assumption 3.2 The feature vector $\phi(s, a)$ in the Critic satisfies the following two conditions: (1) $\|\phi(s, a)\|_2 \le 1,\ \forall\ (s, a) \in \mathcal{S} \times \mathcal{A}$; and (2) the smallest singular value for $\mathbf{E}_{\rho^{\pi_\theta}}[\phi(s, a)\phi(s, a)^\top]$ is lower bounded by a positive constant σ^* for policy π_θ, where θ is the actor parameter obtained from the Actor update.

Assumption 3.2 is widely used in the A-C method to guarantee that the minimization in Eq. (3.5) can be attained by a unique minimizer [35, 107].

Assumption 3.3 The reward $r(s, a)$ satisfies the following two conditions: (1) The reward is upper bounded by a positive constant $r_{\max}$ for all $(s, a) \in \mathcal{S} \times \mathcal{A}$; and (2) the stationary state-action transition matrix $\boldsymbol{P}^{\pi}$ has non-zero spectral gap $1 - \lambda > 0$ for all π.

The first condition in Assumption 3.3 is often used for discounted MDPs to ensure a finite value function (e.g., $Q(s, a) \leq Q_{\max}$) [107, 108, 328]. Moreover, since the samples in the same trajectories are generally correlated, the second condition is adopted to guarantee the concentration properties of the Markov chain, which is generally true for the stationary Markov chain [154, 247].

For any $\lambda \in (-1, 1)$, let $\alpha_1(\lambda) = (1 + \lambda)/(1 - \lambda)$, $\alpha_2(\lambda) = 5/(1 - \lambda)$ where $\alpha_2(0) = 1/3$ and $\alpha_3(\lambda) = \max\{\lambda, 0\}$. Define $\tilde{r}_m = \frac{\sqrt{\alpha_2^2 r_{\max}^2 \alpha_3^2 \ln^2 p - 2m\alpha_1\alpha_3 \ln p - \alpha_2\alpha_3 \ln p}}{m} + r_{\max}$ and then we can have the following main result on the approximation error in the Critic update step.

Proposition 3.1 (Approximation Error in Critic Update) *Under Assumptions 3.1, 3.2, 3.3, the following inequality holds with probability at least* $1 - p$, *for any* $t > 0$, $(s, a) \in \mathcal{S} \times \mathcal{A}$*:*

$$|Q_{\omega_t}(s, a) - Q_{\tilde{\omega}_t}(s, a)| \leq \frac{4((1-\gamma)\tilde{r}_m + \gamma^m R)}{\sqrt{N}(\sigma^*)^2} \cdot \left(-\frac{2}{3N} \log \frac{p}{4d} + \sqrt{\frac{4}{9N^2} \log^2 \frac{p}{4d} - \frac{2}{N} \log \frac{p}{4d}} \right) := \varepsilon_p,$$

where d is the dimension of the Critic parameter ω and R is the radius of Critic parameter space Ω as in Eq. (3.8).

Proposition 3.1 establishes the upper bound for the approximation error in the Critic update, which encapsulates the impact of the finite sample size and the finite-step rollout with Bellman evaluation operator T^{π}. It can be seen from Proposition 3.1 that in order to obtain an accurate evaluation of the policy, we can increase the sample size N in the update Eq. (3.8) and have more steps of rollout with Bellman evaluation operator T^{π}. We remark that Proposition 3.1 considers the correlation across samples, and we appeal to the recent advances in Bernstein's Inequality for Markovian samples [98, 154] to tackle this challenge. In what follows, we first provide the proof the Bernstein's inequality with general Makovian samples and then derive the details of Proposition 3.1.

Proof of Bernstein's Inequality with General Makovian samples In this section, we provide the proof of Bernstein's Inequality with General Makovian samples following the proof in Theorem 2 [154].

With a bit abuse of notation,let π denote the stationary distribution of the Markov chain $\{X_i\}_{i \geq 1}$. We define $\pi(h) := \int h(x)\pi(dx)$ to be the integral of function h with respect to π. Let $\mathcal{L}_2(\pi) = \{h : \pi(h^2) < \infty\}$ be the Hilbert space of square-integrable functions and

$\mathcal{L}_2^0(\pi) = \{h \in \mathcal{L}_2(\pi) : \pi(h) = 0\}$ be the subspace of mean zero functions. Let P be the Markov transition matrix of its underlying (state space) graph and P^* be its adjoint in the Hilbert space. Let $\lambda(P) \in [0, 1]$ be the operator norm of P on $\mathcal{L}_2^0(\pi)$ and $\lambda_r(P) \in [-1, 1]$ be the rightmost spectral value of $(P + P^*)/2$. Then the right spectral gap of P is defined as $1 - \lambda_r$ [199] (We remark that in Assumption 3.3, we assume the absolute spectral gap is non-zero, which implies the right spectral gap is also non-zero. This is true since $-1 \leq \lambda_r \leq \lambda \leq 1$.). Let E^h denote the multiplication operator of function $e^h : x \mapsto e^{h(x)}$. In the Hilbert space $\mathcal{L}_2(\pi)$, we define the norm of a function h to be $\|h\|_\pi = \sqrt{\langle h, h\rangle_\pi}$. Furthermore, we introduce the norm of a linear operator T on $\mathcal{L}_2(\pi)$ as $|||T|||_\pi = \sup\{\|Th\|_\pi : \|h\|_\pi = 1\}$.

We first restate Bernstein's Inequality with General Makovian Samples [154] in the following theorem. Let $\alpha_1(\lambda) = (1+\lambda)/(1-\lambda)$, $\alpha_2(\lambda) = 5/(1-\lambda)$ and $\alpha_2(0) = 1/3$.

Theorem 3.1 (Bernstein's Inequality with General Makovian Samples) *Suppose* $\{X_i\}_{i\geq 1}$ *is a stationary Markov chain with invariant distribution* π *and non-zero right spectral gap* $1 - \lambda_r > 0$, *and* $f :\mapsto x[-c, c]$ *is a function with* $\pi(f) = 0$. *Let* $\sigma^2 = \pi(f^2)$. *Then, for any* $0 \leq t < (1 - \max\{\lambda_r, 0\})/5c$ *and any* $\varepsilon > 0$,

$$\mathbf{P}_\pi\left(\frac{1}{n}\sum_{i=1}^{n} f(X_i) > \varepsilon\right) \leq \exp\left(-\frac{n\varepsilon^2/2}{\alpha_1(\max\{\lambda_r, 0\})\cdot\sigma^2 + \alpha_2(\max\{\lambda_r, 0\})\cdot c\varepsilon}\right). \tag{3.9}$$

Proof **Step1**. Establish the upper bound of $\mathbf{E}\left[e^{t\sum_i^n f_i(X_i)}\right]$.

Let $I : x \mapsto 1$ be the function mapping x to 1 and let Π be the projection operator onto 1, i.e., $\Pi : g \mapsto \langle h, I\rangle_\pi I = \pi(h)I$. Define the León-Perron operator to be $\widehat{P}_\gamma = \gamma I + (1-\gamma)\Pi$, $\gamma \in [0, 1)$. Then we recall the following lemma (Lemma 2, [154]) on the stationary Markov chain [98]. □

Lemma 3.3 *Let* $\{X_i\}$ *be a stationary Markov chain with invariant measure* π *and non-zero right spectral gap* $1 - \lambda_r > 0$. *For any bounded function* f *and any* $t \in \mathbb{R}$,

$$\mathbf{E}_\pi\left[e^{t\sum_{i=1}^n f(X_i)}\right] \leq \left|\left|\left|E^{tf/2}\widehat{P}_{\max\{\lambda_r,0\}}E^{tf/2}\right|\right|\right|_\pi^n.$$

Lemma 3.4 indicates that it is sufficient to prove the upper bound of $\mathbf{E}\left[e^{t\sum_i^n f_i(X_i)}\right]$ by proving the upper bound of $|||E^{tf/2}\widehat{P}_{\max\{\lambda_r,0\}}E^{tf/2}|||_\pi^n$.

To this end, we first invoke the following lemma (Lemma 6, [154]) to construct $\widehat{f}_k \approx f$ such that for any $\lambda \in [0, 1)$, $|||E^{tf/2}\widehat{P}_\lambda E^{tf/2}|||_\pi = \lim_{k\to\infty}\left|\left|\left|E^{t\widehat{f}_k/2}\widehat{P}_\lambda E^{t\widehat{f}_k/2}\right|\right|\right|_\pi$.

Lemma 3.4 *For function $f : x \in \mathcal{X} \mapsto [-c, c]$ such that $\pi(f) = c$, $\pi(f^2) = \sigma^2$. Let $\lceil \cdot \rceil$ be the ceiling function and $\widetilde{f}_k(x) = \left\lceil \frac{f(x)+c}{c/3k} \right\rceil \times \frac{c}{3k} - c$. Let $\widehat{f}_k = \frac{\widetilde{f}_k - \pi(\widetilde{f}_k)}{1+1/3k}$. Then $\widetilde{f}_k$ takes at most $6k + 1$ possible values and satisfies that for any bounded linear operator T acting on the Hilbert Space $\mathcal{L}_2(\pi)$ and any $t \in \mathbb{R}$,*

$$\left|\left|\left| E^{tf/2} T E^{tf/2} \right|\right|\right|_{\pi} = \lim_{k \to \infty} \left|\left|\left| E^{t\widehat{f}_k/2} T E^{t\widehat{f}_k/2} \right|\right|\right|_{\pi}.$$

Assume that the Markov chain $\{\widehat{X}_i\}_{i \geq 1}$, $\widehat{X}_i \in \mathcal{X}$ is generated by the León-Perron operator $\hat{P}_\lambda$. It follows that $\{\widehat{Y}_i\}_{i \geq 1} = \{\hat{f}_k(\widehat{X}_i)\}_{i \geq 1}$ is a Markov chain in the state space $\mathcal{Y} = \hat{f}_k(\mathcal{X})$. We recall the following lemma (Lemma 7, [154]) on the relation between the two Markov chains.

Lemma 3.5 *Let $\widehat{P}_\lambda$ be the León-Perron operator with $\lambda \in [0, 1)$ on state space $\mathcal{X}$. Let f be a function on $\mathcal{X}$. On the finite state space $\mathcal{Y} = \{y \in f(\mathcal{X}) : \pi(\{x : f(x) = y\}) > 0\}$, define a transition matrix $\widehat{Q}_\lambda = \lambda I + (1 - \lambda) I \mu^\top$, with transition vector μ consisting of elements $\pi(\{x : f(x) = y\})$ for $y in \mathcal{Y}$. Let $E^{t\mathcal{Y}}$ denote the diagonal matrix with elements $e^{ty} : y \in \mathcal{Y}$. Then we have,*

$$\left|\left|\left| E^{tf/2} \widehat{P}_\lambda E^{tf/2} \right|\right|\right|_{\pi} = \left|\left|\left| E^{t\mathcal{Y}/2} \widehat{Q}_\lambda E^{t\mathcal{Y}/2} \right|\right|\right|_{\mu}.$$

Next, we bound the term $\left|\left|\left| E^{t\mathcal{Y}/2} \widehat{Q}_\lambda E^{t\mathcal{Y}/2} \right|\right|\right|_{\mu}$ by the expansion of the largest eigenvalue of the perturbed Markov operator $E^{tf/2} P E^{tf/2}$ as a series in t. Specifically, we recall the following result [202].

Lemma 3.6 *Consider a reversible, irreducible Markov chain on finite state space $\mathcal{X}$. Let D be the diagonal matrix with $\{f(x) : x \in \mathcal{X}\}$ and $T^{(m)} = P D^m / m!$ for any $m \geq 0$ with $D^0 = I$. Assume the invariant distribution of the Markov chain is π and the second largest eigenvalue of the transition matrix P is $\lambda_r < 1$. Let $t_0 = \left(2 \left|\left|\left| T^{(1)} \right|\right|\right|_{\pi} (1 - \lambda_r)^{-1} + c_0\right)^{-1}$ for some c_0 such that*

$$\left|\left|\left| T^{(m)} \right|\right|\right|_{\pi} \leq \left|\left|\left| T^{(1)} \right|\right|\right|_{\pi} c_0^{m-1}, \forall m \geq 1.$$

Denote the largest eigenvalue of $P E^{tf}$ by $\beta(t)$ and $Z = (I - P + \Pi)^{-1} - \Pi$. Let $Z^0 = -\Pi$, $Z^{(j)} = Z^j$, $j \geq 1$, $\beta(0) = 1$ and $\beta(m)$, $m \geq 1$ be

$$\beta^{(m)} = \sum_{p=1}^{m} \frac{-1}{p} \sum_{v_1 + \cdots + v_p = m, v_i \geq 1, k_1 + \cdots + k_p = p-1, k_j \geq 0} \text{trace}\left(T^{(v_1)} Z^{(k_1)} \ldots T^{(v_p)} Z^{(k_p)}\right),$$

Then we have the following expansion on $\beta(t)$,

$$\beta(t) = \sum_{m=0}^{\infty} \beta^{(m)} t^m, \quad |t| < t_0.$$

Follow the same line as in [202] (pp. 854–856), denote $\sigma^2 = \|f\|_\pi^2$ and $c = c_0 \geq |||D|||_\pi$ (defined in Lemma 3.6), then we have the following upper bound of $\beta(t)$.

$$\begin{aligned}
\beta(t) &= \beta^{(0)} + \beta^{(1)} t + \sum_{m=2} \beta^{(m)} t^m \\
&\leq 1 + 0 + \sum_{m=2}^{\infty} \frac{\pi\left(f^m\right) t^m}{m!} + \sum_{m=2}^{\infty} \frac{\sigma^2 \lambda t}{5c} \left(\frac{5ct}{1-\lambda_r}\right)^{m-1} \\
&\leq \exp\left(\sum_{m=2}^{\infty} \frac{\pi\left(f^m\right) t^m}{m!} + \sum_{m=2}^{\infty} \frac{\sigma^2 \lambda t}{5c} \left(\frac{5ct}{1-\lambda_r}\right)^{m-1}\right) \\
&\leq \exp\left(\frac{\sigma^2}{c^2}\left(e^{tc} - 1 - tc\right) + \frac{\sigma^2 \lambda t^2}{1-\lambda_r - 5ct}\right) \\
&:= \exp(g_1(t) + g_2(t))
\end{aligned} \tag{3.10}$$

Now we are ready to derive the bound for the term $\mathbf{E}\left[e^{t \sum_i^n f_i(X_i)}\right]$. Following the results in Lemma 3.4, we consider a sequence of f_k such that,

$$\left|\left|\left| E^{tf/2} \widehat{P}_\lambda E^{tf/2} \right|\right|\right|_\pi = \lim_{k\to\infty} \left|\left|\left| E^{t\widehat{f}_k/2} \widehat{P}_\lambda E^{t\widehat{f}_k/2} \right|\right|\right|_\pi.$$

Next, we construct the finite state space counterpart of each pair of $E^{t\widehat{f}_k/2} \widehat{P}_\lambda E^{t\widehat{f}_k/2}$ and π by Lemma 3.5, i.e.,

$$\left|\left|\left| E^{t\widehat{f}_k/2} \widehat{P}_\lambda E^{t\widehat{f}_k/2} \right|\right|\right|_\pi := \left|\left|\left| E^{t\mathcal{Y}_k/2} \widehat{Q}_\lambda E^{t\mathcal{Y}_k/2} \right|\right|\right|_{\mu_k}$$

Let the random variable in the state space $\mathcal{Y}_k$ be Y_k, then the mean and variance of Y_k is $\sum_{y\in\mathcal{Y}_k} \pi\left(\{x : \widehat{f}_k(x) = \mathcal{Y}\}\right) y = \pi\left(\widehat{f}_k\right) = 0$ and $\sum_{y\in\mathcal{Y}_k} \pi\left(\{x : \widehat{f}_k(x) = y\}\right) y^2 = \pi\left(\widehat{f}_k^2\right)$.

For each k, applying Eq. (3.10) gives us,

$$\left|\left|\left| E^{t\mathcal{Y}_k/2} \widehat{Q}_\lambda E^{t\mathcal{Y}_k/2} \right|\right|\right|_{\mu_k} \leq \exp\left(\frac{\pi\left(\widehat{f}_k^2\right)}{c^2}\left(e^{tc} - 1 - tc\right) + \frac{\pi\left(\widehat{f}_k^2\right) \lambda t^2}{1-\lambda_r - 5ct}\right)$$

Note that as $k \to \infty$, we have $\pi\left(\widehat{f_k^2}\right) \to \pi(f^2) = \sigma^2$. Then we have the upper bound for each operator $\left|\left|\left|E^{tf_i/2} P E^{tf_i/2}\right|\right|\right|_\pi$, i.e., for any $\lambda \in [0, 1)$,

$$\left|\left|\left|E^{tf/2} P_\lambda E^{tf/2}\right|\right|\right|_\pi \leq \exp(g_1(t) + g_2(t))$$

where g_1 and g_2 are defined in Eq. (3.10).

Consequently, we obtain the upper bound for $\mathbf{E}\left[e^{t \sum_i^n f_i(X_i)}\right]$ as follows, $\mathbf{E}\left[e^{t \sum_i^n f_i(X_i)}\right]$,

$$\mathbf{E}_\pi\left[e^{t \sum_{i=1}^n f_i(X_i)}\right] \leq \exp\left(\frac{n\sigma^2}{c^2}\left(e^{tc} - 1 - tc\right) + \frac{n\sigma^2 \max\{\lambda_r, 0\} t^2}{1 - \max\{\lambda_r, 0\} - 5ct}\right)$$

Step 2 Use the convex analysis argument to derive the Bernstein's Inequality.

We first restate the following lemma (Lemma 9, [154]) on the terms g_1 and g_2.

Lemma 3.7 *For* $\lambda \in [0, 1)$, *let* $g_1(t) = \frac{n\sigma^2}{c^2}\left(e^{tc} - 1 - tc\right)$ *and* $g_2(t) = \frac{n\sigma^2 \max\{\lambda_r, 0\} t^2}{1 - \max\{\lambda_r, 0\} - 5ct}$, *then for any* $0 \leq t < (1 - \gamma)/5c$, *the Frechet conjugates* $(g_1 + g_2)^*$ *satisfy the following inequalities.*

$$\text{if } \lambda \in (0, 1): \quad (g_1 + g_2)^*(\varepsilon) := \sup_{0 \leq t < (1-\lambda)/5c} \{t\varepsilon - g_1(t) - g_2(t)\} \geq \frac{\varepsilon^2}{2}\left(\frac{1+\lambda}{1-\lambda}\sigma^2 + \frac{5c\varepsilon}{1-\lambda}\right)^{-1}$$

$$\text{if } \lambda = 0: \quad (g_1 + g_2)^*(\varepsilon) = g_1^*(\varepsilon) \geq \frac{\varepsilon^2}{2}\left(\sigma^2 + \frac{c\varepsilon}{3}\right)^{-1}.$$

By the Chernoff bound, we have,

$$-\log \mathbf{P}\left(\frac{1}{n}\sum_{i=1}^n f_i(X_i) > \varepsilon\right) \geq n \times \sup_{t \in \mathbb{R}} \{t\varepsilon - g_1(t) - g_2(t)\}$$

Notice that $g_1(t) = O(t^2)$ and $g_2(t) = O(t^2)$ as $t \to 0$, then for some $t > 0$, we have $t\varepsilon - g_1(t) - g_2(t) > 0$. Meanwhile, when $t \leq 0$, we have $t\varepsilon - g_1(t) - g_2(t) \leq 0$. Thus, we can obtain that,

$$\sup\{t\varepsilon - g_1(t) - g_2(t) : t > 0\} = \sup\{t\varepsilon - g_1(t) - g_2(t) : t \in \mathbb{R}\} = (g_1 + g_2)^*(\varepsilon).$$

Letting $\lambda = \max\{\lambda_r, 0\}$, $\alpha_1(\lambda) = (1 + \lambda)/(1 - \lambda)$, $\alpha_2(\lambda) = 5/(1 - \lambda)$ and $\alpha_2(0) = 1/3$ yields,

$$\mathbf{P}_\pi\left(\frac{1}{n}\sum_{i=1}^n f(X_i) > \varepsilon\right) \leq \exp\left(-\frac{n\varepsilon^2/2}{\alpha_1(\max\{\lambda_r, 0\}) \cdot \sigma^2 + \alpha_2(\max\{\lambda_r, 0\}) \cdot c\varepsilon}\right). \tag{3.11}$$

This concludes the proof. □

Proof of Proposition 3.1 Let $\bar{\omega}_{t+1} = \Gamma_R(\tilde{\omega}_{t+1})$, and assume $\|\phi(s,a)\| \le 1$ uniformly (see Assumption 3.1). Based on the approach in [107], it suffices to upper bound $\|\omega_{t+1} - \tilde{\omega}_{t+1}\|_2$. Observe that

$$\|\omega_{t+1} - \bar{\omega}_{t+1}\|_2 \le \|\widehat{\Phi}\widehat{v} - \Phi v\|_2 \le \|\Phi\|_2 \cdot \|\widehat{v} - v\|_2 + \|\widehat{\Phi} - \Phi\|_2 \cdot \|\widehat{v}\|_2,$$

where Φ and v are given as follows:

$$\widehat{\Phi} = \left(\frac{1}{N}\sum_{l=1}^{N} \phi(s_l, a_l)\,\phi(s_l, a_l)^\top\right)^{-1},$$

$$\Phi = \left(\mathbf{E}_{\rho_{t+1}}\left[\phi(s,a)\phi(s,a)^\top\right]\right)^{-1},$$

$$\widehat{v} = \frac{1}{N}\sum_{l=1}^{N}\left((1-\gamma)\sum_{i=0}^{m-1}\gamma^i r_{l,i} + \gamma^m Q_{\omega_t}\left(s_{l,m}, a_{l,m}\right)\right)\cdot \phi\left(s_{l,m}, a_{l,m}\right),$$

$$v = \mathbf{E}_{\rho_{t+1}}\left[(1-\gamma)\sum_{i=0}^{m-1}\left(\gamma^i r_{l,i} + \gamma^m \boldsymbol{P}_{\pi_{\theta_{t+1}}} Q_{\omega_t}(s_m, a_m)\right)\cdot \phi(s_m, a_m)\right].$$

Recall that the following assumptions are in place: (1) Spectral norm $\|\phi(s,a)\|_2 \le 1$, $\phi(s,a) \in \mathbb{R}^d$; (2) $|r(s,a)| \le r_{\max}$ and $\bar{r} = \mathbf{E}_{s,a} r(s,a)$; (3) $\|\omega_t\|_2 \le R$ and (4) the minimum singular value of the matrix $\mathbf{E}_{\rho_t}[\phi(s,a)\phi(s,a)^\top]$, $t \ge 1$ is uniformly lower bounded by σ^*. It can be shown that $\|\Phi\|_2 \le \frac{1}{\sigma^*}$.

Next, we derive the bound by appealing to Bernstein's Inequality with General Makovian samples. Following Theorem 2 [154], let π_r be the invariant distribution (which is relevant to the current policy π_k) of the stationary Markov chain $\{r_t\}_{t=1}^m$. Suppose that it has non-zero right spectral gap $1 - \lambda_r > 0$. Let $\sigma_r^2 = \int (r - \bar{r})^2 \pi_r(dr)$. Then, we have that for any $\varepsilon > 0$:

$$\mathbf{P}_{\pi_r}\left(\frac{1}{m}\sum_{i=1}^{m}(r_i - \bar{r}) > \varepsilon\right) \le \exp\left(-\frac{m\varepsilon^2/2}{\alpha_1(\max\{\lambda_r, 0\})\cdot\sigma^2 + \alpha_2(\max\{\lambda_r, 0\})\cdot r_{\max}\varepsilon}\right),$$

where $\alpha_1(\lambda) = \frac{1+\lambda}{1-\lambda}$, $\alpha_2(\lambda) = \begin{cases} \frac{1}{3} & \text{if } \lambda = 0 \\ \frac{5}{1-\lambda} & \text{if } \lambda \in (0,1) \end{cases}$.

We conclude that with probability at least $1-p$,

$$\sum_{i=0}^{m-1} r_i \le \frac{\sqrt{\alpha_2^2(\max\{\lambda_r, 0\})^2 \ln p^2 - 2m\alpha_1(\max\{\lambda_r, 0\})\ln p} - \alpha_2(\max\{\lambda_r, 0\})\ln p}{m} + \bar{r} := \tilde{r}_m.$$

It follows that with probability at least $1-p$,

$$\|\hat{v}\|_2 \le (1-\gamma)\tilde{r}_m + \gamma^m R,$$

Further, note that

$$\|v\|_2 \leq (1-\gamma)\bar{r} + \gamma^m R,$$

Since the minimum singular value of $\hat{\Phi}^{-1}$ is no less than $\frac{\sigma^*}{2}$ w.h.p. when N is large enough, we have that

$$\|\hat{\Phi}\|_2 \leq \frac{2}{\sigma^*}.$$

For convenience, define

$$\hat{X} \triangleq \left(\frac{1}{N}\sum_{l=1}^{N} \phi\left(s_l, a_l\right) \phi\left(s_l, a_l\right)^\top\right), \quad X \triangleq \left(\mathbf{E}_{\rho_{t+1}}\left[\phi(s,a)\phi(s,a)^\top\right]\right),$$

and define

$$Z \triangleq \hat{X} - X = \sum_{k=1}^{N} S_k, \tag{3.12}$$

$$S_k \triangleq \frac{1}{N}(\phi_k\phi_k^\top - X), \tag{3.13}$$

where $S_k, k = 1, \ldots, N$ are independent.

Next, we derive the uniform bound on the spectral norm of each summand as follows:

$$\|S_k\|_2 = \frac{1}{N}\|\phi_k\phi_k^\top - X\| \leq \frac{1}{N}(\|\phi_k\phi_k^\top\| + \|X\|) \leq \frac{2}{N}.$$

To this end, we bound the matrix variance statistic $V(Z)$:

$$V(Z) := \|\mathbf{E}[Z^2]\| = \|\sum_{k=1}^{N}\mathbf{E}[S_k^2]\|.$$

Note that the variance of each summand is given by

$$\begin{aligned}
\mathbf{E}[S_k^2] =& \frac{1}{N^2}\mathbf{E}[(\phi_k\phi_k^\top - X)^2] \\
=& \frac{1}{N^2}\mathbf{E}[\|\phi_k\|^2 \cdot \phi\phi^\top - \phi\phi^\top X - X\phi\phi^\top + X^2] \\
\preccurlyeq& \frac{1}{N^2}[\mathbf{E}[\phi\phi^\top] - X^2] \\
\preccurlyeq& \frac{1}{N^2}X.
\end{aligned}$$

Combining the above, we conclude that

$$0 \preccurlyeq \sum_{k=1}^{N} \mathbf{E}[S_k^2] \preccurlyeq \frac{1}{N} X.$$

Observe that

$$\|X\| = \|\mathbf{E}[\phi\phi^\top]\|_2 \le \mathbf{E}[\|\phi\phi^\top\|] = \mathbf{E}\|\phi\|^2 \le 1.$$

Since the spectral norm is the variance statistic given by

$$V(Z) \le \frac{1}{N}\|X\|,$$

appealing to Bernstein's Inequality, we have that

$$\begin{aligned}
\mathbf{P}\{\|Z\| \ge t\} \le& 2d \exp\left(\frac{\frac{-t^2}{2}}{\frac{1}{N}\|X\| + \frac{2t}{3N}}\right), \\
\mathbf{E}[\|Z\|] \le& \sqrt{\frac{2}{N}\|X\| \log(2d)} + \frac{2}{3N}\log(2d) \\
\le& \sqrt{\frac{2}{N}\log(2d)} + \frac{2}{3N}\log(2d).
\end{aligned}$$

This is to say, with probability at least $1 - p/2$, the following holds:

$$\|X - \hat{X}\| \le -\frac{2}{3N}\log\frac{p}{4d} + \sqrt{\frac{4}{9N^2}\log^2\frac{p}{4d} - \frac{2}{N}\log\frac{p}{4d}}.$$

In a nutshell, we have that

$$\begin{aligned}
\|\widehat{\Phi} - \Phi\|_2 =& \|\hat{X}^{-1} - X^{-1}\|_2 \\
=& \|\hat{X}^{-1}(\hat{X} - X)X^{-1}\|_2 \\
=& \|\hat{\Phi}(\hat{X} - X)\Phi\|_2 \\
\le& \frac{2}{(\sigma^*)^2}\|\hat{X} - X\|_2 \\
\le& \frac{4}{\sqrt{N}\,(\sigma^*)^2} \cdot \left(-\frac{2}{3N}\log\frac{p}{4d} + \sqrt{\frac{4}{9N^2}\log^2\frac{p}{4d} - \frac{2}{N}\log\frac{p}{4d}}\right).
\end{aligned}$$

Similarly, the following inequality holds with probability at least $1 - p/2$:

$$\|\widehat{v} - v\|_2 \le -\frac{\delta_1}{3}\log\frac{p}{2(d+1)} + \sqrt{\frac{\delta_1^2}{9}\log^2\frac{p}{2(d+1)} - 2\delta_2\log\frac{p}{2(d+1)}},$$

where d is the dimension of vector φ, $\delta_1 = \frac{1}{N}\left((1-\gamma)(\tilde{r}_m + \bar{r}) + 2\gamma^m R\right)$ and $\delta_2 = \|\mathbf{E}[\hat{v} - v]\|_2$ satisfying

$$\begin{aligned}\delta_2 \leq & \frac{1}{N}\left[(1-\gamma)(|\tilde{r}_m|(|\tilde{r}_m - \bar{r}| + \gamma^m R|\tilde{r}_m - \bar{r}|))\right] \\ \leq & \frac{1-\gamma}{N}[r_{\max} + \gamma^m R]|\tilde{r}_m - \bar{r}|.\end{aligned}$$

Summarizing, we have that

$$\begin{aligned}\left\|\omega_{t+1} - \bar{\omega}_{t+1}\right\|_2 \leq & \|\Phi\|_2 \cdot \|\widehat{v} - v\|_2 + \|\widehat{\Phi} - \Phi\|_2 \cdot \|\widehat{v}\|_2 \\ \leq & -\frac{\delta_1}{3\sigma^*}\log\frac{p}{2(d+1)} + \sqrt{\frac{\delta_1^2}{9}\log^2\frac{p}{2(d+1)} - 2\delta_2\log\frac{p}{2(d+1)}} \\ & + \frac{4((1-\gamma)\tilde{r}_m + \gamma^m R)}{\sqrt{N}(\sigma^*)^2}\left(-\frac{2}{3N}\log\frac{p}{4d} + \sqrt{\frac{4}{9N^2}\log^2\frac{p}{4d} - \frac{2}{N}\log\frac{p}{4d}}\right),\end{aligned}$$

which indicates that with probability at least $1-p$,

$$\begin{aligned}|Q_{\omega_{t+1}} - Q_{\bar{\omega}_{t+1}}| \leq & \left(\frac{4((1-\gamma)\tilde{r}_m + \gamma^m R)}{\sqrt{N}(\sigma^*)^2}\left(-\frac{2}{3N}\log\frac{p}{4d} + \sqrt{\frac{4}{9N^2}\log^2\frac{p}{4d} - \frac{2}{N}\log\frac{p}{4d}}\right)\right) \\ \triangleq & \varepsilon_Q. \end{aligned} \tag{3.14}$$

Remark In the case when Assumption 3.1 does not hold, i.e., we have

$$\inf_{\bar{\omega}\in\Omega} \mathbf{E}_{\rho^{\pi_\theta}}\left[\left((T^{\pi_\theta})^m Q_\omega - \bar{\omega}^\top \phi\right)(s,a)\right] = c_1,$$

where $c_1 > 0$ is a constant. Let $\bar{\omega}_{t+1} = \Gamma_R(\tilde{\omega}_{t+1})$, recall that $\tilde{\omega}$ denotes the solution of Eq. (3.5) and ω denotes the sample-based solution, then we have

$$|Q_{\bar{\omega}_{t+1}} - Q_{\tilde{\omega}_{t+1}}| = c_1$$

From Eq. 3.14, we obtain that,

$$|Q_{\omega_{t+1}} - Q_{\bar{\omega}_{t+1}}| \leq \varepsilon_Q$$

Then the difference between the sample-based solution and the underlying true solution of Eq. 3.5 is,

$$|Q_{\omega_{t+1}} - Q_{\tilde{\omega}_{t+1}}| \leq \varepsilon_Q + c_1.$$

Note that when Assumption 3.1 holds,

$$Q_{\tilde{\omega}_{t+1}} = Q_{\bar{\omega}_{t+1}}.$$

3.3.2 Approximation Error in the Actor Update

In practice, the greedy search step for solving Eq. (3.6) is generally approximated by multiple (e.g., N_a) steps of policy gradient. Based on the policy gradient theorem [302, 322], we can have the following update at gradient step $k \in [1, N_a]$ in the t-th Actor update:

$$\begin{aligned} &\theta_{t,k+1} = \theta_{t,k} + \alpha \mathbf{E}_{(s,a)\sim\rho^{\pi_{\theta_{t,k}}}}[Q_{\omega_{t+1}}(s,a)\nabla_\theta \pi_{\theta_{t,k}}(a|s)], \\ &\theta_{t,1} = \theta_t, \quad \theta_{t,N_a} = \theta_{t+1}, \end{aligned} \tag{3.15}$$

where α is the learning rate. For simplicity, we drop the subscript t in $\theta_{t,k}$ when no confusion will arise and denote $\rho^k := \rho^{\pi_{\theta_k}}$. As in the Critic update, we sample a trajectory with length l by following the current policy π_{θ_k}, i.e., $\{s_1, a_1, s_2, a_2, \ldots, s_l, a_l\}$, to approximate the expectation in Eq. (3.15). Then we can have that

$$\begin{aligned} \theta_{k+1} &= \theta_k + \alpha \frac{1}{l}\sum_{i=1}^{l}[Q_{\omega_{t+1}}(s_i,a_i)\nabla_\theta \pi_{\theta_k}(a_i|s_i)] \\ &:= \theta_k + \alpha(C_{k,t,1} + C_{k,t,2}) + \alpha f_{k,t}, \end{aligned} \tag{3.16}$$

where $C_{k,t,1}$, $C_{k,t,2}$ and $f_{k,t}$ are defined as follows

$$\begin{aligned} C_{k,t,1} &:= 1/l\sum_{i=1}^{l}(Q_{\omega_{t+1}} - Q_{\tilde{\omega}_{t+1}})(s_i,a_i)\nabla_\theta \pi_{\theta_k}(a_i|s_i), \\ C_{k,t,2} &:= 1/l\sum_{i=1}^{l}(Q_{\tilde{\omega}_{t+1}} - Q^{\pi_{\theta_t}})(s_i,a_i)\nabla_\theta \pi_{\theta_k}(a_i|s_i), \\ f_{k,t} &:= 1/l\sum_{i=1}^{l} Q^{\pi_{\theta_t}}(s_i,a_i)\nabla_\theta \pi_{\theta_k}(a_i|s_i). \end{aligned}$$

Here $C_{k,t,1}$ captures the error resulted from using samples to estimate expectation in the Critic update. Based on our result in Proposition 3.1, with high probability, this term will go to 0 when we have infinite samples or infinite rollout length m. Note that $(T^{\pi_{\theta_t}})^m Q_{\omega_t} = Q_{\tilde{\omega}_{t+1}}$ (Critic update) and $\lim_{m\to\infty}(T^{\pi_{\theta_t}})^m Q_{\omega_t} = Q^{\pi_{\theta_t}}$. And $C_{k,t,2}$ implies the approximation error when applying the Bellman operator limited (m) times. This term will go to 0 when $m \to \infty$. $f_{k,t}$ is an unbiased estimation of the gradient of $\mathbf{E}_{(s,a)\sim\rho^k}[Q^{\pi_{\theta_t}}(s,a)]$, i.e., $\mathbf{E}[f_{k,t}] = \mathbf{E}_{(s,a)\sim\rho^k}[Q^{\pi_{\theta_t}}(s,a)\nabla_\theta \pi_{\theta_k}(a|s)]$.

Based on Eq. 3.16, it is clear that the Actor update with the approximation error resulted from the Critic update can be viewed as a stochastic gradient update with some perturbation $C_{k,t} = C_{k,t,1} + C_{k,t,2}$. For convenience, we define

$$h(\omega, \theta) := \mathbf{E}_{(s,a)\sim\rho^{\pi_\theta}} [Q_\omega(s, a)] = \mathbf{E}_{s\sim d^{\pi_\theta}} [v^{\pi_\omega}(s)].$$

Note that in the Actor update, the Critic parameter ω is fixed, and the Actor parameter θ is updated. Let θ^*_{t+1} denote the solution to Eq. 3.6.

Denote the score function $\psi_\theta(a|s) := \nabla_\theta \pi_\theta(a|s)$. We have the following assumptions on ψ_θ.

Assumption 3.4 For any $\theta, \theta' \in \mathbb{R}^d$ and state-action pair $(s, a) \in \mathcal{S} \times \mathcal{A}$, there exist positive constants L_ψ, C_ψ and C_π such that the following holds: (1) $\|\psi_\theta - \psi_{\theta'}\| \le L_\psi \|\theta - \theta'\|$; (2) $\|\psi_\theta\| \le C_\psi$ and (3) $\|\pi_\theta(\cdot|s) - \pi_{\theta'}(\cdot|s)\|_{\mathrm{TV}} \le C_\pi \|\theta - \theta'\|$, where $\|\cdot\|_{TV}$ is the total-variation distance.

The smoothness and bounded property of the score function as stated in the (1) and (2) in Assumption 3.4 are widely adopted in the literature [6, 183, 367, 410], and it has been shown [366] that (3) in Assumption 3.4 can be satisfied for any smooth policy with bounded action space.

Let $L_0 = Q_{\max} L_\psi$, $\alpha \le \frac{1}{2L_0}$, $\kappa = C_\psi \frac{r_{\max}}{1-\gamma}$, $\sigma = 3\kappa$, and $\mu = \frac{g_{\min}}{h^*_{\max} - h_{\max}}$. Here $h_{\max} = \max_{\theta \ne \theta^*} h(\theta, \omega)$, $h^*_{\max} = \max_{\theta = \theta^*} h(\theta, \omega)$, $g_{\min} = \min_{\theta \ne \theta^*} \|\nabla h(\omega, \theta)\|$. Denote $\Upsilon = (1 - \alpha\mu)^{N_a}$. Finally, we present the upper bound of the approximation error in the Actor update.

Proposition 3.2 (Approximation Error in Actor Update) *Given Actor parameter θ_{t-1}, the following inequality holds:*

$$\begin{aligned} &\mathbf{E}_{\theta_t}[h(\omega, \theta_t^*) - h(\omega, \theta_t)|\theta_{t-1}] \\ &\le \Upsilon(h(\omega, \theta_t^*) - h(\omega, \theta_{t-1})) + \Xi_p, \end{aligned}$$

where $\Xi_p = ((C_\psi \varepsilon_p + 2\kappa)^2 + 2\alpha L \sigma^2)/2\mu$.

It can be seen in Proposition 3.2 that the Critic approximation error has direct impact on the Actor update through Ξ_p. Proposition 3.2 reveals that due to the bias and noise induced by the Critic approximation error, running more gradient iterations (the first term on the RHS) do not necessarily guarantee the convergence to the optimal policy $\pi_{\theta_t^*}$.

Proof Observe that the Actor updates use the biased stochastic gradient methods (SGD). For simplicity, we adopt the following notations to study the Actor update:

$$\theta_{k+1} = \theta_k + \alpha(\nabla h(\omega, \theta_k) + b(t) + \beta(t)). \tag{3.17}$$

where $b(t) = \mathbf{E}[C_{k,t}]$ is the bias, α is the step size, and

$$\beta = f_{k,t} + C_{k,t} - \mathbf{E}[C_{k,t}] - \mathbf{E}[f_{k,t}]$$

is the zero-mean noise. Note that the objective function $h(\omega, \theta_k)$ is a function of θ. Denote the optimal value (in this iteration of the Actor update) by $h(\omega, \theta^*)$. □

Proof of the Smoothness and PL Condition of h ***in Actor Update.*** For the sake of tractability, we next give the following two lemmas about the smoothness and Polyak-Lojasiewicz Condition on the objective function $h(\cdot, \theta)$.

Lemma 3.8 (L-smoothness) *Suppose Assumption 3.4 hold. Then function $h(\cdot, \theta)$ is bounded from below by an infimum $h^{\inf} \in \mathbb{R}$, differentiable and ∇h is L-Lipschitz, i.e., $\|\nabla h(\omega, \theta) - \nabla h(\omega, \theta')\| \leq L\|\theta - \theta'\|$, $\forall\, \omega, \theta, \theta'$.*

Lemma 3.9 (μ-PL) *If $\nabla h(\omega, \theta) \neq 0$, then we have $\frac{1}{2}\|\nabla h(\omega, \theta)\| \geq \mu(h(\omega, \theta^*) - h(\omega, \theta))$ $\geq 0, \forall\, \theta, \omega$.*

- [Lemma 3.8] Given Critic parameter ω in the objective function, it can be seen that $\|\nabla h(\omega, \theta) - \nabla h(\omega, \theta')\| \leq Q_{\max}\|\nabla \pi_\theta - \nabla \pi_{\theta'}\|$. Since value function is bounded (e.g., $Q_{\max}$) and the score function $\nabla \pi_\theta$ is L_ψ-smooth (ref. Assumption 6), the constant in Assumption 4 can be easily determined by $L = Q_{\max} L_\psi$.
- [Lemma 3.9] Since the objective function is finite, let $h_{\max} = \max_{\theta \neq \theta^*} h(\theta, \omega)$, $h^*_{\max} = \max_{\theta = \theta^*} h(\theta, \omega)$,. In the case when the gradient is non-zero, let $g_{\min} = \min_{\theta \neq \theta^*} \nabla h$, then we can determine $\mu = \frac{g_{\min}}{h^*_{\max} - h_{\max}} \geq 0$.

Next, we prove the following lemma on the modified version of the descent lemma for smooth function (cf. [11, 240]).

Lemma 3.10 *Suppose Lemmas 3.8 and 3.9 hold. Then, for any stepsize $\alpha \leq \frac{1}{(M_n+1)L}$, the following inequality holds:*

$$\mathbf{E}[h(\omega, \theta_{k+1}) - h(\omega, \theta_k)|\theta_k] \leq \frac{\alpha}{2}\zeta^2 + \frac{\alpha^2 L}{2}\sigma^2 - \frac{\alpha}{2}\|\nabla h(\omega, \theta_k)\|^2.$$

Observe that under the PL-condition (Lemma 3.9), we have the following recursion:

$$\mathbf{E}[h(\omega, \theta_{k+1}) - h(\omega, \theta^*)|\theta_k] \leq (1 - \alpha\mu)\mathbf{E}[h(\omega, \theta_k) - h(\omega, \theta^*)] + \frac{\alpha}{2}\zeta_p^2 + \frac{\alpha^2 L}{2}\sigma^2, \tag{3.18}$$

where $\zeta_p = c_\psi \varepsilon_p + 2C_\psi \frac{r_{\max}}{1-\gamma}$ is defined in Lemma 3.2 and depends on p.

By applying Eq. (3.18) recursively, we obtain the desired results in Proposition 3.2.

$$\mathbf{E}_\theta[\|h(\omega,\theta_t^*) - h(\omega,\theta_t)\||\theta_{t-1}] \leq (1-\alpha\mu)^{N_a}(h(\omega,\theta_t^*) - h(\omega,\theta_{t-1})) + \frac{\zeta_p^2 + 2\alpha L\sigma^2}{2\mu},$$

□

3.4 The Impact of Approximation Errors on Warm-Start Actor-Critic

We next quantify the impact of the approximations errors on the sub-optimality gap of the Warm-Start A-C method with inaccurate Actor/Critic updates. We first cast the A-C method as Newton's Method with perturbation, and then present both the finite-time upper bound and lower bound on the finite-time learning performance.

Actor-Critic Method as Newton's Method with Perturbation. As mentioned earlier, the Critic update follows Eq. (3.8) with finite samples and finite step rollout with Bellman evaluation operator T^π and the Actor update follows Eq. (3.16). Given the policy π_t at time t, we denote the resulting policy of one A-C update as $\hat{\pi}_{t+1}$. Recall that we use $\widetilde{\pi}_{t+1}$ to denote the policy attained the max in $T(\boldsymbol{v}^{\pi_t})$ as illustrated in Fig. 3.1. Furthermore, we define the following notations for ease of our discussion: (1) Denote $\mathcal{E}_{v,t} = \boldsymbol{v}^{\hat{\pi}_{t+1}} - \boldsymbol{v}^{\widetilde{\pi}_{t+1}}$ as the approximation error in the Actor update; (2) Denote $\mathcal{E}_{r,t} = \boldsymbol{r}_{\widetilde{\pi}_{t+1}} - \boldsymbol{r}_{\hat{\pi}_{t+1}}$ as the error in the reward vector, which is induced by the approximation error in the Actor update step; (3) Denote $\mathcal{E}_{P,t} = \boldsymbol{P}_{\widetilde{\pi}_{t+1}} - \boldsymbol{P}_{\hat{\pi}_{t+1}}$ as the error in the transition matrix $\boldsymbol{P}$; (4) Denote $\mathcal{E}_{\hat{J},t} = \boldsymbol{J}_{\widetilde{\boldsymbol{v}}_t}^{-1} - \boldsymbol{J}_{\hat{\boldsymbol{v}}_t}^{-1}$ where $\boldsymbol{J}_{\hat{\boldsymbol{v}}_t} = \boldsymbol{I} - \gamma \boldsymbol{P}_{\hat{\pi}_{t+1}}$ and $\boldsymbol{J}_{\widetilde{\boldsymbol{v}}_t} = \boldsymbol{I} - \gamma \boldsymbol{P}_{\widetilde{\pi}_{t+1}}$.

Following the same line as in Sect. 3.2.2, we treat the A-C algorithm as Newton's method with perturbation $\mathcal{E}_t$, i.e.,

$$\boldsymbol{v}^{\hat{\pi}_{t+1}} := \boldsymbol{v}^{\hat{\pi}_t} - \hat{\mathcal{L}}(t), \tag{3.19}$$

where $\hat{\mathcal{L}}(t) = \boldsymbol{J}_{\hat{\boldsymbol{v}}_t}^{-1}(\boldsymbol{v}^{\hat{\pi}_t} - T(\boldsymbol{v}^{\hat{\pi}_t})) - \mathcal{E}_t$ is the stochastic estimator of Newton's update $\mathcal{L}(t) = \boldsymbol{J}_{\hat{\boldsymbol{v}}^{\pi_t}}^{-1}\left(\boldsymbol{v}^{\hat{\pi}_t} - T\left(\boldsymbol{v}^{\hat{\pi}_t}\right)\right)$, and

$$\begin{aligned}\mathcal{E}_t =& \mathcal{E}_{v,t} + \mathcal{E}_{\hat{J},t}(\boldsymbol{v}^{\hat{\pi}_{t+1}} - (\boldsymbol{r}_{\widetilde{\pi}_{t+1}} + \gamma \boldsymbol{P}_{\widetilde{\pi}_{t+1}} \boldsymbol{v}^{\hat{\pi}_{t+1}}))\\ &- \boldsymbol{J}_{\hat{\boldsymbol{v}}_t}^{-1}(\mathcal{E}_{r,t} + \gamma \mathcal{E}_{P,t}\boldsymbol{v}^{\hat{\pi}_t}),\end{aligned}$$

which can be further decomposed into bias and Martingale difference noise as follows:

$$\begin{aligned}\mathcal{B}(t) \triangleq& \mathbf{E}[\hat{\mathcal{L}}(t)] - \mathcal{L}(t) = \mathbf{E}[\mathcal{E}_t],\\ \mathcal{N}(t) \triangleq& \hat{\mathcal{L}}(t) - \mathbf{E}[\hat{\mathcal{L}}(t)] = \mathcal{E}_t - \mathbf{E}[\mathcal{E}_t].\end{aligned}$$

We have a few observations in order. It can be seen that the perturbation $\mathcal{E}_t$ results from both Actor approximation error (e.g., $\mathcal{E}_{r,t}$, $\mathcal{E}_{P,t}$) and Critic approximation error (e.g., $\mathcal{E}_{v,t}$).

Meanwhile, the learnt Q function in the Critic update Eq. (3.8) is biased in general due to finite rollout steps m which further leads to the biased gradients in the Actor update Eq. 3.16 [183]. More importantly, due to the error propagation effect (see Fig. 3.1), the approximation errors from previous step may get amplified. Clearly, the estimation bias plays an important role in affecting the learning performance, especially when deep neural networks are used as function approximations, which has been extensively investigated using empirical studies [92, 108, 333].

Next, we examine the bias $\mathcal{B}(t)$ based on the approximation errors in the Actor/Critic updates. Combining the results in Propositions 3.1 and 3.2 on the approximation error in the Critic/Actor updates, we define

$$H_t \triangleq \sum_{i=0}^{t} \Upsilon^i \Xi_p + \Upsilon^{t+1}(h(\omega, \theta_t^*) - h(\omega, \theta_0)).$$

Then we have the following result on the bias $B(t)$.

Proposition 3.3 (Upper Bound on the Bias) *Suppose Assumption 3.4 holds. Let $S_\varepsilon(\cdot)$ be an open ball of radius ε. There exist positive constants L_b, and ε, such that when $\theta_{t+1} \in S_\varepsilon(\theta_{t+1}^*)$, the following holds for any $t > 0$,*

$$\|\mathcal{B}(t)\| \leq L_b H_t$$

Proof We first prove the following lemma on the relation between Actor parameter θ and the objective function $h(\omega, \theta)$. □

Lemma 3.11 *There exist a contant $L_h > 0$ and an open ball $S_\varepsilon(\theta_t^*)$ such that for any $\theta_t \in B_\varepsilon(\theta_t^*)$ the following holds for any $t > 0$.*

$$\mathbf{E}[\|\pi_{\theta_t} - \pi^*\|_{\mathrm{TV}}] \leq L_h \mathbf{E}[h(\omega, \theta_t^*) - h(\omega, \theta_t)].$$

Proof By Taylor's expansion, we have

$$h(\omega, \theta^*) = h(\omega, \theta_t) + \nabla h(\omega, \theta_t)(\theta_t^* - \theta_t) + o(\|\theta_t^* - \theta_t\|).$$

Since $h(\omega, \cdot)$ satisfies Polyak-Lojasiewicz Condition, it follows that

$$\|\nabla h(\omega, \theta)\| \geq 2\mu(h(\omega, \theta^*) - h(\omega, \theta)) := L_g \text{ for all } \theta.$$

Note that $L_g > 0$ when $\theta \neq \theta^*$. Then we have that

$$\begin{aligned} h(\omega, \theta_t^*) - h(\omega, \theta_t) =& |\nabla h(\omega, \theta_t)(\theta^* - \theta_t) + o(\|\theta^* - \theta_t\|)| \\ \geq& |\nabla h(\omega, \theta_t)(\theta_t^* - \theta_t)| - |o(\|\theta^* - \theta_t\|)| \\ \geq& L_g\|\theta_t^* - \theta_t\| - L_o\|\theta_t^* - \theta_t\| \\ =& (L_g - L_o)\|\theta_t^* - \theta_t\|, \end{aligned}$$

where the last inequality uses the fact that there exists ε such that when $\|\theta_t - \theta_t^*\| \leq \varepsilon$,

$$|o(\|\theta_t^* - \theta_t\|)| \leq L_o\|\theta_t^* - \theta_t\|, \quad L_o < L_g.$$

Taking expectation over both sides gives

$$\begin{aligned} \mathbf{E}[h(\omega, \theta_t^*) - h(\omega, \theta_t)] =& (L_g - L_o)\mathbf{E}[\|\theta_t^* - \theta_t\|] \\ \geq& (L_g - L_o)\|\mathbf{E}[\theta_t^* - \theta_t]\|. \end{aligned}$$

Then we conclude that the parameter of interest L_h,

$$L_h = \frac{C_\pi}{L_g - L_o} > 0.$$

where C_π is defined in Assumption 3.4. □

We are ready to present the proof of Proposition 3.3. Based on the definition of $\mathcal{E}_{\hat{J},t}$ and $\mathcal{E}_{\hat{T},t}$, we derive the upper bound for each term respectively.

$$\begin{aligned} \mathcal{E}_{\hat{J},t} =& (\boldsymbol{I} - \gamma \boldsymbol{P}_{\hat{\pi}_{t+1}})^{-1} - (\boldsymbol{I} - \gamma \boldsymbol{P}_{\widetilde{\pi}_{t+1}})^{-1} \\ =& (\boldsymbol{I} - \gamma \boldsymbol{P}_{\widetilde{\pi}_{t+1}})^{-1} \left(\gamma \boldsymbol{P}_{\widetilde{\pi}_{t+1}} - \gamma \boldsymbol{P}_{\hat{\pi}_{t+1}}\right) (\boldsymbol{I} - \gamma \boldsymbol{P}_{\hat{\pi}_{t+1}})^{-1}. \end{aligned}$$

Observe that value function $\boldsymbol{v}$ is smooth and upper bounded. We denote the smoothness parameter by L_v, the upper bound by $\|\boldsymbol{v}\| \leq V^{\max}$, and the smoothness of the reward function by L_r.

By taking the norm of both sides and applying Assumptions 3.3, 3.4 and 3.5, we obtain

$$\|\mathcal{E}_{\hat{J},t}\| \leq M^2 L_J L_v \|\widetilde{\pi}_{t+1} - \hat{\pi}_{t+1}\|_{\mathrm{TV}}.$$

Further, observe that

$$\begin{aligned} \mathcal{E}_{\hat{T},t} &= \boldsymbol{r}_{\hat{\pi}_{t+1}} + \gamma \boldsymbol{P}_{\hat{\pi}_{t+1}} \boldsymbol{v}^{\hat{\pi}_t} - (\boldsymbol{r}_{\widetilde{\pi}_{t+1}} + \gamma \boldsymbol{P}_{\widetilde{\pi}_{t+1}} \boldsymbol{v}^{\hat{\pi}_t}), \\ &= \boldsymbol{r}_{\hat{\pi}_{t+1}} - \boldsymbol{r}_{\widetilde{\pi}_{t+1}} + \gamma(\boldsymbol{P}_{\hat{\pi}_{t+1}} - \boldsymbol{P}_{\widetilde{\pi}_{t+1}})\boldsymbol{v}^{\hat{\pi}_t}. \end{aligned}$$

By taking the norm of both sides and applying Assumption 3.5, we obtain

$$\begin{aligned}\|\mathcal{E}_{\hat{T},t}\| &= \|\boldsymbol{r}_{\hat{\pi}_{t+1}} - \boldsymbol{r}_{\widetilde{\pi}_{t+1}}\| + \|\gamma(\boldsymbol{P}_{\hat{\pi}_{t+1}} - \boldsymbol{P}_{\widetilde{\pi}_{t+1}})\boldsymbol{v}^{\hat{\pi}_t}\| \\ &\leq (L_r + \gamma V^{\max})\|\widetilde{\pi}_{t+1} - \hat{\pi}_{t+1}\|_{\mathrm{TV}} \\ &:= L_T^{\max}.\end{aligned}$$

Recall the definition of $\mathcal{E}_t$ is given as

$$\mathcal{E}_t = -\left(\mathcal{E}_{\hat{J},t}(\boldsymbol{v}^{\hat{\pi}_t} - T(\boldsymbol{v}^{\hat{\pi}_t})) + \boldsymbol{J}_{\hat{\boldsymbol{v}}_t}^{-1}\mathcal{E}_{\hat{T},t} + \mathcal{E}_{\hat{T},t}\mathcal{E}_{\hat{J},t}\right).$$

Taking the norm and expectation on both sides yields that

$$\begin{aligned}\|\mathbf{E}[\mathcal{E}_t]\| \leq \mathbf{E}[\|\mathcal{E}_t\|] &= \mathbf{E}\left[\|\mathcal{E}_{\hat{J},t}(\boldsymbol{v}^{\hat{\pi}_t} - T(\boldsymbol{v}^{\hat{\pi}_t})) + \boldsymbol{J}_{\hat{\boldsymbol{v}}_t}^{-1}\mathcal{E}_{\hat{T},t} + \mathcal{E}_{\hat{T},t}\mathcal{E}_{\hat{J},t}\|\right] \\ &\leq L_{\mathcal{E}}\mathbf{E}[\|\widetilde{\pi}_{t+1} - \hat{\pi}_{t+1}\|_{\mathrm{TV}}],\end{aligned}$$

where $L_{\mathcal{E}} = (2V^{\max}K + L_T^{\max})M^2 L_v L_J + M(L_r + \gamma V^{\max}) > 0$ is a constant. Since $\widetilde{\pi}_{t+1} = \pi_{t+1}^*$ is the greedy solution, we thereby complete the proof of Proposition 3.3. □

3.4.1 Upper Bound on Sub-Optimality Gap

In order to address the question "*Under what condition online learning can be significantly accelerated by a warm-start policy*?", we derive the upper bound on the sub-optimality gap.

Case 1: Unbiased Case. We first consider the finite-time upper bound in the unbiased case, i.e., $B(t) = 0,\ \forall t$. In this case, we introduce the following standard assumption on the Jacobian $\boldsymbol{J}_{\boldsymbol{v}}$.

Assumption 3.5 (Local Lipschitz Continuity) For some $0 < q < 1$ there exist constant $0 < L_J < +\infty$ and constant $0 < M < +\infty$ such that starting from the warm-start policy π_0, the policies $\{\hat{\pi}_t,\ t = 1, 2, \ldots\}$ generated by the A-C algorithm satisfy

$$\|\boldsymbol{J}_{\boldsymbol{v}^*} - \boldsymbol{J}_{\boldsymbol{v}^{\hat{\pi}_t}}\| \leq L_J\|\boldsymbol{v}^{\hat{\pi}_t} - \boldsymbol{v}^*\|^q,$$

and $\|\boldsymbol{J}_{\boldsymbol{v}^{\hat{\pi}_t}}^{-1}\| \leq M$.

Intuitively, Assumption 3.5 means that the difference of Jacobian $\|\boldsymbol{J}_{\boldsymbol{v}^{\pi_t}} - \boldsymbol{J}_{\boldsymbol{v}^*}\|$ is small whenever the underlying value functions that induces the policies are close. We note that the conditions of this type are commonly used in the convergence analysis of policy iteration algorithms for exact dynamic programming [118, 264]. In particular, we remark that the Jacobian function (of π or $\boldsymbol{v}^{\pi}$) is non-linear and this assumption implies the learned policy initialized with it is essential for the warm-start policy to be reasonably "close" to the optimal policy. Next, we present the finite-time upper bound in the unbiased case.

Proposition 3.4 (Unbiased Case) *In the unbiased case, i.e.,* $\mathcal{B}_t = 0, \ \forall\, t \geq 0$, *we have*

$$\|\mathbf{E}[\boldsymbol{v}^* - \boldsymbol{v}^{\hat{\pi}_{t+1}}]\| \leq L\|\mathbf{E}[\boldsymbol{v}^* - \boldsymbol{v}^{\hat{\pi}_t}]\|^{1+q} \tag{3.20}$$

where $L := ML_J$ *with* M *and* L_J *defined in Assumption 3.5. By applying Eq. 3.20 recursively, we obtain,*

$$\|\mathbf{E}[\boldsymbol{v}^* - \boldsymbol{v}^{\hat{\pi}_{t+1}}]\| \leq L^{\frac{(1+q)^{t+1}-1}{q}} \|\boldsymbol{v}^* - \boldsymbol{v}^{\pi_0}\|^{(1+q)^{(t+1)}}$$

In Proposition 3.4, as the Warm-start policy is close to the optimal policy, we establish the superlinear convergence of $\mathbf{E}[\boldsymbol{v}^* - \boldsymbol{v}^{\hat{\pi}_{t+1}}]$ in the presence of approximation error $\mathcal{E}(t)$ from both Actor update and Critic update. This observation corroborates the most recent empirically finding [34, 303], where the online RL can further improve the warm-start policy by only few adaptation steps.

Case 2: Bounded Bias. Next, we present the finite-time upper bound in the general case when the bias is upper bounded (as given in Proposition 3.3).

Corollary 3.1 *If Assumption 3.5 holds in the biased case, we have that for any* $t > 0$,

$$\|\mathbf{E}[\boldsymbol{v}^* - \boldsymbol{v}^{\hat{\pi}_{t+1}}]\| \leq L\|\mathbf{E}[\boldsymbol{v}^* - \boldsymbol{v}^{\hat{\pi}_t}]\|^{1+q} + L_b H_t. \tag{3.21}$$

By applying Eq. 3.21 recursively, we obtain,

$$\begin{aligned}\|\mathbf{E}[\boldsymbol{v}^* - \boldsymbol{v}^{\hat{\pi}_{t+1}}]\| \leq\ & \|\boldsymbol{v}^* - \boldsymbol{v}^{\pi_0}\|^{(1+q)^{1+t}} \\ & \cdot (L \cdots ((L + u_1)^{1+q} + u_2)^{1+q} \cdots + u_t),\end{aligned}$$

where $u_t := \frac{L_b H_t}{\|\boldsymbol{v}^* - \boldsymbol{v}^{\pi_0}\|^{(1+q)^{(1+t)}}}$ *and* $L_b H_t$ *is the upper bound of the bias as in Proposition 3.3.*

Proof Based on the update rule of the value function, we have

$$\begin{aligned}\boldsymbol{v}^* - \boldsymbol{v}^{\hat{\pi}_{t+1}} =& \boldsymbol{J}_{\hat{\boldsymbol{v}}_t}^{-1} \boldsymbol{J}_{\hat{\boldsymbol{v}}_t}(\boldsymbol{v}^* - \boldsymbol{v}^{\hat{\pi}_t}) + \boldsymbol{J}_{\hat{\boldsymbol{v}}_t}^{-1}\left(\boldsymbol{v}^{\hat{\pi}_t} - T(\boldsymbol{v}^{\hat{\pi}_t})\right) + \mathcal{E}_t \\ \leq& \boldsymbol{J}_{\hat{\boldsymbol{v}}_t}^{-1} \boldsymbol{J}_{\hat{\boldsymbol{v}}_t}(\boldsymbol{v}^* - \boldsymbol{v}^{\hat{\pi}_t}) - \boldsymbol{J}_{\hat{\boldsymbol{v}}_t}^{-1} \boldsymbol{J}_{\boldsymbol{v}^*}(\boldsymbol{v}^* - \boldsymbol{v}^{\hat{\pi}_t}) - \mathcal{E}_t \\ \leq& \boldsymbol{J}_{\hat{\boldsymbol{v}}_t}^{-1}\left[\boldsymbol{J}_{\hat{\boldsymbol{v}}_t} - \boldsymbol{J}_{\boldsymbol{v}^*}\right](\boldsymbol{v}^* - \boldsymbol{v}^{\hat{\pi}_t}) + \mathcal{E}_t,\end{aligned}$$

which implies that

$$\mathbf{E}_{\hat{\pi}_{t+1}}[\boldsymbol{v}^* - \boldsymbol{v}^{\hat{\pi}_{t+1}} | \boldsymbol{v}^{\hat{\pi}_t}] \leq \mathbf{E}_{\hat{\pi}_{t+1}}[\boldsymbol{J}_{\hat{\boldsymbol{v}}_t}^{-1}]\left[\boldsymbol{J}_{\hat{\boldsymbol{v}}_t} - \boldsymbol{J}_{\boldsymbol{v}^*}\right](\boldsymbol{v}^* - \boldsymbol{v}^{\hat{\pi}_t}) + \mathcal{B}(t).$$

Then, taking expectation over $\hat{\pi}_t$ on both sides gives us,

$$\mathbf{E}_{\hat{\pi}_{t+1},\hat{\pi}_t}[\boldsymbol{v}^* - \boldsymbol{v}^{\hat{\pi}_{t+1}}|\boldsymbol{v}^{\hat{\pi}_t}] \leq \mathbf{E}_{\hat{\pi}_{t+1},\hat{\pi}_t}[\boldsymbol{J}_{\hat{\boldsymbol{v}}_t}^{-1}\left[\boldsymbol{J}_{\hat{\boldsymbol{v}}_t} - \boldsymbol{J}_{\boldsymbol{v}^*}\right]]\mathbf{E}_{\hat{\pi}_t}[(\boldsymbol{v}^* - \boldsymbol{v}^{\hat{\pi}_t})] + \mathcal{B}(t) \quad (3.22)$$

Let $J_t := \boldsymbol{J}_{\hat{\boldsymbol{v}}_t}^{-1}\left[\boldsymbol{J}_{\hat{\boldsymbol{v}}_t} - \boldsymbol{J}_{\boldsymbol{v}^*}\right]$. It follows from Assumption 3.5 that

$$\|J_t\| \leq ML_J\|\boldsymbol{v}^{\hat{\pi}_t} - \boldsymbol{v}^*\|^q := L\|\boldsymbol{v}^{\hat{\pi}_t} - \boldsymbol{v}^*\|^q.$$

where $L = ML_J$ and L_J is defined in Assumption 3.5.

Meanwhile, we have,

$$\begin{aligned}\|\mathbf{E}[J_t]\| \leq& \mathbf{E}[\|J_t\|]\\ \leq& L\mathbf{E}[\|\boldsymbol{v}^{\hat{\pi}_t} - \boldsymbol{v}^*\|^q]\\ \leq& L\|\mathbf{E}[\boldsymbol{v}^{\hat{\pi}_t} - \boldsymbol{v}^*]\|^q,\end{aligned}$$

where the last inequality follows Jensen's inequality.

Then, taking norm on both sides of the inequality (3.22) gives

$$\begin{aligned}\|\mathbf{E}_{\hat{\pi}_{t+1},\hat{\pi}_t}[\boldsymbol{v}^* - \boldsymbol{v}^{\hat{\pi}_{t+1}}|\boldsymbol{v}^{\hat{\pi}_t}]\| \leq& \|\mathbf{E}_{\hat{\pi}_{t+1},\hat{\pi}_t}[\boldsymbol{J}_{\hat{\boldsymbol{v}}_t}^{-1}\left[\boldsymbol{J}_{\hat{\boldsymbol{v}}_t} - \boldsymbol{J}_{\boldsymbol{v}^*}\right]]\mathbf{E}_{\hat{\pi}_t}[(\boldsymbol{v}^* - \boldsymbol{v}^{\hat{\pi}_t})] + \mathcal{B}(t)\|\\ \leq& \|\mathbf{E}_{\hat{\pi}_{t+1},\hat{\pi}_t}[\boldsymbol{J}_{\hat{\boldsymbol{v}}_t}^{-1}\left[\boldsymbol{J}_{\hat{\boldsymbol{v}}_t} - \boldsymbol{J}_{\boldsymbol{v}^*}\right]]\|\|\mathbf{E}_{\hat{\pi}_t}[(\boldsymbol{v}^* - \boldsymbol{v}^{\hat{\pi}_t})]\| + \|\mathcal{B}(t)\|\\ =& \|\mathbf{E}_{\hat{\pi}_{t+1},\hat{\pi}_t}[J_t]\|\|\mathbf{E}_{\hat{\pi}_t}[(\boldsymbol{v}^* - \boldsymbol{v}^{\hat{\pi}_t})]\| + \|\mathcal{B}(t)\|\\ \leq& L\|\mathbf{E}_{\hat{\pi}_t}[(\boldsymbol{v}^* - \boldsymbol{v}^{\hat{\pi}_t})]\|^{1+q} + \|\mathcal{B}(t)\|\end{aligned}$$

Let $a_t = \|\mathbf{E}_{\hat{\pi}_t}[(\boldsymbol{v}^* - \boldsymbol{v}^{\hat{\pi}_t})]\|$ and $b_t = \|\mathcal{B}(t)\|$. Then we have the following recursive inequality,

$$a_{t+1} \leq La_t^{1+q} + b_t, \quad t = 0, 1, \ldots \quad (3.23)$$

Staring from $t = 0$, we have,

$$a_1 \leq La_0^{1+q} + b_0$$

Let $b_0 = u_0 a_0^{1+q}$, where $u_0 = \frac{L_b H_t}{a_0^{1+q}}$, then we have,

$$a_1 \leq (L + u_0)a_0^{1+q}$$

Similarly, let $t = 1$ and $b_1 = u_1 a_0^{(1+q)^2}$ with $u_0 = \frac{L_b H_t}{a_0^{(1+q)^2}}$. Then we have,

$$a_2 \leq (L(L + u_0)^{1+q} + u_1)a_0^{(1+q)^2}$$

By applying Eq. 3.23 recursively, we conclude that

$$\|\mathbf{E}[\boldsymbol{v}^* - \boldsymbol{v}^{\hat{\pi}_{t+1}}]\| \leq \|\boldsymbol{v}^* - \boldsymbol{v}^{\pi_0}\|^{(1+q)^{1+t}} \cdot (L \cdots ((L + u_1)^{1+q} + u_2)^{1+q} \cdots + u_t),$$

where $u_t := \frac{L_b H_t}{\|\boldsymbol{v}^* - \boldsymbol{v}^{\pi_0}\|^{(1+q)(1+t)}}$ and $L_b H_t$ is the upper bound of the bias. □

Implication on Reducing the Performance Gap. The upper bound in Corollary 3.1 sheds light on the impact of warm-start policy π_0 (the first term) and the bias $\{\mathcal{B}(t)\}$ (u_t) (the second term), thereby providing guidance on how to achieve desired finite-time learning performance. When the bias $\mathcal{B}(t) \neq \mathbf{0}$ $(u_t \neq 0)$, the upper bound hinges heavily on the biases in the approximation errors, even when the warm-start policy π_0 is close to the optimal policy (see the second term in Eq. 3.21). In this case, recall the result on the upper bound of the bias $\mathcal{B}(t)$ in Proposition 3.3, where we establish the connection between the bias and the approximation error. As expected, in order to reduce the performance gap, it is essential to decrease the bias in the approximation error, which can be achieved by increasing gradient steps, rollout length and sample sizes.

"Wash-out" Phenomenon. In Corollary 3.1, the product structure between the warm-start term and bias term also implies that the imperfections of the Warm-start policy can be "washed out" by online learning when the bias is close to zero. For instance, when the value function $\boldsymbol{v}^{\pi_0}$ induced by the Warm-start policy π_0 is bounded away from $\boldsymbol{v}^*$, e.g., $\varepsilon < \|\boldsymbol{v}^{\pi_0} - \boldsymbol{v}^*\| < L^{-q}$ and the bias is sufficiently small, e.g., $u_t \leq \varepsilon^{-q} - L$, then we have $\|\mathbf{E}[\boldsymbol{v}^{\pi_1} - \boldsymbol{v}^*]\| \leq \|\boldsymbol{v}^{\pi_0} - \boldsymbol{v}^*\|$. We note that this result corroborates with the observation in the very recent literature [34] and this phenomenon has not been formalized by previous works on error propagation [188, 237]. Furthermore, we clarify that the "Wash-out" phenomenon in Corollary 3.1 would not hold in the case when Assumption 3.5 is not satisfied, which may likely yield a policy far away from the optimal during the online learning.

Remark In the case when the bias is pronounced, Assumption 3.5 can be stringent. Nevertheless, it is of more interest to find lower bounds on the sub-optimality gap, which we turn our attention to next.

3.4.2 Lower Bound on Sub-Optimality Gap

Aiming to understand "*whether online learning can be accelerated by a warm-start policy*", we derive a lower bound to quantify the impact of the bias and the error propagation. Let $(\pi_0, \hat{\pi}_1, \ldots, \hat{\pi}_t)$ be the sequence of policies generated by running t-step A-C algorithm in Eqs. 3.8 and 3.16. Fro convenience, let filtration $\mathcal{F}_t$ be the σ-algebra generated by $(\pi_0, \hat{\pi}_1, \ldots, \hat{\pi}_t)$. We obtain the lower bound by unrolling the recursion of the Newton update (with perturbation) Eq. 3.19.

Theorem 3.2 *Conditioned on the filtration* $\mathcal{F}_t = \sigma(\pi_0, \hat{\pi}_1, \ldots, \hat{\pi}_t)$*, the lower bound of* $\|\mathbf{E}[\boldsymbol{v}^* - \boldsymbol{v}^{\hat{\pi}_{t+1}}|\mathcal{F}_t]\|$ *satisfies that*

$$\|\mathbf{E}\left[\boldsymbol{v}^* - \boldsymbol{v}^{\hat{\pi}_{t+1}}|\mathcal{F}_t\right]\| \geq \|\gamma^{t+1}\bar{\boldsymbol{P}}_{t+1}(\boldsymbol{v}^* - \boldsymbol{v}^{\pi_0}) + \sum_{i=1}^{t}\gamma^i\bar{\boldsymbol{P}}_i\mathcal{B}(t-i) + \mathcal{B}(t)\|, \tag{3.24}$$

where $\bar{\boldsymbol{P}}_{t+1} = \mathbf{E}\left[\left(\prod_{i=0}^{t}\boldsymbol{P}_{\pi_{t+1-i}}\right)\right]$.

Proof Following the value function update rule, we have

$$\begin{aligned}
\boldsymbol{v}^{\hat{\pi}_{t+1}} &= \boldsymbol{v}^{\hat{\pi}_t} - \left(\boldsymbol{J}_{\hat{\boldsymbol{v}}_t}^{-1}\left(\boldsymbol{v}^{\hat{\pi}_t} - T(\boldsymbol{v}^{\hat{\pi}_t})\right) + \mathcal{E}_t\right)\\
&= \boldsymbol{v}^{\hat{\pi}_t} - (L(t) + \mathcal{E}_t)\\
&:= \boldsymbol{v}^{\hat{\pi}_t} - \hat{\mathcal{L}}(t).
\end{aligned}$$

Then, the difference between $\boldsymbol{v}^{\hat{\pi}_{t+1}}$ and $\boldsymbol{v}^*$ is given by

$$\boldsymbol{v}^* - \boldsymbol{v}^{\hat{\pi}_{t+1}} = \boldsymbol{v}^* - \boldsymbol{v}^{\hat{\pi}_t} + \boldsymbol{J}_{\hat{\boldsymbol{v}}_t}^{-1}\left(\boldsymbol{v}^{\hat{\pi}_t} - T(\boldsymbol{v}^{\hat{\pi}_t})\right) + \mathcal{E}_t. \tag{3.25}$$

Observe the following result holds for any $\hat{\pi}_t$,

$$(\boldsymbol{v}^{\hat{\pi}_t} - T(\boldsymbol{v}^{\hat{\pi}_t})) - \underbrace{(\boldsymbol{v}^* - T(\boldsymbol{v}^*))}_{=0} \geq \boldsymbol{J}_{\hat{\boldsymbol{v}}_t}^{2}(\boldsymbol{v}^{\hat{\pi}_t} - \boldsymbol{v}^*). \tag{3.26}$$

Recall our decomposition of the value function update is given as

$$\hat{\mathcal{L}}(t) = \mathcal{L}(t) + \underbrace{\hat{\mathcal{L}}(t) - \mathbf{E}[\hat{\mathcal{L}}(t)]}_{\text{Martingale Difference Noise: } \mathcal{N}(t)} + \underbrace{\mathbf{E}[\hat{\mathcal{L}}(t)] - \mathcal{L}(t)}_{\text{Bias: } \mathcal{B}(t)}.$$

Plugging Eq. 3.26 into Eq. 3.25, we obtain

$$\begin{aligned}
\boldsymbol{v}^* - \boldsymbol{v}^{\hat{\pi}_{t+1}} &= \boldsymbol{v}^* - \boldsymbol{v}^{\hat{\pi}_t} + \left(\boldsymbol{J}_{\hat{\boldsymbol{v}}_t}^{-1}\left(\boldsymbol{v}^{\hat{\pi}_t} - T(\boldsymbol{v}^{\hat{\pi}_t})\right) + \mathcal{E}_t\right)\\
&\geq \left(\boldsymbol{I} - \boldsymbol{J}_{\boldsymbol{v}^{\hat{\pi}_t}})\right)(\boldsymbol{v}^* - \boldsymbol{v}^{\hat{\pi}_t}) + \mathcal{B}(t) + \mathcal{N}(t)\\
&= \gamma\boldsymbol{P}_{\tilde{\pi}_{t+1}}(\boldsymbol{v}^* - \boldsymbol{v}^{\hat{\pi}_t}) + \mathcal{B}(t) + \mathcal{N}(t).
\end{aligned}$$

Taking expectation on both sides yields that

$$\mathbf{E}[\boldsymbol{v}^* - \boldsymbol{v}^{\hat{\pi}_{t+1}}|\boldsymbol{v}^{\hat{\pi}_t}] \geq \gamma\boldsymbol{P}_{\tilde{\pi}_{t+1}}(\boldsymbol{v}^* - \boldsymbol{v}^{\hat{\pi}_t}) + \mathcal{B}(t).$$

Applying the above inequality recursively gives that

$$\begin{aligned}\mathbf{E}\left[\boldsymbol{v}^* - \boldsymbol{v}^{\hat{\pi}_{t+1}}\right] \geq & \gamma^{t+1}\mathbf{E}\left[\left(\prod_{i=0}^{t} \boldsymbol{P}_{\widetilde{\pi}_{t+1-i}}\right)\right](\boldsymbol{v}^* - \boldsymbol{v}^{\pi_0}) \\ & + \sum_{i=1}^{t}\gamma^i \mathbf{E}\left[\left(\prod_{j=0}^{i-1} \boldsymbol{P}_{\widetilde{\pi}_{t+1-j}}\right)\right](\mathcal{B}(t-i)) + \mathcal{B}(t) \\ := & \gamma^{t+1}\bar{\boldsymbol{P}}_{t+1}(\boldsymbol{v}^* - \boldsymbol{v}^{\pi_0}) + \sum_{i=1}^{t}\gamma^i \bar{\boldsymbol{P}}_i \mathcal{B}(t-i) + \mathcal{B}(t),\end{aligned} \tag{3.27}$$

with $\bar{\boldsymbol{P}}_{t+1} = \mathbf{E}\left[\left(\prod_{i=0}^{t} \boldsymbol{P}_{\widetilde{\pi}_{t+1-i}}\right)\right]$. Taking norm on both sides of Eq. 3.27 yields the desired results. □

Error Propagation and Accumulation. It can be seen form Theorem 3.2 that the bias terms $\{\mathcal{B}(t)\}$ add up over time, and the propagation effect of the bias terms is encapsulated by the last two terms on the right side of Eq. (3.24). Clearly, the first term on the right side, corresponding to the impact of the warm-start policy π_0, diminishes with A-C updates. To get a more concrete sense of Theorem 3.2, we consider the following special settings. (1) When the bias is always positive, i.e., $\mathcal{B}(t) > \mathbf{0}$ for all $t \geq 0$, the lower bound in Theorem 3.2 is always positive, i.e., $\|\mathbf{E}\left[\boldsymbol{v}^* - \boldsymbol{v}^{\hat{\pi}_{t+1}}\right]\| \geq \|\mathcal{B}(t)\| > 0$. In this case, the sub-optimal gap remains bounded away from zero. Similar conclusion can be made when the bias is always negative. (2) When the bias term can be either positive or negative, the lower bound is shown as Eq. 3.24. In this case, the learning performance of the A-C algorithm largely depends on the behavior of the Bias term. It can be seen from Theorem 3.2 that even when the warm-start policy is near-optimal, it is still challenging to guarantee that online fine-tuning can improve the policy if the approximation error is not handled correctly. We note that this has also been observed empirically [196, 238].

Remark The primary goal of this work is to make a first attempt to quantify the learning performance of Warm-start RL by studying its convergence behavior. It can be seen from Corollary 3.1 and Theorem 3.2 that the bounds are in terms of the biases $\{B(t)\}$, and the structure of $\{B(t)\}$ remains open and is highly nontrivial. Hence, we submit that the convergence rate and the sampling complexity are of great interest but it is beyond the scope of this work.

Remark We clarify the connection between our work and previous works on the "coverage" requirements (e.g., Assumption A [364]). The concentrability condition [364] characterizes the distance between the visitation distributions of the warm-start policy and some optimal policy for *every* state-action pair. Hence, this "coverage" assumption requires the state-action point-wise distance between the optimal policy and the policy to be upper bounded in the worst-case scenario, implying the bias is also bounded above since the worse-case distance is larger than average distance in general. While in our setting, we evaluate the sub-optimality gap in the average sense, i.e., $\mathbf{E}[\boldsymbol{v}^* - \boldsymbol{v}^{\pi_t}]$, by characterizing the upper bound of the bias from the Actor update and Critic update. Meanwhile, the performance requirements for online learning algorithms in the previous work (e.g., Bellman error is upper bounded by [315]) correspond to the second term on the RHS of Proposition 3.4, Corollary 3.1 and Theorem 3.2, where we show that upper bound of the approximation error in the Actor update has direct impact on the sub-optimality.

3.5 Experiments

Empirical Results. We consider experiments over the Gridworld benchmark task. In particular, we consider the following sizes of the grid to represent different problem complexity, i.e., 10×10, 15×15 and 20×20. The goal of the agent is to find a way (policy) to travel from a specified start location, e.g., the red square in Fig. 3.3, to an assigned target location, e.g., the red hexagram in Fig. 3.3, such that the (discounted) accumulative reward along the way is maximized. Specifically, the action space contains 4 discrete actions, namely, up, down, left, right, which are represented as 1, 2, 3, 4 in the algorithm, respectively. The reward in the goal state is defined as 10 and in the bad state, e.g., the black cube in Fig. 3.3, is -6. The rest of the states result in the reward -1. The discounting factor is set as $\gamma = 0.9$. We consider the grid with 10 rows and 10 columns such that the state space has 100 states. The transition properties of the environment is as follows: the agent will transfer to next state following the chosen action with probability 0.7; the agent will go left of the desired action with probability 0.15 and go right with probability 0.15. For each experiment, the shaded area represents a standard deviation of the average evaluation over 5 training seeds.

Specifically, we consider the following A-C algorithm to solve the Gridworld benchmark task,

Critic Update: The Critic updates its value by applying the Bellman evaluation operator (T^{π}) for m-times ($m \geq 1$), i.e., given policy π, at the t-th step A-C update,

$$\boldsymbol{v}(t+1) = (T^{\pi})^m(\boldsymbol{v}(t)). \tag{3.28}$$

Actor Update: The Actor updates the policy by a greedy step to maximize the learnt $\boldsymbol{v}$ value, i.e.,

$$\pi' = \arg\max_{\pi} T^{\pi}(\boldsymbol{v}(t+1)). \tag{3.29}$$

Impact of the Warm-Start Policy. We first consider the impact of the Warm-Start policy in the ideal setting, where both the Critic update and Actor update is nearly accurate as in ADP. In this case, we let m be large enough, e.g., $m = 1000$, in the Critic update Eq. (3.28). As observed in Fig. 3.4, a 'good' Warm-Start policy can efficiently accelerate the

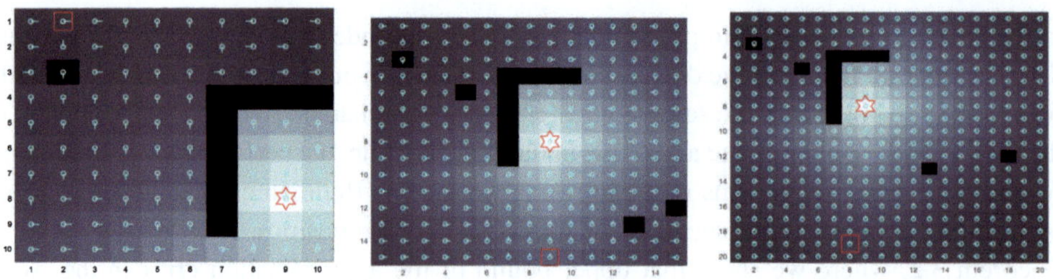

(a) 10× 10 Gridworld. (b) 15× 15 Gridworld. (c) 20× 20 Gridworld.

Fig. 3.3 Gridworld benchmark with different sizes. The colors specify the 'goodness' measure of the state, i.e., the darker color cubes are with lower $v(s)$ value and the agent should avoid those areas. The horizontal lines and vertical lines in each cube point to the direction the agent should take, i.e., policy at every state. Figure 3.3a, b and c show the learning results after 50 iterations of A–C update

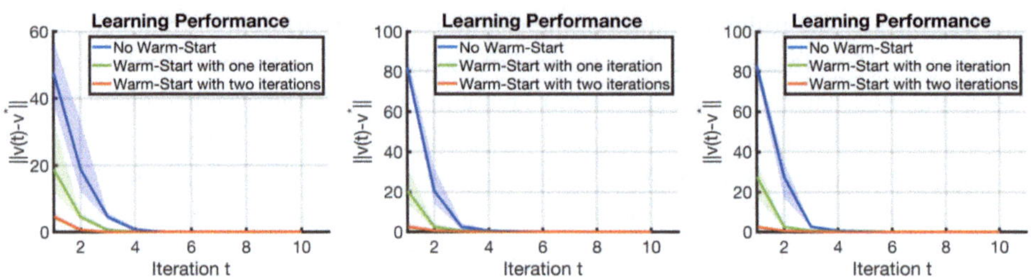

(a) 10 × 10 Gridworld. (b) 15 × 15 Gridworld. (c) 20 × 20 Gridworld.

Fig. 3.4 The impact of the Warm-Start Policy when no approximation errors in Actor update and Critic update. The convergence behavior given different initial policy, i.e., a random policy (no Warm-Start), a Warm-Start policy obtained by running the A-C algorithm for one iteration and two iterations. The x-axis represents the A-C update step and y-axis is the value of the norm $\|\boldsymbol{v}(t) - \boldsymbol{v}^*\|$

learning process, e.g., it only takes two iterations to convergence with a Warm-Start policy. Meanwhile, in all three cases, the performance gap $\|\boldsymbol{v}(t) - \boldsymbol{v}^*\|$ decays over time which reflects our discovery in Corollary 3.1. Specifically, when the Warm-Start policy is not 'good' enough (or even no Warm-Start), the A-C algorithm can still be able to improve the learning performance overtime (see e.g., the first term on the right side of the upper bound in Corollary 3.1).

Impact of the Approximation Error in the Critic Update. We evaluate the impact of the approximation error in the Critic update on the convergence behavior by two approaches. (1) First, we study the Critic update with finite time Bellman evaluation, e.g., $m = 500, 50, 20, 5$. As shown in Fig. 3.5, the inaccurate Critic update impacts the convergence behavior as expected. The case when $m = 5$ shows that the finite time Bellman evaluation may contribute to the slower convergence. (2) Next, we consider the general case when there is approximation error in the Critic update. In particular, we add the uniform noise $e(t)$ in the value

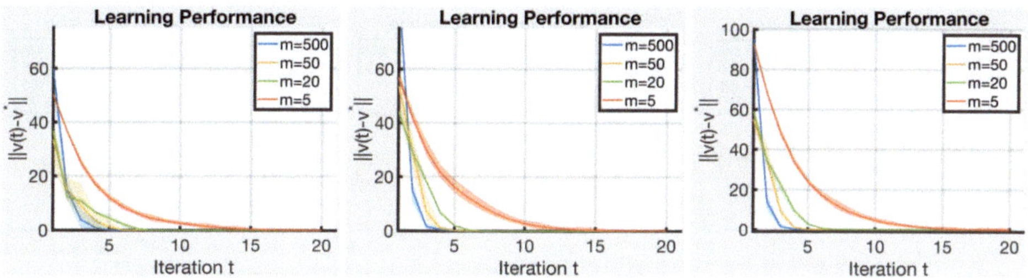

(a) 10 × 10 Gridworld. (b) 15 × 15 Gridworld. (c) 20 × 20 Gridworld.

Fig. 3.5 Learning performance versus rollout length

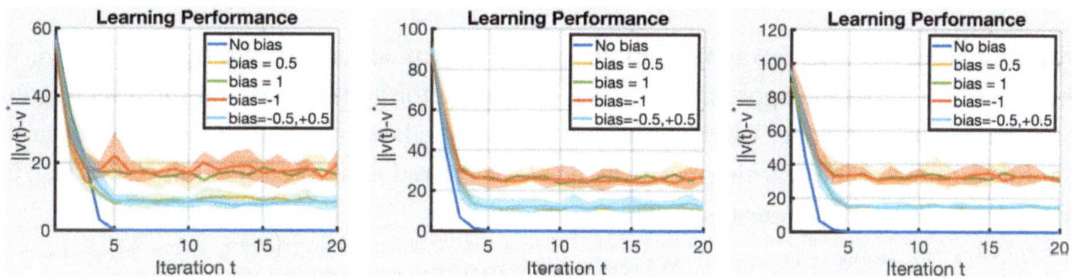

(a) 10 × 10 Gridworld. (b) 15 × 15 Gridworld. (c) 20 × 20 Gridworld.

Fig. 3.6 Illustration of the lower bound in Theorem 3.1

function with different bias, e.g.,$\mathbf{E}[e(t)] = 0, 0.5, 1, -1$. Meanwhile, we also consider the case when the bias can be either $+0.5$ or -0.5 in the learning process, e.g., $\mathbf{E}[e(t)] = 0.5$ with probability 0.5 and $\mathbf{E}[e(t)] = -0.5$ with probability 0.5. The resulting convergence behavior is presented in Fig. 3.6. Notably, it can be clearly seen that both the positive and negative bias may result in an error floor and 'prevent' the algorithm from converging to the optimal (e.g., the last two terms of the lower bound in Theorem 3.2). The experiment results in Fig. 3.6 corroborate our theoretical findings in Proposition 3.1, Corollary 3.1 and Theorem 3.2.

Impact of the Approximation Error in the Actor Update. We investigate the learning performance of the A-C algorithm under inaccurate Actor update. In particular, we add the perturbation on the learnt policy in Eq. (3.29) as follows,

$$\text{Policy}(s) = \begin{cases} \text{Policy}(s), & p, \\ \text{randi}([1, 4]), & 1 - p. \end{cases}$$

where Policy(s) denotes the action should the agent take at the current state s following the learnt policy and randi$([1, 4])$ is a random function to choose the action $1, 2, 3, 4$ uniformly. Thus, with probability p, the agent will choose the action follow the current policy while

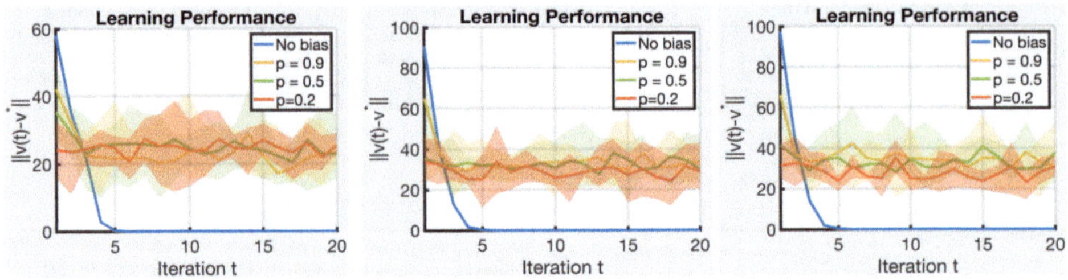

(a) 10×10 Gridworld. (b) 15×15 Gridworld. (c) 20×20 Gridworld.

Fig. 3.7 Convergence behavior versus approximation error in the actor update

with probability $1 - p$, the agent will choose a random action. By setting different p, we show in Fig. 3.7 that the approximation error in the Actor update may significantly degrade the learning performance. Meanwhile, Fig. 3.7 also indicates that decreasing bias can be helpful to improve the learning performance (see the red and green lines in Fig. 3.7). This observation also verifies our results in Theorem 3.2.

3.6 Conclusion

In this chapter, we take a finite-time analysis approach to address the question "*whether and when online learning can be significantly accelerated by a warm-start policy from offline RL?*" in Warm-Start RL. By delving into the intricate coupling between the updates of the Actor and the Critic, we first provide upper bounds on the approximation errors in both the Critic update and Actor update of online adaptation, respectively, where the recent advances on Bernstein's Inequality are leveraged to deal with the sample correlation therein. Based on these results, we next cast the Warm-Start A-C method as Newton's method with perturbation, which serves as the foundation for characterizing the impact of the approximation errors on the finite-time learning performance of Warm-Start A-C. In particular, we provide upper bounds on the sub-optimality gap, which provides guidance on the design of Warm-Start RL for achieving desired finite-time learning performance. And we also derive lower bounds on the sub-optimality gap under biased approximation errors, indicating that the performance gap can be bounded away from zero even with a good prior policy. We note that as the biases structure remains open, the study on the efficiency of Warm-start RL calls for additional work. Finally, it is also worth to explore the setting beyond linear function approximation and further derive the practical warm-start RL algorithm utilizing the theoretical findings in this chapter.

Meta-Learning 4

4.1 Introduction

Transfer learning [256] is a powerful technique that enhances the learning performance of a target task by leveraging knowledge from a related source task. There are two main categories of transfer learning: parameter transfer and sample transfer. In parameter transfer, the learned parameters from the source task are directly copied to the target task's learning model. In sample transfer, training samples from the source task are integrated into the target task's dataset and contribute to its training process. Comparing these two methods, sample transfer can provide additional valuable information and allow for preprocessing of the transferred samples to better align them with the target task, while parameter transfer offers significant savings in training costs and thus is very helpful for models with a large number of parameters such as deep neural networks (DNNs).

As a special case of parameter-based transfer learning, meta-learning, a.k.a., learning to learn, has recently attracted much attention as an effective approach to enable fast learning with a few data samples by learning the valuable knowledge across similar learning tasks. The concept of meta-learning, a.k.a., learning to learn, is not new, which was first proposed by Jürgen Schmidhuber in his diploma thesis [292] in 1987. The primary goal is to learn better learning algorithms and achieve continual self-improving, based on the accumulated experience through the learning process [327, 338]. Different with traditional transfer learning, meta-learning explicitly trains for the ability to learn by uncovering the deep similarity relation among tasks, which enables fast and precise adaptation on unseen tasks. Particularly, meta-learning learns a hypothesis meta-prior among previous tasks such that the learnt meta-prior can be leveraged for quick learning on new related tasks with a small amount of data. The meta-training stage for learning the meta-prior is often cast as a bi-level problem, where the inner loop solves the fast learning problem for a single task given

H. Wang et al., *Continual and Reinforcement Learning for Edge AI*,
Synthesis Lectures on Learning, Networks, and Algorithms,
https://doi.org/10.1007/978-3-031-84363-1_4

the meta-prior, and the outer loop updates the meta-prior such that the solutions to the inner problems improve some overall meta-objective across all tasks. Based on what meta-prior is learnt, meta-learning approaches can be typically divided into three categories: model-based, metric-based, and optimizer-based. (1) The meta-prior in model-based approaches (e.g., [286]) can be a recurrent neural network with self-connected recursive neurons, e.g., a Long-Short Term Memory (LSTM) [144], where the neurons include memory cells to store model experiences and can be updated sequentially to model self-improvement. The model sequentially takes in the dataset and then processes new inputs from the task. Here, the outer loop trains the recurrent network with gradient descent, and the inner loop simply rolls out the recurrent network. (2) The meta-prior in metric-based approaches (e.g., [173, 313, 320, 341]) is a metric space where learning is particularly efficient. The general idea is to learn embedding vectors of input data explicitly and use them to design proper kernel functions for the metric space. The meta-prior can be trained with gradient descent in the outer loop; and the inner loop corresponds to a comparison scheme, e.g., nearest neighbors, in the meta-learned metric space. (3) Optimizer-based approaches learn an optimizer for fast learning in the inner loop, which can be a model initialization [103, 244], a model regularization [403], or hyper-parameters of a learning algorithm [167]. Gradient descent is used not only in the outer loop to train the meta-prior, but also in the inner loop to achieve quick model adaptation based on the meta-prior.

In this chapter, we focus on the optimizer-based meta-learning approach because of its simplicity and generalization ability, which can be very effective for resource-constrained edge nodes. In particular, the idea of learning a model initialization through meta-learning on similar tasks provides us a promising solution to achieve efficient edge learning. For a better understanding of how meta-learning an initialization works, we briefly introduce below the popular algorithm, namely, MAML [103], which learns a model initialization that can be quickly adapted at new tasks. Note that a classical way to learn the initialization is pretraining on a large dataset and fine-tuning on a smaller dataset. However, this classic pre-training approach has no guarantee of learning an initialization that is good for fine-tuning [244], and the learnt model may still need to be heavily adapted to the new task, where severe overfitting can be caused with a small dataset for adaptation [213]. In contrast, MAML directly optimizes the performance with respect to the initialization. More specifically, all the tasks in MAML are assumed to follow a meta-model which is represented by a parameterized function f_θ with parameter θ. At the meta-training time, a set of J tasks $\{\mathcal{T}_j\}_{j=1}^J$ are drawn from the prior distribution $\mathbb{P}(\mathcal{T})$ with the corresponding dataset $\mathcal{D}_j$. For each training task $\mathcal{T}_j$, its dataset $\mathcal{D}_j$ is divided into two separate sets, the training set $\mathcal{D}_j^{train}$ and the testing set $\mathcal{D}_j^{test}$. When adapting to task $\mathcal{T}_j$, the model's parameters θ become ϕ_j. The updated parameter ϕ_j can be computed using one or more gradient descent steps with the training set $\mathcal{D}_j^{train}$, e.g., for a given loss function $\mathcal{L}$ and learning rate α,

$$\phi_j = \theta - \alpha \nabla_\theta \mathcal{L}(\mathcal{D}_j^{train}, \theta). \tag{4.1}$$

MAML explicitly optimizes for the model initialization θ such that the updated parameter ϕ_j with the training dataset $\mathcal{D}_j^{train}$ can achieve good testing performance on the testing dataset $\mathcal{D}_j^{test}$ for the task $\mathcal{T}_j$. Mathematically, this can be formulated as the following optimization problem:

$$\min_{\theta} \sum_{\mathcal{T}_j \sim \mathbb{P}(\mathcal{T})} \mathcal{L}(\mathcal{D}_j^{test}, \theta - \alpha \nabla_\theta \mathcal{L}(\mathcal{D}_j^{train}, \theta)).$$

During meta-testing, the task-specific model ϕ_i for a new testing task $\mathcal{T}_i$ can be quickly adapted from the meta initialization θ based on (4.1) using the small local dataset $\mathcal{D}_i$. Such a fast learning capability with small datasets makes meta-learning naturally an attractive technique to achieve real-time edge intelligence.

4.2 Recent Algorithm Development

Optimizer-based meta-learning. Following the development of MAML, there have been a lot of studies on more advanced optimizer-based meta-learning strategies. To reduce the computational complexity, Nichol et al. [244] introduces a first-order meta-learning algorithm called Reptile, which does not require the computation of the second-order derivatives. To address the task ambiguity problem, a probabilistic meta-learning algorithm is proposed in Finn et al. [104] to incorporate a parameter distribution that is trained via a variational lower bound. Raghu et al. [270] conducts an in-depth empirical study on MAML and finds that feature reuse is the main reason behind the success of MAML. Thus motivated, they further proposes a simplification of MAML by removing the inner loop for the model update except the task-specific head of the underlying neural network. An Hessian-free framework is further proposed in Song et al. [314] to solve the meta-learning problem based on Evolution Strategies. Collins et al. [65] introduces the notion of task-robustness by minimizing the maximum loss over the meta-training tasks and proposes a task robust meta-learning algorithm. Along a different line, Fallah et al. [96] establishes the convergence of one-step MAML for non-convex loss functions, and then proposes a Hessian-free MAML to reduce the computational cost with theoretical guarantee. The convergence for multi-step MAML is studied in Ji et al. [153]. Wang et al. [348] further characterizes the gap between the stationary point and the global optimum of MAML in a general non-convex setting. There are also several studies investigating data-free meta-learning for data privacy protection, by learning a model initialization from a collection of pre-trained models without the original data. Wang et al. [353] proposes a model fusion method based on distributionally robust optimization to diversify the learned task embeddings. A framework PURER is proposed in Hu et al. [146] to distill training data from the pretrained models progressively through a sequence of pseudo episodes. Wei et al. [357] introduces a faster and better data-free meta-learning framework, by leveraging a meta-generator for task data recovery for pretrained models and a meta-learner trained with implicit gradient alignment for better generalization.

Federated meta-learning. The marriage of meta-learning and federated learning [233] has recently garnered much attention, giving rise to a new research direction, namely federated meta-learning. In particular, the empirical successes of such an integration have been corroborated in Chen et al. [51] and Jiang et al. [155]. Lin et al. [213] establishes the convergence of federated meta-learning for strongly convex functions and investigates the impact of task similarity. Another work [97] studies the case of non-convex functions with stochastic gradient descent. A different federated meta-learning approach is proposed in Dinh et al. [86], based on a proximal meta-learning method with moreau envelopes. Lin et al. [216] studies the adaptability and computational efficiency together and proposes a federated meta-learning approach to jointly learn a backbone network and a channel gating module. By leveraging ADMM to decompose the original problem into multiple subproblems, Yue et al. [387] develops ADMM-FedMeta to reduce the computational and the communication costs. [223] proposes an asynchronous federated meta-learning mechanism by incorporating the staleness of local models into the model aggregation. To relax the assumption that data distributions among different clients are similar, a group-based federated meta-learning framework is proposed in Yang et al. [371] to adaptively divide the clients into groups based on the similarity of their data distribution and learn separate models for each group.

4.3 Online Meta-Learning

For learning on a single edge device, the ML model needs to be quickly adapted in a lifelong manner for learning new tasks with only a few data samples. To achieve this, note that two key aspects of human intelligence are the abilities to quickly learn complex tasks and continually update their knowledge base for faster learning of future tasks. Meta-learning [103, 173, 274] and online learning [48, 132, 297] are two main research directions that try to equip learning agents with these abilities. In particular, meta-learning aims to facilitate quick learning of new unseen tasks by building a prior over model parameters based on the knowledge of related tasks, whereas online learning deals with the problem where the task data is sequentially revealed to a learning agent. Combining these two directions leads to a promising approach, namely online meta-learning [103, 134, 375], for fast adaptation of new tasks. In this part, we will introduce a memory- and computation-efficient online meta-learning approach for enabling efficient on-device continual learning.

Considering the setup where online tasks arrive one at a time, the objective of online meta-learning is to continuously update the meta prior based on which the new task can be learnt more quickly after the agent encounters more tasks. In online meta-learning, the agent typically maintains two separate models, i.e., the *meta-model* to capture the underlying common knowledge across tasks and the *online task model* for solving the current task in hand. Most of the existing studies [4, 103] in online meta-learning follow a "resetting" strategy: quickly adapt the online task model from the meta model using the current data, update the meta model and reset the online task model back to the updated meta model at the beginning of the next task. This strategy generally works well when the task boundaries are known and the task distribution remains stationary. However, in many real-world data streams the task boundaries are not directly visible to the agent [45, 134, 271], and the task distributions can dynamically change during the online learning stage. Therefore, in this work we seek to solve the online meta-learning problem in such more realistic settings.

Needless to say, how to efficiently solve the online meta-learning problem without knowing the task boundaries in the non-stationary environments is nontrivial due to the following key questions: (1) *How to update the meta model and the online task model?* Clearly, the "resetting" strategy at the moment of new data arriving is not desirable, as adapting from the previous task model is preferred when the new data belongs to the same task with the previous data. On the other hand, the meta model update should be distinct between in-distribution (IND) tasks, where the current knowledge should be preserved, and out-of-distribution tasks (OOD), where the new knowledge should be learnt quickly. (2) *How to make the system lightweight for fast online learning?* The nature of online meta-learning precludes sophisticated learning algorithms, as the agent should be able to quickly adapt to different tasks typically without access to the previous data. And dealing with the environment non-stationarity should not significantly increase the computational cost, considering that the environment could change fast during online learning.

In this chapter, we first propose a novel online meta-learning algorithm in non-stationary environments without knowing the task boundaries, which appropriately addresses the problems above. More specifically, we first propose two simple but effective mechanisms to detect the task switches using the classification loss and detect the distribution shift using the Helmholtz free energy [224], respectively, as motivated by empirical observations. Based on these detection mechanisms, our algorithm provides a finer treatment on the online model updates, which brings in the following benefits: (1) (*task knowledge reuse*) The detection of task switches enables our algorithm to reuse the best model available for each task, avoiding the "resetting" to the meta model at each step as in most previous studies; (2) (*judicious meta model update*) The detection of distribution shift allows our algorithm to update the meta model in a way that the new knowledge can be quickly learnt for out-of-distribution tasks whereas the previous knowledge can be preserved for in-distribution tasks; (3) (*efficient memory usage*) Our algorithm does not reuse/store any of the previous data and updates the meta model at each online episode based only on the current data, which clearly differs from most existing studies [105, 271, 375] in online meta-learning. Besides, we conduct extensive experiments in three different standard benchmarks for online meta-learning. As indicated by the experimental results, our algorithm significantly outperforms the related baselines methods on all benchmarks. The ablation study also verifies the effectiveness of the proposed detection mechanisms. We also provide a regret analysis of the proposed algorithm by taking task boundary detection into account, where a sublinear task-averaged regret can be achieved under mild conditions. In particular, our result captures a trade-off between the impact of task similarity on the performance of standard online meta-learning with known task boundaries and the performance under task boundary detection uncertainty. Namely, when tasks are more similar, better performance can be achieved due to less task variations over time, but it is harder to detect task switches.

4.3.1 Related Work

Online Learning. In online learning [48, 132, 139], cost functions are sequentially revealed to an agent which is required to select an action before seeing each cost. One of the most studied approach is follow the leader (FTL) [132], which updates the parameters at each step using all previously seen loss functions. Regularized versions of FTL have also been introduced to improve stability [2, 300]. Similar in spirit to our work in terms of computational resources, online gradient descent (OGD) [408] takes a gradient descent step at each round using only the revealed loss. However, traditional online learning methods do not efficiently leverage past experience and optimize for zero-shot performance without any adaptation. In this work, we study the online meta-learning problem, in which the goal is to optimize for quick adaptation on future tasks as the agent continually sees more tasks.

Online Meta-learning. Online meta learning was first introduced in Finn et al. [105]. Pioneering methods [105, 375, 385] follow a FTL-like design approach, which requires storing previous tasks and leads to a linear growth of memory requirement. Follow-the-regularized-leader (FTRL) [300] approach has also been extended to the online meta learning setting in Balcan et al. [25] and Khodak et al. [167], resulting in a better memory requirement. Acar et al. [4] proposed a memory-efficient approach based on summarizing previous task experiences into one state vector. However, these approaches require knowledge of task boundaries and "reset" the task model to the meta model at each online episode [78, 103]. Similar to Acar et al. [4], our algorithm also overcomes the linear memory scaling. But unlike their method, we do not need to know task boundaries and consider dynamic environments. Rajasegaran et al. [271] alleviates the "resetting" issue by updating the online model always starting from its previous state, which however needs to store previous models and has limited performance especially in dynamic environments where successive tasks can be very different.

None of the methods above considered the online meta-learning problem in a dynamic environment setting where the task distributions change substantially over time without knowing the task boundaries. Caccia et al. [45] is the first work that empirically evaluated the proposed algorithm in a dynamic environment, but did not propose a method to quickly learn the knowledge for out-of-distribution tasks. In contrast, we update the meta representations in a way that preserves the in-distribution knowledge while continually improving fast adaptation for out-of-distribution tasks.

4.3.2 Background and Problem Formulation

Background. We first briefly introduce some related concepts about online meta-learning.

Online learning. In the general online learning problem, loss functions are sequentially revealed to a learning agent: at each step t, the agent first selects an action θ_t, and then a cost $f_t(\theta_t)$ is incurred. The goal of the agent is to select a sequence of actions to minimize the following *static* regret

$$\mathrm{R}(T) = \sum_{t=1}^{T} f_t\left(\theta_t\right) - \min_\theta \sum_{t=1}^{T} f_t\left(\theta\right), \tag{4.2}$$

i.e., the gap between the agent's predictions $\{f_t(\theta_t)\}_{t=1}^{T}$ and the performance of the best *static* model in hindsight. A successful agent achieves a regret $\mathrm{R}(T)$ that grows sublinearly in T. Online learning is a well studied field and we refer the readers to Hazan et al. [140] for more information.

Online meta-learning. As a marriage between online learning and meta-learning, online meta-learning [105, 134, 375] aims to achieve the following two features: (i) fast adaptation to the current task (the meta-learning aspect); (ii) learn to adapt even faster as it sees more tasks (the online learning aspect). Specifically, the agent observes a stream of tasks $\mathcal{S} = \{\mathcal{T}_1, \mathcal{T}_2, \ldots, \mathcal{T}_T\}$ sampled from $\mathbb{P}(\mathcal{T})$, where tasks are revealed one at a time. For each task $\mathcal{T}_t$, the agent has access to a support set S_t for task-specific adaptation and a query set Q_t for evaluation. The goal here is to select a sequence of meta models $\{\mathbf{w}_t\}$ for achieving sublinear growth of the following regret

$$\mathrm{R}_{\mathrm{meta}}(T) = \sum_{t=1}^{T} f_t\left(U_t\left(\theta_t\right)\right) - \min_\theta \sum_{t=1}^{T} f_t\left(U_t(\theta)\right) \tag{4.3}$$

where U_t is the task adaptation function depending on the support set S_t, and the cost function f_t is evaluated using the adapted parameters $U_t\left(\mathbf{w}_t\right)$ on the query set Q_t. Intuitively, the agent seeks to learn a better meta model which leads to better task models for future tasks after seeing more tasks.

Online meta-learning in non-stationary environments. Differently from most online meta-learning studies [4, 105, 134, 271, 375, 385], in this work we consider a more realistic scenario:

Pre-trained meta model. In many real applications, there is plenty of data available for pre-training, and it is unrealistic to deploy an agent in complex dynamic environments without any basic knowledge of the tasks at hand [45]. Therefore, following the same line as in Caccia et al. [45], we assume that there is a set of training tasks $\{\mathcal{T}_i^0\}_{i=1}^{M}$ drawn from some unknown distribution $\mathbb{P}^0(\mathcal{T})$. And as standard in meta-learning, each pre-training task $\mathcal{T}_i^0$ has a support dataset S_i^0 and a query dataset Q_i^0. In this work, we employ MAML over the training tasks to learn a pre-trained meta model.

Unknown task boundaries. During the online meta-learning phase, we assume that the task boundaries are unknown, i.e., the so-called task-agnostic setup [45], in the sense that the agent does not know if the new coming data at time t belongs to the previous task or a new task. To model the uncertainty about task boundaries, we assume that at any time t the new data belongs to the previous task with probability $p \in (0, 1)$ or to a new task with probability $1 - p$.

Non-stationary task distributions. During the online meta-learning phase, the agent could encounter new tasks that are sampled from other distributions instead of the pre-training one $\mathbb{P}^0(\mathcal{T})$. To capture this non-stationarity in task distribution, we assume that whenever a new task arrives during online learning, it will be sampled either from $\mathbb{P}^0(\mathcal{T})$ with probability $\eta \in (0, 1)$ or from a new (w.r.t. $\mathbb{P}^0(\mathcal{T})$) distribution with probability $1 - \eta$. Note that we do not restrict the number of new distributions that can be encountered during online learning and the task distributions can be revisited.

4.3.3 Proposed Algorithm Under Distribution Shifts

To address the online meta-learning problem mentioned above for non-stationary environments, we next propose a simple but effective algorithm, called onLine mEta lEarning under Distribution Shifts (LEEDS), based on the detection of task switches and distribution shift to assist fast online learning. Following most studies [45, 271] in online meta-learning, we maintain two separate models during the online learning stage: θ for the meta model and ϕ for the online task model.

Detection of task switches and distribution shift: To enable fast learning of a new task in online learning, the detection mechanisms can not be overly sophisticated, but have to be efficient with high detection accuracy. Towards this end, we propose two different methods for detecting the task switch and the distribution shift, respectively, which work in concert as key components of LEEDS.

Detection of task switches. To understand the learning behaviors under task switches, we evaluate the classification loss of the previous task model using the newly coming data, i.e., $\mathcal{L}(\phi_{t-1}; S_t)$ for time t, where $\mathcal{L}$ is the loss function, ϕ_{t-1} is the previous online model at time $t - 1$, and S_t is the current support set. The left plot in Fig. 4.1 shows the empirical results on an online few-shot image recognition problem. As depicted, the loss value keeps decreasing as the agent receives more data from the same task but suddenly increases whenever a new task arrives. This is clearly reasonable as the learnt online model for the previous task does not fit the new task anymore. Inspired by this empirical observation, we use a simple mechanism based on the value of $\mathcal{L}(\phi_{t-1}; S_t)$ to detect the task boundaries: there is a task switch whenever the loss is above some pre-defined threshold. As demonstrated later in Sect. 4.3.5, such a simple mechanism is indeed quite effective as corroborated by its high detection accuracies on various online meta-learning problems.

Detection of distribution shift. To efficiently determine if a new task is IND or OOD, i.e., sampled from the pre-training task distribution or not, we consider an energy-based OOD detection mechanism with a binary classifier $C_{\mathcal{T}}(\cdot; \theta)$ defined as follows

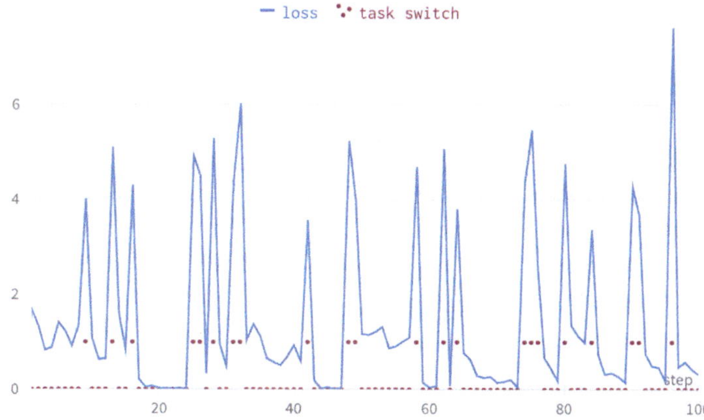

Method	Memory
Ours	$O(1)$
[Rajasegaran et al. 2022]	$O(T)$
[Caccia et al. 2020]	$O(\frac{1}{1-p})$
[Acar et al. 2021]	$O(1)$
[Finn et al. 2019]	$O(T)$

Fig. 4.1 **Left plot**: Variations of the online loss for a pre-trained meta model using MAML which is deployed for online learning. Red dot at 0 means no task switch at that time, and at 1 means the task switched at that time. **Right table:** Comparison of the memory requirements among different methods. T is number of online rounds and $p \in (0, 1)$ is non-stationarity level

$$C_\tau(\mathbf{x}; \theta) = \begin{cases} 1 & \text{if } -\mathrm{E}(\mathbf{x}; \theta) \leq \tau \\ 0 & \text{if } -\mathrm{E}(\mathbf{x}; \theta) > \tau \end{cases} \tag{4.4}$$

where $\mathrm{E}(\mathbf{x}; \theta) = -\delta \log \sum_{k=1}^{K} \exp(-g_k(\mathbf{x}; \theta)/\delta)$ corresponds to the Helmholtz free energy for input $\mathbf{x}$, $g_k(\cdot; \theta)$ is the k-th component of the prediction model's output, and δ is the temperature parameter. The hyperparameter τ is a threshold that can be set using the pre-training tasks. As shown in Liu et al. [224], the negative energy is linearly aligned with the likelihood of the of input $\mathbf{x}$ under model θ, making it a useful score for distinguishing IND and OOD tasks.

Update of meta and online parameters: Based on the two detection schemes, the next question is how to update the meta and task models for fast adaption in dynamic environments.

Without such detection mechanisms, previous online meta-learning algorithms [4, 105] typically adapt the task model from the meta model using a support set, evaluate the adapted model on a query set, and reset the task model to the meta model when new data is received from the online data stream process, and then repeat the process again. However, such a "resetting" scheme can be sub-optimal in realistic scenarios. For instance, if the newly received data belongs to the same task with the previous data, the agent should update the task model by starting from previously adapted parameters instead of from the meta model.

In contrast, the simple but effective detection mechanisms in this work enable a more elegant treatment to the knowledge update during online learning:

(1) If there is a task switch at time t, i.e., $\mathcal{L}(\phi_{t-1}; S_t) > \ell$ where ℓ is the threshold, adapting from the meta model is generally better than adapting from the task model of the previous task. Therefore, we first obtain the online task model ϕ_t from the meta model using the new data:

$$\phi_t = \theta_{\text{adapt}} = \theta_{t-1} - \alpha_1 \nabla_{\theta_{t-1}} \mathcal{L}(\theta; S_t),$$

and then update the meta model no matter if there is a distribution shift, so as to incorporate the knowledge of the new task to the meta model:

$$\theta_t = \theta_{t-1} - \alpha_2 \nabla_\theta \mathcal{L}(\theta_{\text{adapt}}; Q_t).$$

(2) If there is no task switch, i.e., $\mathcal{L}(\phi_{t-1}; S_t) \le \ell$, we continue to update the task model from the previous task model using the new data, different from the "resetting" scheme in the literature:

$$\phi_t = \phi_{t-1} - \alpha_1 \nabla_{\phi_{t-1}} \mathcal{L}(\phi_{t-1}; S_t).$$

To accelerate the knowledge learning for the new domains, we further distinguish the meta model update for IND and OOD tasks. In particular, if the current task is an IND task, we will only update the meta model once at the beginning of this task. That is to say, the meta model will not be further updated within the same task. In stark contrast, if the current task is an OOD task, we continue to update the meta model whenever new data for this task arrives as follows:

$$\theta_{\text{adapt}} = \theta_{t-1} - \alpha_1 \nabla_\theta \mathcal{L}(\theta_{t-1}; S_t), \quad \theta_t = \theta_{t-1} - \alpha_2 \nabla_\theta \mathcal{L}(\theta_{\text{adapt}}; Q_t).$$

The details of LEEDS can be found in Algorithm 1. Note that the performance of LEEDS is indeed robust to the threshold parameters ℓ and τ as shown later in our experimental results.

Memory friendly: One important feature of the proposed algorithm LEEDS is that the meta model update is only based on the current data. In contrast, most of the previous studies store the previous data in the memory for the meta model update. A comparison of the memory requirements among different approaches is summarized in the right table in Fig. 4.1. As will be shown later in the experiments, LEEDS significantly outperforms the related baselines without the need of storing any previous data. This encouraging result suggest that the community could also learn from careful explorations of simpler designs, besides emphasizing algorithmic complexities.

Algorithm 1 onLine mEta lEarning under Distribution Shifts (LEEDS)

Input: Dynamic stream $\mathcal{S}$, pre-training distribution $\mathbb{P}^0(\mathcal{T})$, stepsizes α_1 and α_2, thresholds ℓ and τ.
Perform pre-training phase using MAML method on tasks drawn from $\mathbb{P}^0(\mathcal{T})$
while stream $\mathcal{S}$ is ON **do**
  $D_t \longleftarrow \mathcal{S}$ // receive current data from online data stream S_t, Q_t
  $S_t, Q_t \longleftarrow D_t$ // split data into support and query
  if $\mathcal{L}(\phi_{t-1}; S_t) \leq \ell$ (i.e., no switch) **then**
    $\phi_t = \phi_{t-1} - \alpha_1 \nabla_{\phi_{t-1}} \mathcal{L}(\phi_{t-1}; S_t)$ // adapt starting from previous online model
    Evaluate ϕ_t on query set Q_t
    if $C_\tau(S_t; \theta_{t-1})$ (i.e., covariate shift) **then**
      $\theta_{\text{adapt}} = \theta_{t-1} - \alpha_1 \nabla_\theta \mathcal{L}(\theta_{t-1}; S_t)$, $\theta_t = \theta_{t-1} - \alpha_2 \nabla_\theta \mathcal{L}(\theta_{\text{adapt}}; Q_t)$ // update meta model
    end if
  else
    $\theta_{\text{adapt}} = \theta_{t-1} - \alpha_1 \nabla_{\theta_{t-1}} \mathcal{L}(\theta_{t-1}; S_t)$ // adapt starting from meta model using support set
    Set $\phi_t = \theta_{\text{adapt}}$ and Evaluate ϕ_t on query set Q_t
    $\theta_t = \theta_{t-1} - \alpha_2 \nabla_\theta \mathcal{L}(\theta_{\text{adapt}}; Q_t)$ // update meta model
  end if
end while

4.3.4 Theoretical Results

In the online meta-learning with distribution shifts, it is clear that the *static* comparator in Eq. (4.3) is not sufficient to capture the non-stationarity, as one cannot expect to have a single meta-model for all task distributions. Hence, we consider a task-averaged regret (TAR) by following Balcan et al. [25]

$$\boldsymbol{R}_{\text{avg}} = \frac{1}{T} \sum_{t=1}^{T} \left(\sum_{k=1}^{K_t} f_t^k\left(\phi_t^k\right) - \sum_{k=1}^{K_t} f_t^k\left(\phi_t^*\right) \right) \tag{4.5}$$

where T is the number of tasks, K_t is the number of steps within task t, and ϕ_t^* is the dynamic comparator for each task t. For simplicity, we denote f_t^k as the within-task loss function in step k of task t (evaluated on the query set). As shown in Balcan et al. [25], one cannot expect to achieve a TAR that decreases w.r.t. T because the dynamic comparators $\{\phi_t^*\}_t$ can force a constant loss for each task t. However, the average is still not taken w.r.t. the total number of steps $\sum_{t=1}^{T} K_t$, but only w.r.t. the number of tasks T. Therefore, the objective here is to achieve a TAR that is sublinear in K_t. Note that the within-task loss functions $\{f_t^t\}_{k=1}^{K_t}$ are usually non-adversarial in many practical online meta-learning problems. For example, in few-shot online image classification problem, $\{f_t^t\}_{k=1}^{K_t}$ correspond to different evaluations of the same classification loss function on different query sets for the same task.

In what follows, we can show that a constant TAR is achievable for the non-adversarial case even after taking the detection errors of task boundaries into consideration.

Let the comparator ϕ_t^* be a minimizer of each of the loss functions $f_t^k, k = 1, \ldots, K_t$. To formally characterize the non-adversariality of the within-task loss functions $\{f_t^t\}_{k=1}^{K_t}$ in online meta-learning, we make the following assumptions.

Assumption 4.1 The comparator ϕ_t^* is a fixed point of the adaptation mappings U_t^k at step k within task t, i.e., $U_t^k(\phi_t^*) = \phi_t^*$ for $k = 1, \ldots, K_t$.

Assumption 4.2 The adaptation mapping U_t^k is a contraction, i.e., there exists some ρ_t^k such that for all ϕ_1, ϕ_2, $\|U_t^k(\phi_1) - U_t^k(\phi_2)\| \leq \rho_t^k \|\phi_1 - \phi_2\|$.

Assumption 4.1 characterizes the property of the dynamic comparator ϕ_t^*. It is clear that Assumption 4.1 will hold when there exists a task model ϕ that can perfectly achieve zero loss on all data points sampled from the data distribution for task t. For example, in over-parameterized regime, one can always train a model to perfectly predict all training data points with near-zero loss. Assumption 4.2 can be easily satisfied in online meta-learning [105]. For example, the one step gradient descent mapping, i.e., $U_t^k(\phi) = \phi - \alpha \nabla \hat{f}_t^k(\phi)$, satisfies Assumption 4.2 when the function $\hat{f}_t^k$ is β-smooth and μ-strongly convex, and the stepsize α is chosen in $(0, \frac{2}{\beta})$. Define $\rho = \max_{t,k} \rho_t^k$.

Assumption 4.3 Each function f_t^k is L-smooth.

Note that we do not make any assumption on the convexity of the within-task loss functions f_t^k. For any step r in the online meta-learning process (where each task t includes K_t steps), let P_r be the current data distribution and ϕ_r be the task model obtained after adaptation. To analyze the error probability of the detection mechanisms proposed in this work, we make the following assumption on the single-data loss function ℓ:

Assumption 4.4 Let $\ell(\cdot, \xi)$ be the loss on a data point ξ and $0 \leq \ell(\cdot, \xi) \leq M$. We assume that there exist constants $\ell_m \leq \ell_p$ such that the following holds for any step r and s: (1) $E_{\xi \sim P_r} \ell(\phi_r, \xi) \leq \ell_m$; (2) $E_{\xi \sim P_r} \ell(\phi_s, \xi) \geq \ell_p$ if $P_r \neq P_s$.

Assumption 4.4 characterizes the comparison of the expected loss w.r.t. a certain data distribution between 1) the model adapted on this distribution and 2) the model adapted on another distribution. This essentially means, in expectation, the loss after adaptation is less than the loss of another task model, which is crucial for a threshold-based detection scheme to be applicable.

By updating the meta-model using OGD [408] on the meta loss $\|\phi_t^0 - \phi_t^*\|^2$ after each task t, we can have the following theorem to characterize the expected TAR w.r.t. the uncertainty of the task-boundary detection.

Theorem 4.1 *Suppose Assumptions 4.1, 4.2, 4.3, 4.4 hold. Let* $R = \sum_{t=1}^{T} K_t$ *be the total number of online rounds, and* $S = c \log R$ *be the number of data points used for adaptation at each step, where* c *is some positive constant. Then the expected TAR is bounded as*

$$\mathbb{E}[\boldsymbol{R}_{\text{avg}}] \le O\left(\frac{\sigma_*^2 + \frac{\log T}{T}}{1-\rho^2} + R^{-\left(\frac{c(\ell_p-\ell_m)^2}{2M^2}-2\right)}\right),$$

where the expectation is taken over the task-boundary detection uncertainty. $\sigma_*^2 = \frac{1}{T}\sum_{t=1}^{T}\left\|\phi_t^* - \phi^*\right\|^2$ *and* $\phi^* = \frac{1}{T}\sum_{t=1}^{T}\phi_t^*$ *denote the variance and the mean of the comparators* $\{\phi_t^*\}_{t=1}^{T}$*, respectively. In particular, a constant expected TAR can be achieved by selecting* $c > \frac{4M^2}{(\ell_p-\ell_m)^2}$.

Proof Recall the task-averaged regret:

$$\boldsymbol{R}_{\text{avg}} = \frac{1}{T}\sum_{t=1}^{T}\left(\sum_{k=1}^{K_t} f_t^k\left(\phi_t^k\right) - \sum_{k=1}^{K_t} f_t^k\left(\phi_t^*\right)\right).$$

We consider the setting where at each round the agent can perform some task-specific adaptation before evaluation on the loss function f_t^k [105]. Let U_t^k be the mapping that defines the adaptation procedure at each round. For instance, a popular choice of the mapping function U_t^k is the one step gradient descent mapping: $U_t^k(\phi) = \phi - \alpha\nabla\hat{f}_t^k(\phi)$, where α is a learning rate and $\hat{f}_t^k$ is an approximation of the loss function f_t^k, e.g., computed on a small support set from task t.

Denote $\hat{f}_r(\phi_{r-1}) = \frac{1}{|\mathcal{S}_r|}\sum_{\xi\in\mathcal{S}_r}\ell(\phi_{r-1},\xi) := \hat{f}_{r,r-1}$ as the adaptation loss at round r, where $\ell(\cdot,\xi)$ is a classification loss on data point ξ. We first provide an upper bound on the detection error.

The total probability of error of the threshold-based detection scheme at round r is given by:

$$P_{\text{error}} = P\left(P_r \neq P_{r-1}\right)P\left(\hat{f}_{r,r-1} < \tau \,\middle|\, P_r \neq P_{r-1}\right) + P\left(P_r = P_{r-1}\right)P\left(\hat{f}_{r,r-1} > \tau \,\middle|\, P_r = P_{r-1}\right) \tag{4.6}$$

where the first term characterizes the probability of miss detection of the task boundary, and the second term is the probability of false alarm when the underlying task does not change.

Based on Assumption 4.4, for a data distribution P_r and support set $\mathcal{S}_r$ drawn i.i.d. from P_r, the Hoeffding inequality yields:

$$P\left(\frac{1}{|\mathcal{S}_r|}\sum_{\xi\in\mathcal{S}_r}\ell(\phi_{r-1},\xi) - E_{\xi\sim P_r}\,\ell(\phi_{r-1},\xi) < -\epsilon\right) \le \exp\left(\frac{-2|\mathcal{S}_r|\epsilon^2}{M^2}\right)$$

Hence,

$$
\begin{aligned}
&P\left(\hat{f}_{r,r-1} - \ell_p < -\epsilon \,\middle|\, P_r \neq P_{r-1}\right) \\
\leq &P\left(\frac{1}{|S_r|}\sum_{\xi \in S_r} \ell(\phi_{r-1}, \xi) - E_{\xi \sim P_r}\, \ell(\phi_{r-1}, \xi) < -\epsilon \,\middle|\, P_r \neq P_{r-1}\right) \\
\leq &\exp\left(\frac{-2|S_r|\epsilon^2}{M^2}\right).
\end{aligned}
$$

By setting $\epsilon = \frac{\ell_p - \ell_m}{2}$, we can have

$$
P\left(\hat{f}_{r,r-1} < \frac{\ell_m + \ell_p}{2} \,\middle|\, P_r \neq P_{r-1}\right) \leq \exp\left(\frac{-|S_r|(\ell_p - \ell_m)^2}{2M^2}\right).
$$

Thus, setting the threshold $\tau = \frac{\ell_m + \ell_p}{2}$ yields

$$
P\left(\hat{f}_{r,r-1} < \tau \,\middle|\, P_r \neq P_{r-1}\right) \leq \exp\left(\frac{-|S_r|(\ell_p - \ell_m)^2}{2M^2}\right). \tag{4.7}
$$

Using the other side of the Hoeffding inequality, we have:

$$
\begin{aligned}
&P\left(\hat{f}_{r,r-1} - \ell_m > \epsilon \,\middle|\, P_r = P_{r-1}\right) \\
\leq &P\left(\frac{1}{|S_r|}\sum_{\xi \in S_r} \ell(\phi_{r-1}, \xi) - E_{\xi \sim P_r} \ell(\phi_{r-1}, \xi) > \epsilon \,\middle|\, P_r = P_{r-1}\right) \\
\leq &P\left(\frac{1}{|S_r|}\sum_{\xi \in S_r} \ell(\phi_{r-1}, \xi) - E_{\xi \sim P_r} \ell(\phi_{r-1}, \xi) > \epsilon\right) \\
\leq &\exp\left(\frac{-2|S_r|\epsilon^2}{M^2}\right).
\end{aligned}
$$

Therefore, we have:

$$
P\left(\hat{f}_{r,r-1} > \frac{\ell_m + \ell_p}{2} \,\middle|\, P_r = P_{r-1}\right) \leq \exp\left(\frac{-|S_r|(\ell_p - \ell_m)^2}{2M^2}\right),
$$

because $\epsilon = \frac{\ell_p - \ell_m}{2}$. Thus, we obtain

$$
P\left(\hat{f}_{r,r-1} > \tau \,\middle|\, P_r = P_{r-1}\right) \leq \exp\left(\frac{-|S_r|(\ell_p - \ell_m)^2}{2M^2}\right). \tag{4.8}
$$

Combining Eqs. (4.6), (4.7) and (4.8), and using the fact that $P(P_r \neq P_{r-1}) = 1 - P(P_r = P_{r-1})$ can yield the following upper bound on the probability of error when the threshold is set to $\tau = \frac{\ell_m + \ell_p}{2}$:

$$P_{\text{error}} \leq \exp\left(\frac{-|S_r|(\ell_p - \ell_m)^2}{2M^2}\right). \tag{4.9}$$

Based on Assumption 4.3, we have for any ϕ:

$$f_t^k(\phi_t^k) \leq f_t^k(\phi) + \left\langle \nabla f_t^k(\phi), \phi_t^k - \phi \right\rangle + \frac{L}{2}\|\phi_t^k - \phi\|^2.$$

Therefore, by setting ϕ to be the comparator ϕ_t^* and using Assumption 4.1, we obtain:

$$f_t^k(\phi_t^k) - f_t^k(\phi_t^*) \leq \frac{L}{2}\|\phi_t^k - \phi_t^*\|^2.$$

Thus, summing over $k = 1, \ldots, K_t$ yields

$$\sum_{k=1}^{K_t} f_t^k(\phi_t^k) - \sum_{k=1}^{K_t} f_t^k(\phi_t^*) \leq \frac{L}{2}\sum_{k=1}^{K_t}\|\phi_t^k - \phi_t^*\|^2. \tag{4.10}$$

We next bound the term $\sum_{k=1}^{K_t}\|\phi_t^k - \phi_t^*\|^2$ from above. It follows that

$$\begin{aligned}
\|\phi_t^k - \phi_t^*\|^2 &= \|U_t^k(\phi_t^{k-1}) - U_t^k(\phi_t^*)\|^2 \\
&\leq \rho^2\|\phi_t^{k-1} - \phi_t^*\|^2 \\
&\leq \rho^{2k}\|\phi_t^0 - \phi_t^*\|^2,
\end{aligned} \tag{4.11}$$

where the first inequality uses Assumption 4.2. Therefore, combining Eqs. (4.10) and (4.11) gives that

$$\begin{aligned}
\sum_{k=1}^{K_t} f_t^k(\phi_t^k) - \sum_{k=1}^{K_t} f_t^k(\phi_t^*) &\leq \frac{L}{2}\sum_{k=1}^{K_t}\rho^{2k}\|\phi_t^0 - \phi_t^*\|^2 \\
&\leq \frac{L}{2(1-\rho^2)}\|\phi_t^0 - \phi_t^*\|^2.
\end{aligned} \tag{4.12}$$

Hence, summing Eq. (4.12) over $t = 1, \ldots, T$ yields

$$\sum_{t=1}^{T}\left(\sum_{k=1}^{K_t} f_t^k\left(\phi_t^k\right) - \sum_{k=1}^{K_t} f_t^k\left(\phi_t^*\right)\right) \leq \frac{L}{2(1-\rho^2)}\sum_{t=1}^{T}\|\phi_t^0 - \phi_t^*\|^2.$$

The task-averaged regret is therefore upper bounded as follows

$$\boldsymbol{R}_{\text{avg}} \leq \frac{L}{2(1-\rho^2)T} \sum_{t=1}^{T} \|\phi_t^0 - \phi_t^*\|^2. \tag{4.13}$$

.

Next, we show that the term $\frac{1}{T}\sum_{t=1}^{T} \|\phi_t^0 - \phi_t^*\|^2$ converges as $T \to \infty$. We have,

$$\frac{1}{T}\sum_{t=1}^{T} \|\phi_t^0 - \phi_t^*\|^2 = \frac{1}{T}\sum_{t=1}^{T} \left(\|\phi_t^0 - \phi_t^*\|^2 - \|\phi_t^* - \phi^*\|^2\right) + \frac{1}{T}\sum_{t=1}^{T} \|\phi_t^* - \phi^*\|^2, \tag{4.14}$$

where $\phi_* = \frac{1}{T}\sum_{t=1}^{T} \phi_t^*$ is the mean of the dynamic comparators. Note that ϕ^* is the minimizer of the summation $\sum_{t=1}^{T} \|\phi_t^* - \phi^*\|^2$. Thus, if we use an algorithm such as OGD [408] to update the initialization ϕ_t^0 with loss $\|\phi_t^0 - \phi_t^*\|^2$ then summation in the first term at the right hand side of Eq. (4.14) corresponds to the regret of OGD on a stream of strongly-convex functions, which is well-known to be $O(\log T)$. The second term $\frac{1}{T}\sum_{t=1}^{T} \|\phi_t^* - \phi^*\|^2 = \sigma_*^2$ corresponds to the variance of the dynamic comparators. Hence, combining Eqs. (4.13) and (4.14) we obtain

$$\boldsymbol{R}_{\text{avg}} = O\left(\frac{\sigma_*^2 + \frac{\log T}{T}}{1-\rho^2}\right). \tag{4.15}$$

Using Eq. (4.9) for $R = \sum_{t=1}^{T} K_t$ rounds, we have with probability at least $1 - R\exp\left(\frac{-S(\ell_p-\ell_m)^2}{2M^2}\right)$ the task-average regret is bounded as in Eq. (4.15). Hence, the expected task-averaged regret for the algorithm without exact task-boundary detection is upper bounded as

$$\mathbb{E}\,\boldsymbol{R}_{\text{avg}} \leq O\left(\frac{\sigma_*^2 + \frac{\log T}{T}}{1-\rho^2} + R^2\exp\left(\frac{-S(\ell_p-\ell_m)^2}{2M^2}\right)\right) \leq O\left(\frac{\sigma_*^2 + \frac{\log T}{T}}{1-\rho^2} + R^{-\left(\frac{c(\ell_p-\ell_m)^2}{2M^2}-2\right)}\right), \tag{4.16}$$

where the second inequality follows by selecting support sets of size $S = c\log R$, which completes the proof of Theorem 4.1. Further choosing $c > \frac{4M^2}{(\ell_p-\ell_m)^2}$ ensures a constant expected task-averaged regret. □

Intuitively, the first term in the upper bound in Theorem 4.1 captures the TAR for online meta-learning with known task boundaries, whereas the second term characterizes the impact of the task-boundary detection uncertainty. As shown in Theorem 4.1, when the tasks become more similar, the first term will decrease because the variance σ_*^2 is smaller, whereas the second term can increase because the constants ℓ_p and ℓ_m become closer (i.e., it is harder to detect task switches if tasks are more similar). Therefore, Theorem 4.1 captures a trade-off between the impact of task similarity on the performance of standard online meta-learning and the performance under the task boundary detection uncertainty. In practice, the optimal actions ϕ_t^* are usually not available for the meta updates. Thus, in our algorithm we use the alternative MAML-like updates which has been shown to be effective for learning meta parameters. As demonstrated in the different experiments in the next section, such updates can indeed serve as a good practical alternative.

4.3.5 Experiments

Experimental setup. We investigate the performance of LEEDS on three standard online meta-learning benchmarks, Omniglot-MNIST-FashionMNIST, Tiered-ImageNet and Symbols, compared to multiple related baseline algorithms. Specifically, we pre-train the meta model in one domain and then deploy it in a dynamic environment where tasks can be drawn from new domains. We evaluate all the algorithms using the average of test losses obtained throughout the entire online learning stage. To investigate the impact of the non-stationary level on the learning performance, we further consider two different cases of the environment non-stationarity: A moderately stationary case where the probability of not switching to a new task is set to $p = 0.9$, and a low stationary case where $p = 0.75$. We do not consider the cases where p is very small, as an algorithm that just assumes task switches at each round should perform well in such cases. We compare algorithms over 10000 episodes unless otherwise stated. See more details about datasets and baselines in Sect. 4.3.5.4. More comprehensive results and training curves of different methods are also provided in Sect. 4.3.5.3.

For all the experiments, whenever a new task needs to be revealed, it will be drawn from either the pre-training domain with probability 0.5, or from one of the OOD domains with probability 0.5. For Tiered-ImageNet, because only ood2 is trully OOD w.r.t. the pre-training task distribution, we increase the sampling probability of ood2 to 0.5 which is consistent to the protocol 50−50% for IND and OOD tasks in all our experiments. More details about the experimental setup including the neural network architectures and the hyperparameter search are deferred to Sect. 4.3.5.5.

4.3.5.1 Main Results

Results on Omniglot-MNIST-FashionMNIST (OMF). The online evaluations of the compared methods are shown in Table 4.1 for non-stationary levels $p = 0.9$ and $p = 0.75$. For each setting we report separately the online accuracies on pre-training domain and on the other two OOD domains, to show how our method keeps improving on the OOD domains while also remembering the pre-tarining tasks. Clearly, our method LEEDS achieves superior performance compared to all other baseline algorithms in both settings. More specifically,

Table 4.1 Average accuracy over 10000 online episodes on **Omniglot-MNIST-FashionMNIST** benchmark under different non-stationarity levels. "pre-train" domain: **Omniglot**; "ood1" domain: **MNIST**, "ood2" domain: **FashionMNIST**. The advantage of our algorithm LEEDS over the other baselines is more significant in the ood domains. See Sect. 4.3.5.4 for more details about the datasets

Method	Non-stationarity level $p = 0.9$			Non-stationarity level $p = 0.75$		
	Pre-train	ood1	ood2	pre-train	ood1	ood2
LEEDS	**99.39** ±0.09	**96.44** ±0.11	**82.87** ±0.19	**98.97** ±0.10	**95.68** ±0.12	**81.49** ±0.22
CMAML++	98.78 ±0.12	92.52 ±0.19	76.16 ±0.28	97.39 ±0.11	89.07 ±0.20	73.35 ±0.35
CMAML	89.79 ±0.54	84.06 ±0.80	69.70 ±0.63	75.51 ±0.94	70.41 ±1.22	58.58 ±1.27
FOML	89.20 ±0.61	70.84 ±0.76	64.83 ±0.74	81.68 ±0.59	59.24 ±0.78	58.07 ±0.77
MAML	95.07 ±0.10	62.02 ±0.14	54.67 ±0.17	95.51 ±0.11	62.31 ±0.13	54.83 ±0.13
ANIL	96.54 ±0.12	42.14 ±0.16	40.12 ±0.13	96.88 ±0.11	42.08 ±0.11	40.00 ±0.13
MetaOGD	84.05 ±1.66	73.73 ±1.39	60.03 ±1.60	85.67 ±1.57	75.09 ±1.43	60.51 ±1.65
BGD	63.58 ±2.25	46.12 ±2.10	44.89 ±1.41	23.86 ±2.36	17.97 ±2.17	19.81 ±1.63
MetaBGD	77.73 ±1.26	59.11 ±0.84	54.67 ±0.82	43.95 ±1.31	27.87 ±0.94	30.14 ±0.98

on the IND domain all methods pre-trained using MAML perform similarly, but are outperformed by LEEDS and CMAML++ which can detect task boundaries. However, on the OOD domains our algorithm significantly outperforms all other baselines, including CMAML++. This is due to the key OOD adaptation module that allows LEEDS to dynamically adapt the meta model based on the task distribution. Interestingly, comparing the performance for MAML and ANIL provides some insights on the limitations of re-using pre-trained representations in non-stationary environments. In fact, the ANIL baseline, which does not adapt its inner representations, performs poorly compared to MAML on the OOD domains, but achieves similar results on the pre-training domain. Also, the results highlight some limitations of the recently introduced FOML [271] method, which achieves lower performance than other competitive baselines. This is because FOML requires the tasks to be not mutually exclusive, which may not hold for the standard few-shot benchmarks considered in our experiments (Table 4.2).

Results on Tiered-ImageNet (TI) and Synbols (SB). We report the online accuracies on all domains and on OOD domains for these two benchmarks in Table 4.3. Because the distribution of the pre-training tasks is similar to the OOD ones for the Tiered-ImageNet benchmark, methods such as MAML can perform reasonably well. In fact, in the lower non-stationary case ($p = 0.75$), MAML is able to outperform the more complex CMAML++ baseline. However, our algorithm still achieves the best performance under both non-stationary levels and in both benchmarks. Note that in the larger **TI** dataset case, the FOML algorithm, which stores all previously seen tasks, runs out of memory after around 6500 online episodes. Again because of similarity between OOD and IND tasks in the **TI** benchmark, static representations learned by ANIL are useful for all domains.

Table 4.2 Average accuracy over 20000 and 10000 online episodes on **Tiered-ImageNet** benchmark and **Synbols** benchmark, respectively, under different non-stationarity levels. The different domains are distinct splits of the original Tiered-ImageNet dataset (please see experimental setup in Sect. 4.3.5.4 for details on how these splits are obtained). In **Synbols**, the "pre-train" domain corresponds to 3 different alphabets from Synbols dataset, "ood1" corresponds to a new (w.r.t. the pre-training one) alphabet from Synbols dataset, and "ood2" contains font classification tasks. Full table with variance is in Sect. 4.3.5.3

Method	Tiered-ImageNet				Synbols			
	Non-stationarity $p = 0.9$		Non-stationarity $p = 0.75$		Non-stationarity $p = 0.9$		Non-stationarity $p = 0.75$	
	All	ood	All	ood	All	ood	All	ood
LEEDS	**66.07**	**67.43**	**64.52**	**65.80**	**85.12**	**67.48**	**82.22**	**63.68**
CMAML++	63.83	63.75	61.28	61.96	81.14	62.39	79.74	60.70
FOML	35.90	35.87	32.02	31.61	46.40	41.73	37.46	34.13
MAML	62.37	61.00	62.54	60.88	76.25	42.70	74.87	43.84
ANIL	59.78	57.61	59.57	57.38	64.58	34.51	72.69	35.66
MetaOGD	57.01	57.32	56.80	56.94	72.04	46.69	67.93	42.66
BGD	40.95	41.44	35.48	35.97	25.63	25.61	27.53	27.17
MetaBGD	49.21	50.01	44.58	45.30	53.74	42.25	40.79	34.63

Table 4.3 Average accuracy over 20000 online episodes on **Tiered-ImageNet** benchmark under different non-stationarity levels. The different domains are distinct splits of the original Tiered-ImageNet dataset (please see experimental setup in Sect. 4.3.5 for details on how these splits are obtained)

Method	Non-stationarity $p = 0.9$		Non-stationarity $p = 0.75$	
	All domains	ood domains	All domains	ood domains
LEEDS	**66.07** ± 0.24	**67.43** ± 0.38	**64.52** ± 0.17	**65.80** ± 0.31
CMAML++	63.83 ± 0.27	63.75 ± 0.55	61.28 ± 0.23	61.96 ± 0.41
FOML	35.90 ± 0.56	35.87 ± 0.83	32.02 ± 0.42	31.61 ± 0.69
MAML	62.37 ± 0.46	61.00 ± 0.72	62.54 ± 0.37	60.88 ± 0.65
ANIL	59.78 ± 0.21	57.61 ± 0.38	59.57 ± 0.22	57.38 ± 0.36
MetaOGD	57.01 ± 0.28	57.32 ± 0.66	56.80 ± 0.25	56.94 ± 0.42
BGD	40.95 ± 0.85	41.44 ± 1.15	35.48 ± 0.76	35.97 ± 1.09
MetaBGD	49.21 ± 1.05	50.01 ± 1.25	44.58 ± 1.12	45.30 ± 1.20

4.3.5.2 Ablation Studies

Task boundaries detection. The table in the right of Fig. 4.2 provides the precision and recall scores of the task switch detection schemes for our method and CMAML++. Our detection scheme outperforms that of CMAML++ in all metrics. This is because, the detection scheme in CMAML++ is based on comparing successive losses, which could lead to over detection

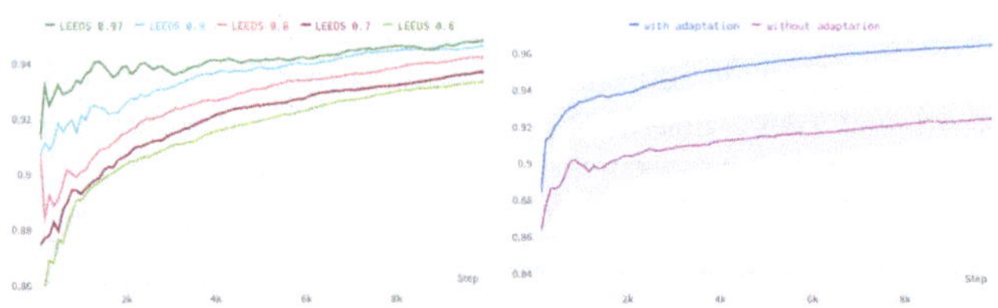

	Method	Prec.	Rec.
TI	LEEDS	**96.6**	**99.27**
	CMAML++	90.93	92.98
SB	LEEDS	**91.81**	**99.37**
	CMAML++	87.09	93.89

Fig. 4.2 Left: LEEDS under different p. **Center**: LEEDS with and without domain adaptation. **Right**: Task boundaries detection on **Tiered-ImageNet (TI)** and **Synbols (SB)**

of task boundaries, especially when the task loss is too high at the first time the task is revealed to the online algorithm.

Importance of domain adaptation module. We investigate the importance of the distribution shift detection module that allows our algorithm LEEDS to update the meta model differently for in-distribution and out-of-distribution tasks. Figure 4.2 shows the performance of our algorithm with and without the distribution shift detection module. The performance of the algorithm significantly improves ($\sim 4.3\%$ improvement) with this module. This shows that such a simple mechanism can effectively boost the online learning performance by allowing the agent to learn more from OOD data while also remembering pre-training knowledge.

Sensitivity to frequency of task switches. Figure 4.2 shows the performance of our algorithm for different values of the probability p of task switches. The performance increases with p, which shows that our algorithm LEEDS can successfully re-use previous task knowledge to increase performance (Table 4.4).

Table 4.4 Average accuracy over 10000 online episodes on **Synbols** benchmark under different non-stationarity levels. The "pre-train" domain corresponds to 3 different alphabets from Synbols dataset, "ood1" corresponds to a new (w.r.t. the pre-training one) alphabet from Synbols dataset, and "ood2" contains font classification tasks

Method	Non-stationarity $p = 0.9$		Non-stationarity $p = 0.75$	
	All domains	ood domains	All domains	ood domains
LEEDS	**85.12** ±0.91	**67.48** ±0.97	**82.22** ±0.32	**63.68** ±0.36
CMAML++	81.14 ±1.05	62.39 ±1.00	79.74 ±1.07	60.70 ±1.12
FOML	46.40 ±0.61	41.73 ±0.73	37.46 ±0.27	34.13 ±0.31
MAML	76.25 ±0.63	42.70 ±0.68	74.87 ±0.42	43.84 ±0.45
ANIL	64.58 ±0.32	34.51 ±0.54	72.69 ±0.30	35.66 ±0.49
MetaOGD	72.04 ±0.67	46.69 ±0.77	67.93 ±0.59	42.66 ±0.62
BGD	25.63 ±0.07	25.61 ±0.09	27.53 ±0.08	27.17 ±0.11
MetaBGD	53.74 ±0.41	42.25 ±0.52	40.79 ±0.23	34.63 ±0.33

4.3.5.3 More Experimental Results

Accuracy tables with variances for Tiered-ImageNet and Synbols datasets.

Sensitivity to thresholds ℓ and τ and temperature δ.

Figures 4.3 and 4.4 illustrate the sensitivity of our algorithm LEEDS with respect to the thresholds ℓ and τ and the temperature parameter δ in the energy-based detection module. As depicted in Fig. 4.3, when the threshold ℓ is too small, the algorithm tends to over detect task switches (as indicated by low Recall for $\ell = 0.5$ in the table), which results in inferior performance of the algorithm due to ineffective reuse of task knowledge. On the other hand when ℓ is too large, the high misdetection rate (e.g., indicated by low Precision for $\ell = 5$) results in the algorithm mostly fine-tuning the online task model ϕ_t with the current task support data. As expected, this results in a failure mode (the algorithm diverges) due to the adversariality of different tasks. We find that values of ℓ in the rage [1.5, 2.3] yield the best performance of our algorithm.

Figure 4.4a shows that larger values of τ, which collapse to updating the meta-model at each step (even for pretraining task distribution), does not substantially improve the performance. This demonstrates the advantage of the distinct meta-update scheme proposed for in- and out-of-distribution tasks, which avoids unnecessary frequent meta-updates for the pretraining tasks and thus allows a more judicious usage of computational budget. Lower values of τ (e.g. $\tau = 15$) tend to detect all task distributions as the pertaining one, and thus corresponds to eliminating the domain adaptation component of our algorithm. We also find that simply setting the temperature $\delta = 1$ in the energy expression yields the best performance, and large values of δ also eliminates the effectiveness of the energy-based detection module (Fig. 4.4b). This is in fact in accordance to the finding in Liu et al. [224], which also suggests setting $\delta = 1$.

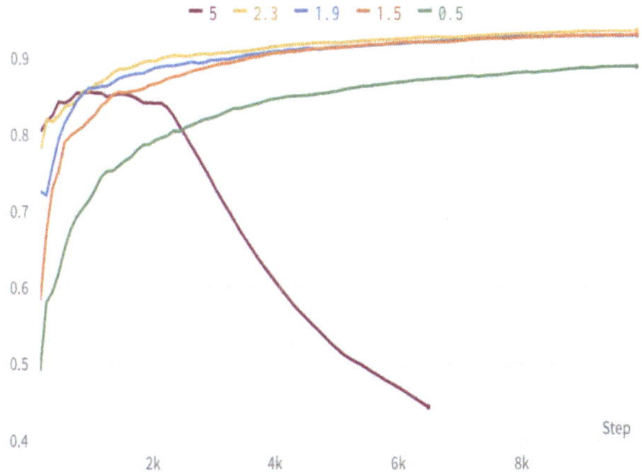

ℓ	Precision	Recall
$\ell = 0.5$	99.9	68.0
$\ell = 1.5$	99.9	98.4
$\ell = 1.9$	99.9	99.3
$\ell = 2.3$	99.9	99.7
$\ell = 5$	38.13	99.9

Fig. 4.3 Performance of LEEDS for different values of the threshold ℓ. **Left plot**: Performance on all encountered domains during online learning. **Right table**: Task boundaries detection for different values of ℓ. Experiments are conducted on the Omniglot-MNIST-FashionMNIST benchmark

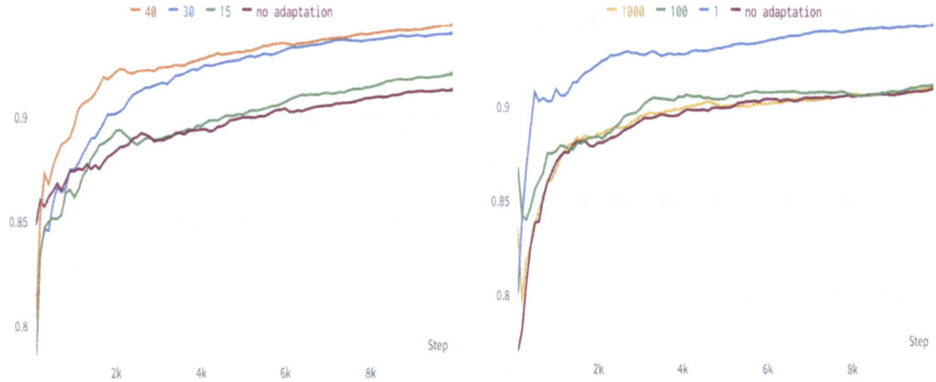

(a) Sensitivity to energy threshold τ (b) Sensitivity to temperature scale

Fig. 4.4 Performance of LEEDS for: **a** different values of the energy threshold τ and **b** different scales of the temperature δ. For both plots we report the performance on all encountered domains during online learning. Experiments are conducted on the Omniglot-MNIST-FashionMNIST benchmark

Figures 4.5, 4.6, 4.7 and 4.8 show the online evaluation curves of the different methods for for all settings. The different plots further show that our algorithm LEEDS outperforms other baseline algorithms in the OOD domains and at the same time also retains its performance on the pre-training tasks. The performance of the methods that do not adapt meta-parameters during online learning phase (such as MAML and ANIL) drops drastically when OOD tasks are far away from the pre-training ones (as shown in Table 4.1 for **Omniglot-MNIST-FashionMNIST**). For settings in which OOD tasks are close to the pre-training ones (such as in the **Tiered-ImageNet** dataset), MAML can perform similarly to CMAML++.

The superior performance of LEEDS even for the pre-training domain particularly shows that the re-use of task knowledge is beneficial for online meta-learning, as opposed to the usual practice of "resetting" to meta-parameters at each step (Fig. 4.9).

By comparing the two plots in Fig. 4.10, it can be seen that the advantage of our domain adaptation module is more significant when the OOD domains are far away from the pre-training one, as is the case for the FashionMNIST OOD domain compared to the Omniglot pre-training domain.

4.3.5.4 Further Descriptions About Datasets and Baseline Methods

Datasets We study dynamic online meta-learning on the following benchmarks:

Omniglot-MNIST-FashionMNIST dataset. For this dataset, we consider 10-ways 5-shots classification tasks. We pre-train the meta model on a subset of the Omniglot dataset and then deploy it in the online learning environment where tasks are sampled from either the full Omniglot dataset, or from one of the OOD datasets, i.e., the MNIST or FashionMNIST datasets.

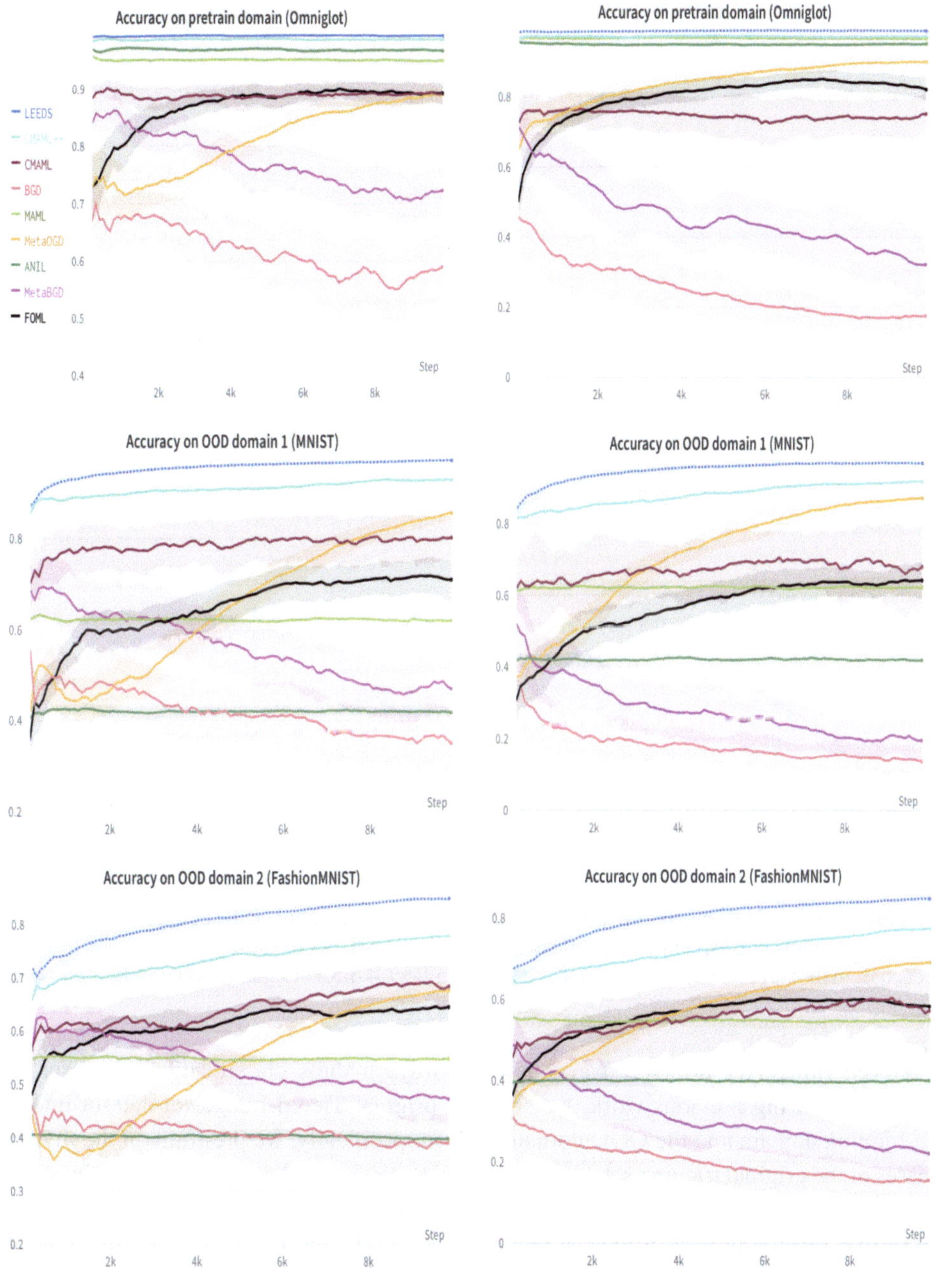

Fig. 4.5 Online evaluations in each of the encountered domains during online learning phase for the **Omniglot-MNIST-FashionMNIST** benchmark. First column corresponds to non-stationarity level $p = 0.9$. In second column $p = 0.75$. LEEDS is the only method that is able to preserve pre-training knowledge while substantially increasing performance in OOD domains. Legend in first plot only

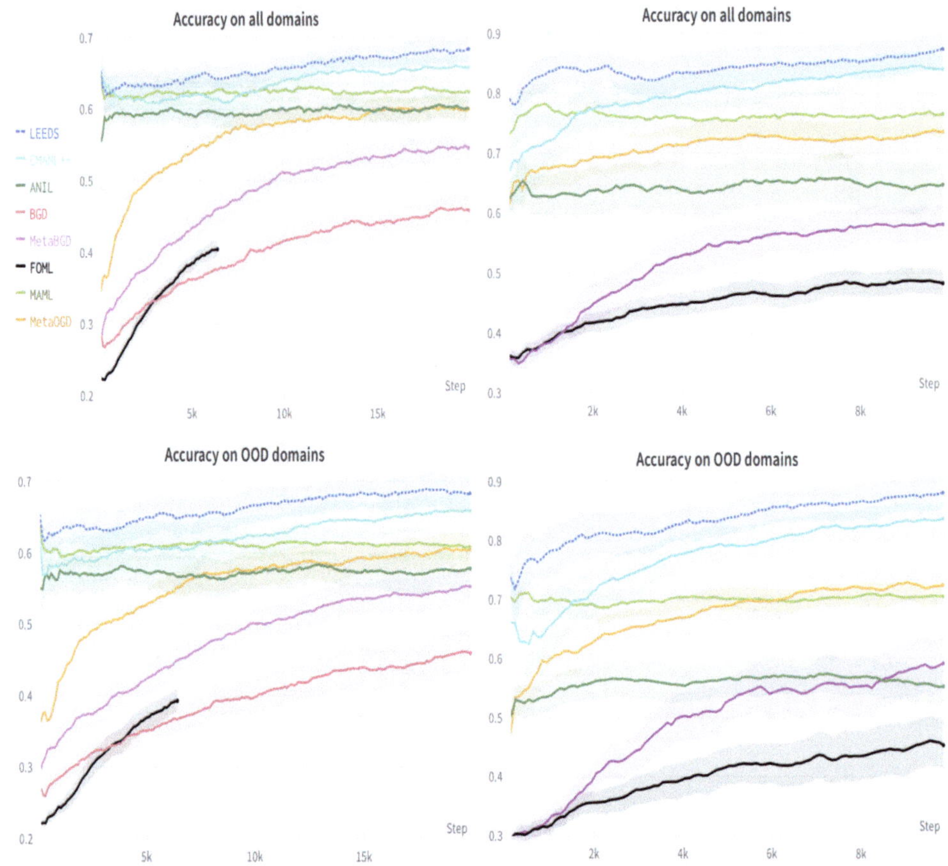

Fig. 4.6 Online evaluations for the **Tiered-ImageNet (TI)** and **Synbols (SB)** benchmarks under $p = 0.9$. First columns correspond to **TI** and second column to **SB**. More results including performance in each domain and under different p can be found in Sect. 4.3.5.3. Legend shown in first plot only

Tiered-ImageNet dataset. We consider 5-ways 5-shots classification tasks for this dataset. Following previous work, we split the original Tiered-ImageNet dataset into the pre-training domain and the OOD domains. More specifically, we pre-train on the first 200 classes of the original training set, use the remaining classes as the first OOD domain (ood1), and set the original test classes as second OOD domain (ood2). During online learning phase, we set the full training set to be IND domain, and ood1 and ood2 as OOD domains. We evaluate algorithms over 20000 online learning episodes.

Synbols dataset. We consider 4-ways 4-shots classification tasks in this dataset. The meta model is pre-trained on characters from 3 different alphabets and deployed on characters from a new alphabet (ood1). We also consider font classification tasks as additional OOD tasks (ood2).

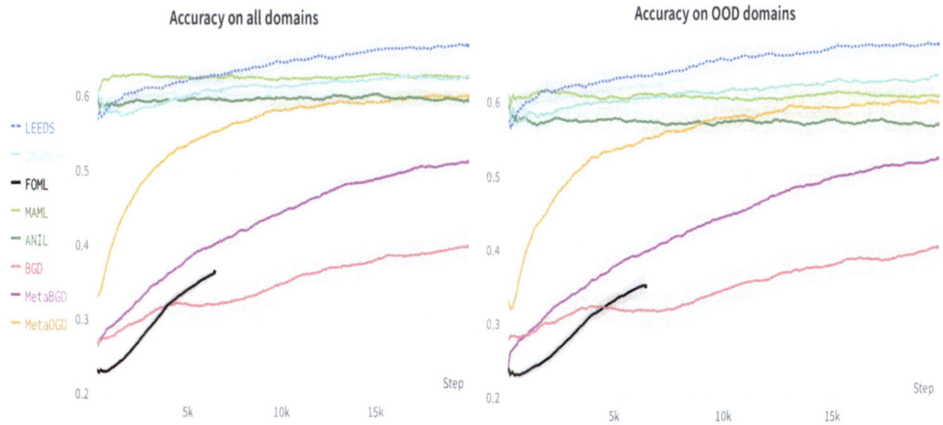

Fig. 4.7 Online evaluations for the **Tiered-ImageNet (TI)** benchmark under $\mathbf{p = 0.75}$. **Left**: Accuracies on all encountered domains during online learning. **Right**: Accuracies on all encountered OOD domains during online learning. We compare all baselines on a 16GB GPU memory budget and FOML runs out of memory for this benchmark due to its linear growth in memory requirement

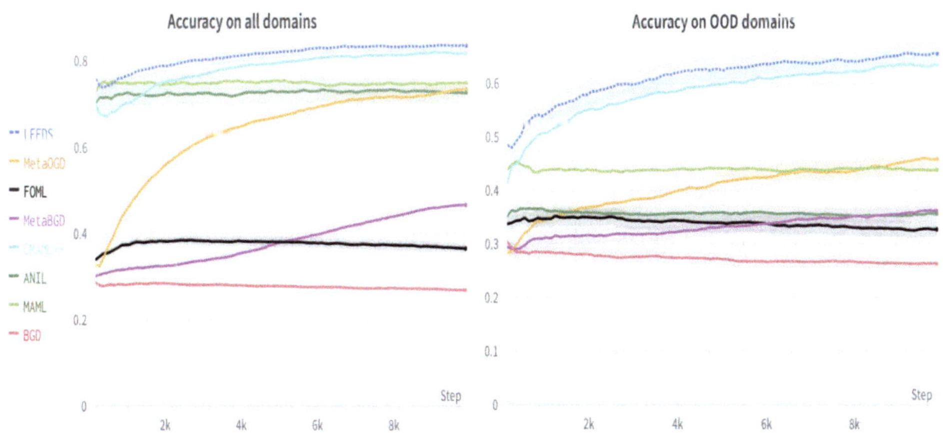

Fig. 4.8 Online evaluations for the **Synbols (SB)** benchmark under $\mathbf{p = 0.75}$. **Left**: Accuracies on all encountered domains during online learning. **Right**: Accuracies on all encountered OOD domains during online learning

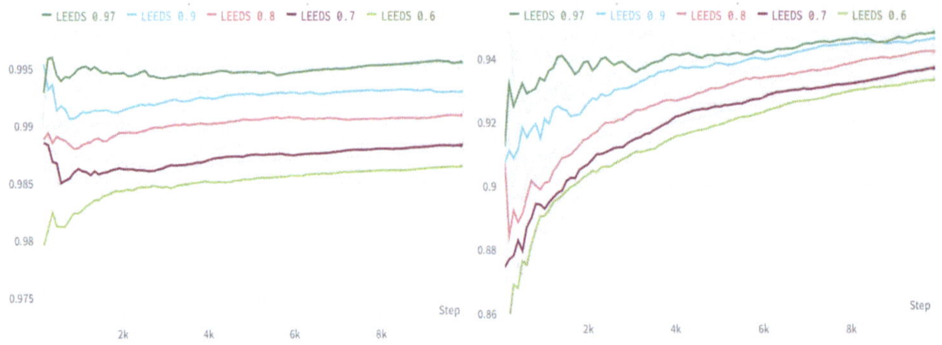

Fig. 4.9 Performance of our algorithm LEEDS under different p. **Left plot**: Evaluations in pre-training domain. **Right plot**: Evaluations in all domains

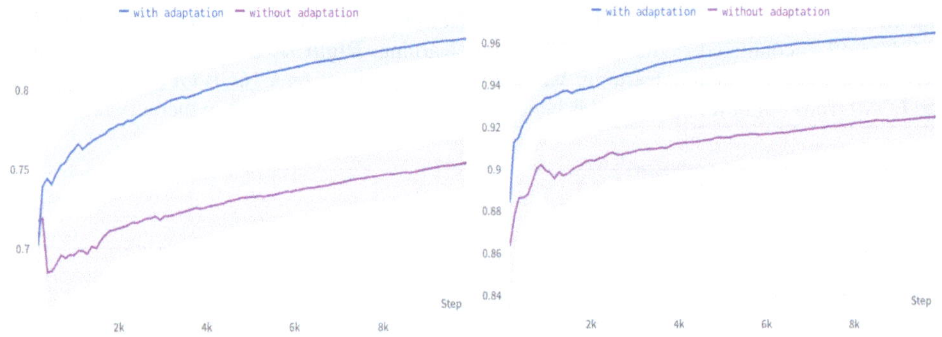

Fig. 4.10 Performance of LEEDS with and without energy-based domain adaptation module. **Left plot**: Evaluations in OOD domain **FashionMINIST**. **Right plot**: Evaluations in OOD domain **MNIST**

Baseline Methods We compare our algorithm with the following baseline methods for online meta-learning.

(1) **C-MAML** [45] **and C-MAML++**: the continual MAML approach (C-MAML) pre-trains the meta model using MAML and employs an online learning strategy based on task boundaries detection. Since C-MAML does not evaluate the task models on separate query sets, for a fair comparison we adapt it to do so and call the resulting algorithm C-MAML++.
(2) **FOML** [271]: the fully online meta-learning method updates online parameters using the latest online data and maintains a concurrent meta-training process to guide the online updates regularized by the meta model.
(3) **MAML** [103] **and ANIL** [270]: the MAML baseline consists of an offline pre-training phase and an online deployment phase. During offline pre-training the method learns

a common meta-initialization that will be used for all tasks, and the meta-initialization will never be updated at the online stage. The task model is adapted from the meta-initialization using the support set and evaluated on the query set. ANIL is similar to MAML but with partial parameter adaption, i.e., only the last layer is adapted for each task.

(4) **MetaBGD** [45] **and BGD** [392]: the baseline MetaBGD combines MAML and the Bayesian gradient descent method during online learning.

(5) **MetaOGD** [408]: the meta online gradient descent method simply updates the meta-model at each step using a MAML-like meta-objective evaluated on the current task data.

4.3.5.5 Further Experimental Details and Hyperameter Search

Further Experimental Specifications In all our experiments, we consider classification tasks. The cross-entropy loss between predictions and true labels is used to train all models. We use the same convolutional neural network (CNN) architectures widely adopted in few-shot learning literature [45, 103, 105], which include four convolutional blocks followed by a linear classification layer. Each convolutional block is a stack of one 3×3 convolution layer followed by BatchNormalization, ReLU, and 2×2 MaxPooling layers. For the **Omniglot-MNIST-FashionMNIST** benchmark, we use 64 filters in each convolutional layer and downsample the gray-scale images to 28×28 spacial resolution so as to have 64-dimensional feature vectors before classification. For **Tiered-ImageNet** and **Synbols** datasets, the inputs are respectively $3 \times 64 \times 64$ and $3 \times 32 \times 32$ RGB images, resulting respectively in 1024- and 256-dimensional feature vectors for 64 filters in each convolution layer.

Further Implementation Specifications The implementation of FOML [271] method is not released yet, and hence we compare with our own implementation of their algorithm. For all other baselines, we used their publicly available implementations. Codes for our algorithm LEEDS and all other baselines are provided in the supplementary materials of our submission. All codes are tested with Python 3.6 and Pytorch 1.2.

For example to run our algorithm LEEDS with the best hyperparameters that we obtained for the Omniglot-MNIST-FashionMNIST dataset under $p = 0.9$, one can run the following command:

```
python main.py —algo leeds —use_best 1
```

The experiment setting (e.g., the dataset to use) can be changed in configurations.yaml file. We run all methods on a single NVIDIA Tesla P100 GPU. All compared algorithms except FOML were able to run on 16GB GPU memory. FOML requires at least 32GB to reach 12000 online episodes for the Tiered-ImageNet dataset.

4.3.5.6 Heuristics for Setting the Thresholds

For the energy threshold τ, we follow the strategy in Liu et al. [224], i.e., we set the threshold τ using the pre-training tasks. More specifically, we set τ so that 95% of the pre-training inputs are correctly detected as pre-training data. In standard few-shot learning experimental setups, the true labeling for each individual task is usually randomly chosen. When there is a task switch, the task specific model learnt from the previous task generally does not fit the new task anymore, where the learning performance could be similar to that of a random model. Thus motivated, we find that a good heuristic to choose a starting value for the threshold ℓ is to use the loss value evaluated on a random model. For example, for 10-ways classification tasks that value would be $\ell_r = -\log(1/10) = 2.3$.

4.4 Conclusion

In this chapter, we introduced the technique meta-learning which can enable fast learning of new tasks at the network edge with a few data points. We first presented a brief overview of the recent algorithm development in terms of optimizer-based meta-learning and federated meta-learning, which served the algorithm foundations for continual edge-AI from the perspective of online meta-learning. To better demonstrate the development in online meta-learning, we showcased our recent study about online meta-learning in non-stationary environments without knowing the task boundaries. To address the problems therein, we proposed LEEDS for efficient meta model and online task model updates. In particular, based two simple but effective detection mechanisms of the task switches and the distribution shift, LEEDS can efficiently reuse the best task model available without resetting to the meta model and distinguish the meta model updates for in-distribution tasks and out-of-distribution tasks so as to quickly learn the new knowledge for new distributions while preserving the old knowledge of the pre-training distribution. In particular, the meta model update in LEEDS was based on the current data only, eliminating the need of storing previous data. Extensive experiments corroborated the superior performance of LEEDS over related baseline methods on multiple benchmarks.

Part II
Efficient Algorithm Design for Edge AI

The second part of this book will focus on efficient algorithm design that tailors the machine learning techniques introduced in Part I to meet the practical constraints in edge AI. In particular, the algorithms will be developed by following the three frameworks proposed earlier, i.e., edge-only, cloud-edge collaboration, and edge-edge collaboration.

Edge-Only Learning via Continual Learning with Enhanced Knowledge Transfer

5

To enable edge-only continual learning on a single edge device, it is important to reflect on the remarkable learning capability of human beings during their lifespan. In particular, with more tasks being learnt, the edge device is expected to be able to learn a new task more easily by leveraging the accumulated knowledge from old tasks (forward knowledge transfer), and also further improve the learning performance of old tasks based on the gained knowledge of related new tasks (backward knowledge transfer). Therefore, in this chapter, we will explore CL algorithm designs for more efficient on-device learning by enhancing both forward and backward knowledge transfer across different tasks.

5.1 Introduction

Most existing CL methods focus on addressing the catastrophic forgetting problem [231], i.e., the neural network may easily forget the knowledge of old tasks when learning a new task. The main strategy therein is to avoid the model change of old tasks when learning the new task. For example, regularization-based methods (e.g., Kirkpatrick et al. [171], Aljundi et al. [12] and Liu and Liu [220]) penalize the modification of important weights of old tasks; parameter-isolation based methods (e.g., Fernando et al. [102], Serra et al. [294], Yoon et al. [380] and Hung et al. [147]) fix the model learnt for old tasks; and memory-based methods (e.g., Chaudhry et al. [49], Farajtabar et al. [100] and Saha et al. [283]) aim to update the model with minimal interference introduced to old tasks. While effectively mitigating catastrophic forgetting, such a model-freezing strategy inevitably limits the backward knowledge transfer, by implicitly and conservatively treating the model update of the new task as interference to old tasks. Intuitively, careful modifications of the learnt model

H. Wang et al., *Continual and Reinforcement Learning for Edge AI*,
Synthesis Lectures on Learning, Networks, and Algorithms,
https://doi.org/10.1007/978-3-031-84363-1_5

of old tasks may further improve the learning performance especially when the new task shares similar knowledge to the old tasks.

On the other hand, little attention has been put on the knowledge transfer across tasks in CL with neural networks. Progressive Network [282] considers the forward knowledge transfer but learns one model for each task. Lin et al. [218] proposes trust region gradient projection (TRGP) to facilitate forward knowledge transfer from correlated old tasks to the new task through a scaling matrix, but the model is still updated along the direction orthogonal to the input subspaces of old tasks to mitigate forgetting. Ke et al. [162] studies the knowledge transfer in a mixed sequence of similar and dissimilar tasks, but employs a complicated task similarity detection mechanism where two additional networks need to be trained first for each task before learning the new task.

In this chapter, we seek to explore CL from a new perspective by going beyond merely addressing forgetting: *When and how could we improve the learnt model of old tasks to facilitate backward knowledge transfer?* In particular, we consider a fixed neural network and the scenario where the data of old tasks is not accessible when learning a new task, which naturally rules out the expansion-based methods (e.g., Yoon et al. [380] and Hung et al. [147]) and the experience-replay methods (e.g., Riemer et al. [278] and Chaudhry et al. [50]). Clearly, achieving backward knowledge transfer is very challenging in this case as mitigating forgetting herein is already nontrivial. To tackle this challenge, note that the orthogonal-projection based CL methods (e.g., Saha et al. [283] and Lin et al. [218]), which update the model with gradients orthogonal to the input subspaces of old tasks, have demonstrated the state-of-the-art performance under the setting above. Thus motivated, we propose a new orthogonal-projection based CL method to carefully modify the learnt model of old tasks for better backward knowledge transfer.

More specifically, we first introduce notions of 'sufficient projection' and 'positive correlation' based on the gradient projection onto the subspaces of old tasks to characterize the task correlation. We theoretically show that when the task gradients are sufficiently aligned in the old task subspace, appropriately updating the learnt model of old tasks when learning the new task could improve the learning performance and possibly lead to a better model for the old tasks. Based on this analysis, we next propose an orthogonal-projection based ContinUal learning method with Backward knowlEdge tRansfer (CUBER), which 1) first identifies the positively correlated old tasks for the new task and 2) carefully updates the learnt model of selected old tasks along with the model learning for the new task. The experimental results on standard CL benchmarks show that CUBER can achieve the best backward knowledge transfer compared to all the considered baseline CL methods, which consequently improves the overall learning performance. In particular, to the best of our knowledge, CUBER is the first to demonstrate positive backward knowledge transfer on these benchmarks using a fixed capacity network without experience replay, whereas all the compared baselines still suffer from catastrophic forgetting with negative backward knowledge transfer.

5.2 Conditions on Improving the Learnt Model of Old Tasks

In continual learning, a sequence of tasks $\mathbb{T} = \{t\}_{t=1}^{T}$ arrives sequentially. For each task t, there is a dataset $\mathbb{D}^t = \{(\boldsymbol{x}_i^t, \boldsymbol{y}_i^t)\}_{i=1}^{N^t}$ with N^t sample pairs, which is sampled from some unknown distribution $\mathbb{P}^t$. In this work, we consider a fixed capacity neural network with weights $\boldsymbol{w}$. When learning a new task t, we only have access to the dataset $\mathbb{D}^t$ and no data-points of old tasks are available. We further denote $\mathcal{L}(\boldsymbol{w}, \mathbb{D}^t) = \mathcal{L}_t(\boldsymbol{w})$ as the loss function for training, e.g., mean squared and cross-entropy loss, and $\boldsymbol{w}^t$ as the model after learning task t.

Intuitively, if the new task t has strong similarity with some old tasks, appropriate model update of the new task would not introduce forgetting of the similar old tasks, but can lead to a better model for these old tasks due to the backward knowledge transfer. To formally characterize this task similarity, we introduce the following conditions on the task gradients.

Definition 5.1 (*Sufficient Projection*) For any new task $t \in [1, T]$, we say it has sufficient gradient projection on the input subspace of old task $j \in [0, t-1]$ if for some $\epsilon_1 \in (0, 1)$

$$\|\text{proj}_{S^j}(\nabla \mathcal{L}_t(\boldsymbol{w}^{t-1}))\|_2 \geq \epsilon_1 \|\nabla \mathcal{L}_t(\boldsymbol{w}^{t-1})\|_2.$$

Here proj_{S^j} defines the projection on the input subspace S^j of task j: $\text{proj}_{S^j}(A) = AB^j(B^j)'$ for some matrix A and B^j is the bases for S^j. This definition of sufficient projection follows the same line of the trust region defined in Lin et al. [218], which implies that task t and j may have sufficient common bases between their input subspaces and hence are strongly correlated, because the gradient lies in the span of the input [394].

Definition 5.2 *(Positive Correlation)* For any new task $t \in [1, T]$, we say it has positive correlation with old task $j \in [0, t-1]$ if for some $\epsilon_2 \in (0, 1)$

$$\langle \nabla \mathcal{L}_j(\boldsymbol{w}^j), \nabla \mathcal{L}_t(\boldsymbol{w}^{t-1}) \rangle \geq \epsilon_2 \|\nabla \mathcal{L}_j(\boldsymbol{w}^j)\|_2 \|\nabla \mathcal{L}_t(\boldsymbol{w}^{t-1})\|_2.$$

While sufficient projection suggests the possibly strong correlation between two tasks t and j, Definition 5.2 goes one step further by introducing the positive correlation between tasks, in the sense that the initial model gradient of task t is not conflicting with the model gradient when learning task j. In fact, it can be shown that the positive correlation implies the sufficient projection in Definition 5.1 for some $\epsilon_1 \geq \epsilon_2$. Note that both Definitions 5.1 and 5.2 characterize the correlation based on the initial model $\boldsymbol{w}^{t-1}$, which allows the task correlation detection before learning the new task t.

$$\mathcal{F}(\mathbb{W}) = \mathcal{L}(\mathbb{W}, \mathcal{D}_1) + \mathcal{L}(\mathbb{W}, \mathcal{D}_2),\ \boldsymbol{g}_1(\mathbb{W}) = \nabla_{\mathbb{W}} \mathcal{L}(\mathbb{W}, \mathcal{D}_1),\ \boldsymbol{g}_2(\mathbb{W}) = \nabla_{\mathbb{W}} \mathcal{L}(\mathbb{W}, \mathcal{D}_2)$$

- (Rule #1): $\mathbb{W}^{k+1} = \mathbb{W}^k - \alpha[\boldsymbol{g}_2(\mathbb{W}^k) - \text{proj}_{S^1}(\boldsymbol{g}_2(\mathbb{W}^k))]$,
- (Rule #2): $\mathbb{W}^{k+1} = \mathbb{W}^k - \alpha \boldsymbol{g}_2(\mathbb{W}^k)$.

For ease of exposition, consider the scenario with a sequence of two tasks 1 and 2. Let $\mathcal{F}(\boldsymbol{w}) = \mathcal{L}(\boldsymbol{w}, \mathcal{D}_1) + \mathcal{L}(\boldsymbol{w}, \mathcal{D}_2)$, $\boldsymbol{g}_1(\boldsymbol{w}) = \nabla_{\boldsymbol{w}} \mathcal{L}(\boldsymbol{w}, \mathcal{D}_1)$ and $\boldsymbol{g}_2(\boldsymbol{w}) = \nabla_{\boldsymbol{w}} \mathcal{L}(\boldsymbol{w}, \mathcal{D}_2)$. Given the model learnt for task 1 as $\boldsymbol{w}^1$, we consider the following two model update rules for learning task 2, *where only the data of task 2 is available*:

- (Rule #1): $\boldsymbol{w}_{k+1} = \boldsymbol{w}_k - \alpha[\boldsymbol{g}_2(\boldsymbol{w}_k) - \text{proj}_{S^1}(\boldsymbol{g}_2(\boldsymbol{w}_k))]$,
- (Rule #2): $\boldsymbol{w}_{k+1} = \boldsymbol{w}_k - \alpha \boldsymbol{g}_2(\boldsymbol{w}_k)$.

Here $\boldsymbol{w}_0 = \boldsymbol{w}^1$ and $k \in [0, K-1]$. Rule #1 characterizes the model update of the orthogonal-projection based methods, where no interference is introduced to task 1 as the model change is orthogonal to the input subspace S^1 of task 1 [218] (and hence the learnt model of task 1 will not be updated). In contrast, the vanilla gradient descent Rule #2 for learning task 2 will inevitably modify the learnt model $\boldsymbol{w}^1$ for task 1 unless $\text{proj}_{S^1}(\boldsymbol{g}_2(\boldsymbol{w}_k)) = 0$, i.e., the extreme case where both forgetting and knowledge transfer do not occur. In what follows, we evaluate the performance of these update rules.

Theorem 5.1 *Suppose that the loss $\mathcal{L}$ is B-Lipschitz and $\frac{H}{2}$-smooth. Let* $\alpha < \min \left\{ \frac{1}{H}, \frac{\gamma \|\boldsymbol{g}_1(\boldsymbol{w}_0)\|}{HBK} \right\}$ *and* $\epsilon_2 \geq \frac{(2+\gamma^2)\|\boldsymbol{g}_1(\boldsymbol{w}_0)\|}{4\|\boldsymbol{g}_2(\boldsymbol{w}_0)\|}$ *for some* $\gamma \in (0, 1)$. *We have the following results:*
(1) If $\mathcal{L}$ is convex, Rule #2 for task 2 converges to the optimal model $\boldsymbol{w}^* = \arg\min \mathcal{F}(\boldsymbol{w})$*;*
(2) If $\mathcal{L}$ is nonconvex, Rule #2 for task 2 converges to the first order stationary point, i.e.,

$$\min_k \|\nabla \mathcal{F}(\boldsymbol{w}_k)\|^2 < \frac{2}{\alpha K}[\mathcal{F}(\boldsymbol{w}_0) - \mathcal{F}(\boldsymbol{w}^*)] + \frac{4+\gamma^2}{2}\|\boldsymbol{g}_1(\boldsymbol{w}_0)\|^2.$$

Proof For a $H/2$-smooth loss function $\mathcal{L}$, it can be easily shown that $\mathcal{F}$ is H-smooth.
(1) For any $k \in [0, K]$, we can have

$$\begin{aligned}
\mathcal{F}(\boldsymbol{w}_{k+1}) \leq & \mathcal{F}(\boldsymbol{w}_k) + \nabla \mathcal{F}(\boldsymbol{w}_k)^T(\boldsymbol{w}_{k+1} - \boldsymbol{w}_k) + \frac{H}{2}\|\boldsymbol{w}_{k+1} - \boldsymbol{w}_k\|^2 \\
= & \mathcal{F}(\boldsymbol{w}_k) + (\boldsymbol{g}_1(\boldsymbol{w}_k) + \boldsymbol{g}_2(\boldsymbol{w}_k))^T(-\alpha \boldsymbol{g}_2(\boldsymbol{w}_k)) + \frac{\alpha^2 H}{2}\|\boldsymbol{g}_2(\boldsymbol{w}_k)\|^2 \\
= & \mathcal{F}(\boldsymbol{w}_k) - \left[\alpha - \frac{\alpha^2 H}{2}\right]\|\boldsymbol{g}_2(\boldsymbol{w}_k)\|^2 - \alpha\langle \boldsymbol{g}_1(\boldsymbol{w}_k), \boldsymbol{g}_2(\boldsymbol{w}_k)\rangle.
\end{aligned} \tag{5.1}$$

For the term $\langle \boldsymbol{g}_1(\boldsymbol{w}_k), \boldsymbol{g}_2(\boldsymbol{w}_k)\rangle$, it follows that

$$\begin{aligned}
& \langle \boldsymbol{g}_1(\boldsymbol{w}_k), \boldsymbol{g}_2(\boldsymbol{w}_k)\rangle \\
= & \langle \boldsymbol{g}_1(\boldsymbol{w}_k) - \boldsymbol{g}_1(\boldsymbol{w}_0) + \boldsymbol{g}_1(\boldsymbol{w}_0), \boldsymbol{g}_2(\boldsymbol{w}_k)\rangle \\
= & \langle \boldsymbol{g}_1(\boldsymbol{w}_k) - \boldsymbol{g}_1(\boldsymbol{w}_0), \boldsymbol{g}_2(\boldsymbol{w}_k)\rangle + \langle \boldsymbol{g}_1(\boldsymbol{w}_0), \boldsymbol{g}_2(\boldsymbol{w}_k)\rangle \\
= & \langle \boldsymbol{g}_1(\boldsymbol{w}_k) - \boldsymbol{g}_1(\boldsymbol{w}_0), \boldsymbol{g}_2(\boldsymbol{w}_k)\rangle + \langle \boldsymbol{g}_1(\boldsymbol{w}_0), \boldsymbol{g}_2(\boldsymbol{w}_k) - \boldsymbol{g}_2(\boldsymbol{w}_0)\rangle + \langle \boldsymbol{g}_1(\boldsymbol{w}_0), \boldsymbol{g}_2(\boldsymbol{w}_0)\rangle.
\end{aligned} \tag{5.2}$$

Because

$$
\begin{aligned}
&2\langle g_1(w_k) - g_1(w_O), g_2(w_k)\rangle + \|g_1(w_k) - g_1(w_O)\|^2 + \|g_2(w_k)\|^2 \\
=&\|g_1(w_k) - g_1(w_O) + g_2(w_k)\|^2 \geq 0,
\end{aligned}
$$

we have

$$
\langle g_1(w_k) - g_1(w_O), g_2(w_k)\rangle \geq -\frac{1}{2}\|g_1(w_k) - g_1(w_O)\|^2 - \frac{1}{2}\|g_2(w_k)\|^2. \tag{5.3}
$$

Following the same line, it can be shown that

$$
\langle g_1(w_O), g_2(w_k) - g_2(w_O)\rangle \geq -\frac{1}{2}\|g_2(w_k) - g_2(w_O)\|^2 - \frac{1}{2}\|g_1(w_O)\|^2. \tag{5.4}
$$

Combining Eqs. 5.2, 5.3 and 5.4 gives a lower bound on $\langle g_1(w_k), g_2(w_k)\rangle$, i.e.,

$$
\begin{aligned}
&\langle g_1(w_k), g_2(w_k)\rangle \\
\geq& -\frac{1}{2}\|g_1(w_k) - g_1(w_O)\|^2 - \frac{1}{2}\|g_2(w_k)\|^2 \\
&-\frac{1}{2}\|g_2(w_k) - g_2(w_O)\|^2 - \frac{1}{2}\|g_1(w_O)\|^2 + \langle g_1(w_O), g_2(w_O)\rangle \\
\geq& -\frac{H^2}{8}\|w_k - w_O\|^2 - \frac{1}{2}\|g_2(w_k)\|^2 \\
&-\frac{H^2}{8}\|w_k - w_O\|^2 - \frac{1}{2}\|g_1(w_O)\|^2 + \langle g_1(w_O), g_2(w_O)\rangle \\
=& -\frac{H^2}{4}\|w_k - w_O\|^2 - \frac{1}{2}\|g_2(w_k)\|^2 - \frac{1}{2}\|g_1(w_O)\|^2 + \langle g_1(w_O), g_2(w_O)\rangle,
\end{aligned} \tag{5.5}
$$

where the second inequality is true because of the smoothness of the loss function.

Based on the update Rule #2, it can be seen that

$$
w_k = w_O - \alpha\sum_{i=0}^{k-1} g_2(w_i). \tag{5.6}
$$

Therefore, continuing with Eq. 5.1, we can have

$$
\begin{aligned}
&\mathcal{F}(w_{k+1}) \\
\leq&\mathcal{F}(w_k) - \left[\alpha - \frac{\alpha^2 H}{2}\right]\|g_2(w_k)\|^2 - \alpha\langle g_1(w_k), g_2(w_k)\rangle \\
\leq&\mathcal{F}(w_k) - \left[\alpha - \frac{\alpha^2 H}{2}\right]\|g_2(w_k)\|^2 + \frac{\alpha^3 H^2}{4}\|\sum_{i=0}^{k-1} g_2(w_i)\|^2 + \frac{\alpha}{2}\|g_2(w_k)\|^2 \\
&+\frac{\alpha}{2}\|g_1(w_O)\|^2 - \alpha\langle g_1(w_O), g_2(w_O)\rangle
\end{aligned}
$$

$$
\begin{aligned}
=&\mathcal{F}(\boldsymbol{w}_k) - \left[\frac{\alpha}{2} - \frac{\alpha^2 H}{2}\right] \|\boldsymbol{g}_2(\boldsymbol{w}_k)\|^2 + \frac{\alpha^3 H^2}{4}\|\sum_{i=0}^{k-1} \boldsymbol{g}_2(\boldsymbol{w}_i)\|^2 + \frac{\alpha}{2}\|\boldsymbol{g}_1(\boldsymbol{w}_0)\|^2 - \alpha\langle \boldsymbol{g}_1(\boldsymbol{w}_0), \boldsymbol{g}_2(\boldsymbol{w}_0)\rangle \\
\leq&\mathcal{F}(\boldsymbol{w}_k) - \left[\frac{\alpha}{2} - \frac{\alpha^2 H}{2}\right] \|\boldsymbol{g}_2(\boldsymbol{w}_k)\|^2 + \frac{\alpha^3 H^2}{4}\|\sum_{i=0}^{k-1} \boldsymbol{g}_2(\boldsymbol{w}_i)\|^2 + \frac{\alpha}{2}\|\boldsymbol{g}_1(\boldsymbol{w}_0)\|^2 \\
&- \alpha\epsilon_2\|\boldsymbol{g}_1(\boldsymbol{w}_0)\|\|\boldsymbol{g}_2(\boldsymbol{w}_0)\|,
\end{aligned}
$$

where the last inequality is based on Definition 5.2.

Next, it can be shown that

$$
\alpha \leq \frac{\gamma\|\boldsymbol{g}_1(\boldsymbol{w}_0)\|}{HBK} \leq \frac{\gamma\|\boldsymbol{g}_1(\boldsymbol{w}_0)\|}{H\|\sum_{i=0}^{k-1} \boldsymbol{g}_2(\boldsymbol{w}_i)\|}.
$$

It then follows that

$$
\begin{aligned}
&\frac{1}{2}\|\boldsymbol{g}_1(\boldsymbol{w}_0)\|^2 + \frac{\alpha^2 H^2}{4}\|\sum_{i=0}^{k-1} \boldsymbol{g}_2(\boldsymbol{w}_i)\|^2 \\
\leq&\frac{1}{2}\|\boldsymbol{g}_1(\boldsymbol{w}_0)\|^2 + \frac{\gamma^2\|\boldsymbol{g}_1(\boldsymbol{w}_0)\|^2}{4H^2\|\sum_{i=0}^{k-1} \boldsymbol{g}_2(\boldsymbol{w}_i)\|^2} H^2\|\sum_{i=0}^{k-1} \boldsymbol{g}_2(\boldsymbol{w}_i)\|^2 \\
=&\frac{2+\gamma^2}{4}\|\boldsymbol{g}_1(\boldsymbol{w}_0)\|^2.
\end{aligned} \tag{5.7}
$$

Therefore, we can obtain that

$$
\begin{aligned}
\mathcal{F}(\boldsymbol{w}_{k+1}) \leq&\mathcal{F}(\boldsymbol{w}_k) - \left[\frac{\alpha}{2} - \frac{\alpha^2 H}{2}\right] \|\boldsymbol{g}_2(\boldsymbol{w}_k)\|^2 + \frac{\alpha(2+\gamma^2)}{4}\|\boldsymbol{g}_1(\boldsymbol{w}_0)\|^2 - \alpha\epsilon_2\|\boldsymbol{g}_1(\boldsymbol{w}_0)\|\|\boldsymbol{g}_2(\boldsymbol{w}_0)\| \\
\leq&\mathcal{F}(\boldsymbol{w}_k) - \left[\frac{\alpha}{2} - \frac{\alpha^2 H}{2}\right] \|\boldsymbol{g}_2(\boldsymbol{w}_k)\|^2 \\
<&\mathcal{F}(\boldsymbol{w}_k),
\end{aligned}
$$

where the second inequality is true because $\epsilon_2 \geq \frac{(2+\gamma^2)\|\boldsymbol{g}_1(\boldsymbol{w}_0)\|}{4\|\boldsymbol{g}_2(\boldsymbol{w}_0)\|}$. This sufficient decrease of the objective function value indicates that the optimal $\mathcal{F}(w^*)$ can be obtained eventually for convex loss functions.

(2) For a non-convex loss function $\mathcal{L}$, we can have the following as in Eq. 5.1:

$$
\begin{aligned}
\mathcal{F}(w_{k+1}^r) \leq&\mathcal{F}(\boldsymbol{w}_k) - \left[\alpha - \frac{\alpha^2 H}{2}\right] \|\boldsymbol{g}_2(\boldsymbol{w}_k)\|^2 - \alpha\langle \boldsymbol{g}_1(\boldsymbol{w}_k), \boldsymbol{g}_2(\boldsymbol{w}_k)\rangle \\
\stackrel{(a)}{=}&\mathcal{F}(\boldsymbol{w}_k) - \left[\alpha - \frac{\alpha^2 H}{2}\right] \|\boldsymbol{g}_2(\boldsymbol{w}_k)\|^2 - \frac{\alpha}{2}[\|\nabla\mathcal{F}(\boldsymbol{w}_k)\|^2 - \|\boldsymbol{g}_1(\boldsymbol{w}_k)\|^2 - \|\boldsymbol{g}_2(\boldsymbol{w}_k)\|^2] \\
=&\mathcal{F}(\boldsymbol{w}_k) - \left[\frac{\alpha}{2} - \frac{\alpha^2 H}{2}\right] \|\boldsymbol{g}_2(\boldsymbol{w}_k)\|^2 - \frac{\alpha}{2}\|\nabla\mathcal{F}(\boldsymbol{w}_k)\|^2 + \frac{\alpha}{2}\|\boldsymbol{g}_1(\boldsymbol{w}_k)\|^2 \\
=&\mathcal{F}(\boldsymbol{w}_k) - \left[\frac{\alpha}{2} - \frac{\alpha^2 H}{2}\right] \|\boldsymbol{g}_2(\boldsymbol{w}_k)\|^2 - \frac{\alpha}{2}\|\nabla\mathcal{F}(\boldsymbol{w}_k)\|^2 + \frac{\alpha}{2}\|\boldsymbol{g}_1(\boldsymbol{w}_k) - \boldsymbol{g}_1(\boldsymbol{w}_0) + \boldsymbol{g}_1(\boldsymbol{w}_0)\|^2
\end{aligned}
$$

$$
\begin{aligned}
&\leq \mathcal{F}(\boldsymbol{w}_k) - \left[\frac{\alpha}{2} - \frac{\alpha^2 H}{2}\right] \|\boldsymbol{g}_2(\boldsymbol{w}_k)\|^2 - \frac{\alpha}{2}\|\nabla \mathcal{F}(\boldsymbol{w}_k)\|^2 + \alpha\|\boldsymbol{g}_1(\boldsymbol{w}_k) - \boldsymbol{g}_1(\boldsymbol{w}_0)\|^2 \\
&\quad + \alpha\|\boldsymbol{g}_1(\boldsymbol{w}_0)\|^2 \\
&\overset{(b)}{\leq} \mathcal{F}(\boldsymbol{w}_k) - \left[\frac{\alpha}{2} - \frac{\alpha^2 H}{2}\right] \|\boldsymbol{g}_2(\boldsymbol{w}_k)\|^2 - \frac{\alpha}{2}\|\nabla \mathcal{F}(\boldsymbol{w}_k)\|^2 + \frac{H^2\alpha^3}{4}\|\sum_{i=0}^{k-1} \boldsymbol{g}_2(\boldsymbol{w}_i)\|^2 \\
&\quad + \alpha\|\boldsymbol{g}_1(\boldsymbol{w}_0)\|^2,
\end{aligned}
$$

where (a) is because $\nabla \mathcal{F}(\boldsymbol{w}_k) = \boldsymbol{g}_1(\boldsymbol{w}_k) + \boldsymbol{g}_2(\boldsymbol{w}_k)$, and (b) is because of the smoothness of $\mathcal{L}$ and Eq. 5.6.

Therefore,

$$
\begin{aligned}
&\min_k \|\nabla \mathcal{F}(\boldsymbol{w}_k)\|^2 \\
&\leq \frac{1}{K}\sum_{k=0}^{K-1} \|\nabla \mathcal{F}(\boldsymbol{w}_k)\|^2 \\
&\leq \frac{2}{\alpha K}\sum_{k=0}^{K-1}\left[\mathcal{F}(\boldsymbol{w}_k) - \mathcal{F}(\boldsymbol{w}_{k+1}) + \frac{H^2\alpha^3}{4}\|\sum_{i=0}^{k-1} \boldsymbol{g}_2(\boldsymbol{w}_i)\|^2 + \alpha\|\boldsymbol{g}_1(\boldsymbol{w}_0)\|^2 - \left[\frac{\alpha}{2} - \frac{\alpha^2 H}{2}\right]\|\boldsymbol{g}_2(\boldsymbol{w}_k)\|^2\right] \\
&\leq \frac{2}{\alpha K}[\mathcal{F}(\boldsymbol{w}_0) - \mathcal{F}(\boldsymbol{w}_K)] + \frac{H^2\alpha^2}{2(K \quad 1)}\sum_{k=1}^{K-1}\|\sum_{i=0}^{k-1} \boldsymbol{g}_2(\boldsymbol{w}_i)\|^2 + 2\|\boldsymbol{g}_1(\boldsymbol{w}_0)\|^2 - \frac{1-\alpha H}{K}\sum_{k=0}^{K-1}\|\boldsymbol{g}_2(\boldsymbol{w}_k)\|^2 \\
&\overset{(a)}{\leq} \frac{2}{\alpha K}[\mathcal{F}(\boldsymbol{w}_0) - \mathcal{F}(\boldsymbol{w}_K)] + \frac{\gamma^2}{2}\|\boldsymbol{g}_1(\boldsymbol{w}_0)\|^2 + 2\|\boldsymbol{g}_1(\boldsymbol{w}_0)\|^2 - \frac{1-\alpha H}{K}\sum_{k=0}^{K-1}\|\boldsymbol{g}_2(\boldsymbol{w}_k)\|^2 \\
&\leq \frac{2}{\alpha K}[\mathcal{F}(\boldsymbol{w}_0) - \mathcal{F}(\boldsymbol{w}^*)] + \frac{4+\gamma^2}{2}\|\boldsymbol{g}_1(\boldsymbol{w}_0)\|^2 - \frac{1-\alpha H}{K}\sum_{k=0}^{K-1}\|\boldsymbol{g}_2(\boldsymbol{w}_k)\|^2 \\
&\leq \frac{2}{\alpha K}[\mathcal{F}(\boldsymbol{w}_0) - \mathcal{F}(\boldsymbol{w}^*)] + \frac{4+\gamma^2}{2}\|\boldsymbol{g}_1(\boldsymbol{w}_0)\|^2
\end{aligned}
$$

where (a) holds due to $\mathcal{F}(\boldsymbol{w}^*) \leq \mathcal{F}(\boldsymbol{w}_K)$ and $\|\sum_{i=0}^{k-1} \boldsymbol{g}_2(\boldsymbol{w}_i)\|^2 \leq \frac{\gamma^2}{\alpha^2 H^2}\|\boldsymbol{g}_1(\boldsymbol{w}_0)\|^2$ based on Eq. 5.7.

Theorem 5.1 indicates that updating the model with Rule #2 will lead to the convergence to the minimizer of the joint objective function $\mathcal{F}(\boldsymbol{w})$ in the convex case, and the convergence to the first order stationary point in the nonconvex case, when task 1 and 2 satisfy the positive correlation with $\epsilon_2 \geq \frac{(2+\gamma^2)\|\boldsymbol{g}_1(\boldsymbol{w}_0)\|}{4\|\boldsymbol{g}_2(\boldsymbol{w}_0)\|}$. That is to say, Rule #2 not only results in a good model for task 2, but can also be beneficial for the joint learning of task 1 and 2. Note that since $\boldsymbol{w}_0$ is the learnt model of task 1, in general we have $\|\boldsymbol{g}_1(\boldsymbol{w}_0)\| < \|\boldsymbol{g}_2(\boldsymbol{w}_0)\|$.

Theorem 5.2 *Suppose that the loss $\mathcal{L}$ is B-Lipschitz and $\frac{H}{2}$-smooth. We have the following results:*

(1) Let $\boldsymbol{w}^c$ *and* $\boldsymbol{w}^r$ *be the model parameters after applying one update to some initial model* $\boldsymbol{w}$ *by using Rule #1 and Rule #2, respectively. Suppose* $\alpha < \min\left\{\frac{1}{H}, \frac{\gamma\|\boldsymbol{g}_1(\boldsymbol{w}_0)\|}{HBK}\right\}$*,* $\epsilon_1 \geq \sqrt{\frac{1+2\alpha H}{2+\alpha H}}$ *and* $\epsilon_2 \geq \frac{(2+\gamma^2)\|\boldsymbol{g}_1(\boldsymbol{w}_0)\|}{4\|\boldsymbol{g}_2(\boldsymbol{w}_0)\|}$ *for some* $\gamma \in (0, 1)$*. It follows that* $\mathcal{F}(\boldsymbol{w}^r) \leq \mathcal{F}(\boldsymbol{w}^c)$*;*

(2) Let $\boldsymbol{w}_k$ *be the k-th iterate for task 2 with Rule #2. Suppose that* $\langle \boldsymbol{g}_1(\boldsymbol{w}_0), \boldsymbol{g}_2(\boldsymbol{w}_i)\rangle \geq \epsilon_2\|\boldsymbol{g}_1(\boldsymbol{w}_0)\|\|\boldsymbol{g}_2(\boldsymbol{w}_i)\|$ *for* $i \in [0, k-1]$ *and* $\alpha \leq \frac{4\epsilon_2\|\boldsymbol{g}_1(\boldsymbol{w}_0)\|}{HBk^{1.5}}$*. It follows that* $\mathcal{L}_1(\boldsymbol{w}_k) \leq \mathcal{L}_1(\boldsymbol{w}^1)$.

Proof (1) For Rule #1, we have

$$\boldsymbol{w}^c = \boldsymbol{w} - \alpha[\boldsymbol{g}_2(\boldsymbol{w}) - \text{proj}_{S^1}(\boldsymbol{g}_2(\boldsymbol{w}))] = \boldsymbol{w} - \alpha\tilde{\boldsymbol{g}_2}(\boldsymbol{w}). \tag{5.8}$$

For Rule #2, we have

$$\boldsymbol{w}^r = \boldsymbol{w} - \alpha\boldsymbol{g}_2(\boldsymbol{w}). \tag{5.9}$$

Based on the smoothness of the objective function, we can have an upper bound on $\mathcal{F}(\boldsymbol{w}^r)$:

$$\mathcal{F}(\boldsymbol{w}^r) \leq \mathcal{F}(\boldsymbol{w}) - \left[\alpha - \frac{\alpha^2 H}{2}\right]\|\boldsymbol{g}_2(\boldsymbol{w})\|^2 - \alpha\langle \boldsymbol{g}_1(\boldsymbol{w}), \boldsymbol{g}_2(\boldsymbol{w})\rangle, \tag{5.10}$$

and a lower bound on $\mathcal{F}(\boldsymbol{w}^c)$:

$$\mathcal{F}(\boldsymbol{w}^c) \geq \mathcal{F}(\boldsymbol{w}) + \langle \nabla\mathcal{F}(\boldsymbol{w}), \boldsymbol{w}^c - \boldsymbol{w}\rangle - \frac{H}{2}\|\boldsymbol{w}^c - \boldsymbol{w}\|^2. \tag{5.11}$$

Combining Eqs. 5.10 and 5.11, it can be shown that

$$\begin{aligned}
&\mathcal{F}(\boldsymbol{w}^r)\\
\leq&\mathcal{F}(\boldsymbol{w}^c) - \langle \nabla\mathcal{F}(\boldsymbol{w}), \boldsymbol{w}^c - \boldsymbol{w}\rangle + \frac{H}{2}\|\boldsymbol{w}^c - \boldsymbol{w}\|^2 - \left[\alpha - \frac{\alpha^2 H}{2}\right]\|\boldsymbol{g}_2(\boldsymbol{w})\|^2 - \alpha\langle \boldsymbol{g}_1(\boldsymbol{w}), \boldsymbol{g}_2(\boldsymbol{w})\rangle\\
=&\mathcal{F}(\boldsymbol{w}^c) - \langle \boldsymbol{g}_1(\boldsymbol{w}) + \boldsymbol{g}_2(\boldsymbol{w}), -\alpha\tilde{\boldsymbol{g}_2}(\boldsymbol{w})\rangle + \frac{H\alpha^2}{2}\|\tilde{\boldsymbol{g}_2}(\boldsymbol{w})\|^2 - \left[\alpha - \frac{\alpha^2 H}{2}\right]\|\boldsymbol{g}_2(\boldsymbol{w})\|^2\\
&- \alpha\langle \boldsymbol{g}_1(\boldsymbol{w}), \boldsymbol{g}_2(\boldsymbol{w})\rangle\\
=&\mathcal{F}(\boldsymbol{w}^c) + \alpha\langle \boldsymbol{g}_1(\boldsymbol{w}), \tilde{\boldsymbol{g}_2}(\boldsymbol{w})\rangle + \alpha\langle \boldsymbol{g}_2(\boldsymbol{w}), \tilde{\boldsymbol{g}_2}(\boldsymbol{w})\rangle + \frac{H\alpha^2}{2}\|\tilde{\boldsymbol{g}_2}(\boldsymbol{w})\|^2 - \left[\alpha - \frac{\alpha^2 H}{2}\right]\|\boldsymbol{g}_2(\boldsymbol{w})\|^2\\
&- \alpha\langle \boldsymbol{g}_1(\boldsymbol{w}), \boldsymbol{g}_2(\boldsymbol{w})\rangle\\
=&\mathcal{F}(\boldsymbol{w}^c) + \alpha\|\tilde{\boldsymbol{g}_2}(\boldsymbol{w})\|^2 + \frac{H\alpha^2}{2}\|\tilde{\boldsymbol{g}_2}(\boldsymbol{w})\|^2 - \left[\alpha - \frac{\alpha^2 H}{2}\right]\|\boldsymbol{g}_2(\boldsymbol{w})\|^2 - \alpha\langle \boldsymbol{g}_1(\boldsymbol{w}), \boldsymbol{g}_2(\boldsymbol{w})\rangle
\end{aligned} \tag{5.12}$$

where the last equality is true because both $\boldsymbol{g}_1(\boldsymbol{w})$ and $\text{proj}_1(\boldsymbol{g}_2(\boldsymbol{w}))$ are orthogonal to $\tilde{\boldsymbol{g}_2}(\boldsymbol{w})$.

For the term $\langle g_1(w), g_2(w)\rangle$, based on Eq. 5.5, it follows that

$$\begin{aligned}
&\langle g_1(w), g_2(w)\rangle \\
\geq& -\frac{H^2}{4}\|w - w_0\|^2 - \frac{1}{2}\|g_2(w)\|^2 - \frac{1}{2}\|g_1(w_0)\|^2 + \langle g_1(w_0), g_2(w_0)\rangle \\
\geq& -\frac{H^2}{4}\|w - w_0\|^2 - \frac{1}{2}\|g_2(w)\|^2 - \frac{1}{2}\|g_1(w_0)\|^2 + \epsilon_2\|g_1(w_0)\|\|g_2(w_0)\|. \quad (5.13)
\end{aligned}$$

Suppose that w is the model update at n-th iteration where $n \leq K$. For Rule #1,

$$\begin{aligned}
\|w - w_0\|^2 =&\alpha^2\|\sum_{i=0}^{n}\tilde{g_2}(w_i)\|^2 \\
\leq&\frac{\gamma^2\|g_1(w_0)\|^2}{H^2B^2K^2}n\sum_{i=0}^{n}\|\tilde{g_2}(w_i)\|^2 \\
\leq&\frac{\gamma^2 n^2\|g_1(w_0)\|^2}{H^2K^2} \\
\leq&\frac{\gamma^2\|g_1(w_0)\|^2}{H^2}.
\end{aligned}$$

Similarly for Rule #2, we can also obtain that

$$\|w - w_0\|^2 \leq \frac{\gamma^2\|g_1(w_0)\|^2}{H^2}.$$

Therefore, continuing with Eq. 5.13, we can have

$$\begin{aligned}
&\langle g_1(w), g_2(w)\rangle \\
\geq& -\frac{2+\gamma^2}{4}\|g_1(w_0)\|^2 + \epsilon_2\|g_1(w_0)\|\|g_2(w_0)\| - \frac{1}{2}\|g_2(w)\|^2 \\
\geq& -\frac{1}{2}\|g_2(w)\|^2
\end{aligned}$$

where the last inequality holds because $\epsilon_2 \geq \frac{(2+\gamma^2)\|g_1(w_0)\|}{4\|g_2(w_0)\|}$.

Based on Eq. 5.12, it follows that

$$\begin{aligned}
\mathcal{F}(w^r) \leq&\mathcal{F}(w^c) + \alpha\|\tilde{g_2}(w)\|^2 + \frac{H\alpha^2}{2}\|\tilde{g_2}(w)\|^2 - \left[\alpha - \frac{\alpha^2 H}{2}\right]\|g_2(w)\|^2 + \frac{\alpha}{2}\|g_2(w)\|^2 \\
=&\mathcal{F}(w^c) - \left[\frac{\alpha}{2} - \frac{\alpha^2 H}{2}\right]\|g_2(w)\|^2 + \left[\alpha + \frac{\alpha^2 H}{2}\right]\|\tilde{g_2}(w)\|^2 \\
\overset{(a)}{\leq}&\mathcal{F}(w^c) - \left[\frac{\alpha}{2} - \frac{\alpha^2 H}{2}\right]\|g_2(w)\|^2 + (1-\epsilon_1^2)\left[\alpha + \frac{\alpha^2 H}{2}\right]\|g_2(w)\|^2 \\
\overset{(b)}{\leq}&\mathcal{F}(w^c)
\end{aligned}$$

where (a) holds because

$$\begin{aligned}\|\boldsymbol{g}_2(\boldsymbol{w})\|^2 =&\|\text{proj}_1(\boldsymbol{g}_2(\boldsymbol{w})) + \tilde{\boldsymbol{g}_2}(\boldsymbol{w})\|^2\\ =&\|\text{proj}_1(\boldsymbol{g}_2(\boldsymbol{w}))\|^2 + \|\tilde{\boldsymbol{g}_2}(\boldsymbol{w})\|^2\\ \geq&\epsilon_1^2\|\boldsymbol{g}_2(\boldsymbol{w})\|^2 + \|\tilde{\boldsymbol{g}_2}(\boldsymbol{w})\|^2,\end{aligned}$$

and (b) is true because $\epsilon_1 \geq \sqrt{\frac{1+2\alpha H}{2+\alpha H}}$.

(2) It can be seen that

$$\begin{aligned}\mathcal{L}_1(\boldsymbol{w}_k) \leq&\mathcal{L}_1(\boldsymbol{w}_0) + \langle \boldsymbol{g}_1(\boldsymbol{w}_0), \boldsymbol{w}_k - \boldsymbol{w}_0\rangle + \frac{H}{4}\|\boldsymbol{w}_k - \boldsymbol{w}_0\|^2\\ =&\mathcal{L}_1(\boldsymbol{w}_0) + \langle \boldsymbol{g}_1(\boldsymbol{w}_0), -\alpha\sum_{i=0}^{k-1}\boldsymbol{g}_2(\boldsymbol{w}_i)\rangle + \frac{\alpha^2 H}{4}\|\sum_{i=0}^{k-1}\boldsymbol{g}_2(\boldsymbol{w}_i)\|^2\\ =&\mathcal{L}_1(\boldsymbol{w}_0) - \alpha\sum_{i=0}^{k-1}\langle \boldsymbol{g}_1(\boldsymbol{w}_0), \boldsymbol{g}_2(\boldsymbol{w}_i)\rangle + \frac{\alpha^2 H}{4}\|\sum_{i=0}^{k-1}\boldsymbol{g}_2(\boldsymbol{w}_i)\|^2\\ \leq&\mathcal{L}_1(\boldsymbol{w}_0) - \alpha\epsilon_2\|\boldsymbol{g}_1(\boldsymbol{w}_0)\|[\sum_{i=0}^{k-1}\|\boldsymbol{g}_2(\boldsymbol{w}_i)\|] + \frac{\alpha^2 Hk}{4}\sum_{i=0}^{k-1}\|\boldsymbol{g}_2(\boldsymbol{w}_i)\|^2.\end{aligned}$$

For $\alpha \leq \frac{4\epsilon_2\|\boldsymbol{g}_1(\boldsymbol{w}_0)\|}{HBk^{1.5}}$, we can have

$$\begin{aligned}\frac{\alpha Hk}{4}\sum_{i=0}^{k-1}\|\boldsymbol{g}_2(\boldsymbol{w}_i)\|^2 \leq&\frac{\epsilon_2\|\boldsymbol{g}_1(\boldsymbol{w}_0)\|}{B\sqrt{k}}\sum_{i=0}^{k-1}\|\boldsymbol{g}_2(\boldsymbol{w}_i)\|^2\\ \leq&\frac{\epsilon_2\|\boldsymbol{g}_1(\boldsymbol{w}_0)\|\left(\sum_{i=0}^{k-1}\|\boldsymbol{g}_2(\boldsymbol{w}_i)\|^2\right)}{\sqrt{\sum_{i=0}^{k-1}\|\boldsymbol{g}_2(\boldsymbol{w}_i)\|^2}}\\ \leq&\epsilon_2\|\boldsymbol{g}_1(\boldsymbol{w}_0)\|\sqrt{\sum_{i=0}^{k-1}\|\boldsymbol{g}_2(\boldsymbol{w}_i)\|^2}\\ \leq&\epsilon_2\|\boldsymbol{g}_1(\boldsymbol{w}_0)\|[\sum_{i=0}^{k-1}\|\boldsymbol{g}_2(\boldsymbol{w}_i)\|].\end{aligned}$$

Therefore, it follows that $\mathcal{L}_1(\boldsymbol{w}_k) \leq \mathcal{L}_1(\boldsymbol{w}_0)$.

Intuitively, the first part of Theorem 5.2 shows that updating with Rule #2 can achieve lower loss value compared to Rule #1 after one step gradient update when task 1 and 2 satisfy the sufficient projection with $\epsilon_1 \geq \sqrt{\frac{1+2\alpha H}{2+\alpha H}}$ and the positive correlation with $\epsilon_2 \geq \frac{(2+\gamma^2)\|\boldsymbol{g}_1(\boldsymbol{w}_0)\|}{4\|\boldsymbol{g}_2(\boldsymbol{w}_0)\|}$. If the positive correlation condition also holds for iterates of Rule #2 when learning the task 2, the second part of Theorem 5.2 indicates that updating the model indeed leads to a better model for task 1 with respect to $\mathcal{L}_1$. In a nutshell, when (1) the task 2 has sufficient gradient projection onto the subspace of task 1 and (2) the projected gradient is also aligned well with the gradient of task 1 in the subspace of task 1, updating the model along $\boldsymbol{g}_2$ will modify the learnt model $\boldsymbol{w}^1$ towards a favorable direction for CL and enable the backward knowledge transfer to task 1.

It is worth noting that the conditions for Theorems 5.1 and 5.2 depend only on the initial model gradient $\boldsymbol{g}_2(\boldsymbol{w}^1)$ before learning task 2 and the gradient $\boldsymbol{g}_1(\boldsymbol{w}^1)$ when learning task 1, which are both easily accessible and calculated. Particularly, the positive correlation can be evaluated by only storing the gradient $\boldsymbol{g}_1(\boldsymbol{w}^1)$ instead of the data of task 1. In stark contrast, the task gradient correlation characterization in Chaudhry et al. [49] and Riemer et al. [278] involves the gradient evaluation of old tasks with respect to the current model weight, and hence requires the data of old tasks when learning the new task.

5.3 Continual Learning with Enhanced Knowledge Transfer

Based on the theoretical analysis above, we next propose a continual learning method with backward knowledge transfer (CUBER), by selectively updating the learnt model of old tasks when learning the new task. In particular, CUBER works in a layer-wise manner: Given a L-layer network, CUBER first characterizes the task correlation for each layer, and then employs different strategies to learn the new task depending on the task correlation.

More specifically, denote the set of weights as $\boldsymbol{w} = \{\boldsymbol{w}_l\}_{l=1}^{L}$, where $\boldsymbol{w}_l$ is the layer-wise weight for layer l. Given the data input $\boldsymbol{x}_i^t$ for task t, let $\boldsymbol{x}_{l,i}^t$ be the input of layer l and $\boldsymbol{x}_{1,i}^t = \boldsymbol{x}_i^t$. Denote f as the operation of the network layer. The output $\boldsymbol{x}_{l+1,i}^t$ for layer l can be then computed as $\boldsymbol{x}_{l+1,i}^t = f(\boldsymbol{w}_l, \boldsymbol{x}_{l,i}^t)$. Here we denote $\boldsymbol{x}_{l,i}^t$ as the representations of the input $\boldsymbol{x}_i^t$ at layer l. Given a new task $t \geq 2$, we characterize its task correlation with some old task $j \in [1, t-1]$ for layer l into three different regimes based on Definitions 5.1 and 5.2.

Regime 1 (no forgetting): We say that task $j \in Reg_{l,1}^t$ for layer l if the following holds:

$$\|\text{proj}_{S_l^j}(\nabla \mathcal{L}_t(\boldsymbol{w}_l^{t-1}))\| \leq \epsilon_1 \|\nabla \mathcal{L}_t(\boldsymbol{w}_l^{t-1}))\|.$$

In this case, the layer-wise input subspace S_l^j for task j and S_l^t for task t are treated as nearly orthogonal ((a) in Fig. 5.1). As a result, there is little knowledge transfer between these two tasks, and updating the model along with $\nabla \mathcal{L}_t(\boldsymbol{w}_l)$ would not introduce much interference to task j. To reinforce the knowledge protection for task j, the model will be updated based on orthogonal projection:

$$\nabla \mathcal{L}_t(\boldsymbol{w}_l) \longleftarrow \nabla \mathcal{L}_t(\boldsymbol{w}_l) - \text{proj}_{S_l^j}(\nabla \mathcal{L}_t(\boldsymbol{w}_l)).$$

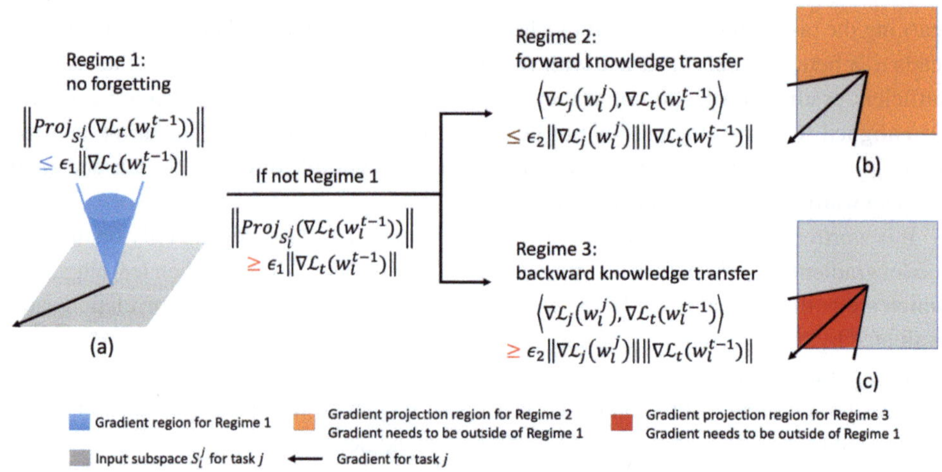

Fig. 5.1 A simple illustration of the layer-wise task correlation detection. Given the new task t, an old task j belongs to (1) Regime 1 if the initial model gradient $\nabla\mathcal{L}_t(\boldsymbol{w}_l^{t-1})$ of task t has small projection onto the subspace S_l^j of task j; (2) Regime 2 if the strong projection condition is satisfied while the projection of $\nabla\mathcal{L}_t(\boldsymbol{w}_l^{t-1})$ onto S_l^j is not aligned well with the gradient of task j; (3) Regime 3 if both the strong projection condition and the positive correlation condition are satisfied

Regime 2 (forward knowledge transfer): We say that task $j \in Reg_{l,2}^t$ for layer l if the following holds:

$$\|\text{proj}_{S_l^j}(\nabla\mathcal{L}_t(\boldsymbol{w}_l^{t-1}))\| \geq \epsilon_1\|\nabla\mathcal{L}_t(\boldsymbol{w}_l^{t-1}))\|,$$
$$\langle\nabla\mathcal{L}_j(\boldsymbol{w}_l^j), \nabla\mathcal{L}_t(\boldsymbol{w}_l^{t-1})\rangle \leq \epsilon_2\|\nabla\mathcal{L}_j(\boldsymbol{w}_l^j)\|\|\nabla\mathcal{L}_t(\boldsymbol{w}_l^{t-1})\|.$$

In this case, task j and task t can be strongly correlated for layer l but possibly with 'negative' correlation ((b) in Fig. 5.1), in the sense that updating the model along with $\nabla\mathcal{L}_t(\boldsymbol{w}_l)$ would substantially modify the learnt model $\text{proj}_{S_l^j}(\boldsymbol{w}^{t-1})$ for task j in an unfavorable way and lead to the forgetting of task j. As there will be better forward knowledge transfer from the old task j to the new task t for layer l, we leverage the scaled weight projection in Lin et al. [218] to facilitate the forward knowledge transfer through a scaling matrix $Q_l^{j,t}$, whereas the model is still updated using orthogonal projection to protect the knowledge of old tasks:

$$\begin{aligned}\nabla\mathcal{L}_t(\boldsymbol{w}_l) &\longleftarrow \nabla\mathcal{L}_t(\boldsymbol{w}_l) - \text{proj}_{S_l^j}(\nabla\mathcal{L}_t(\boldsymbol{w}_l)),\\ Q_l^{j,t} &\longleftarrow Q_l^{j,t} - \beta\nabla_Q\mathcal{L}_t(\boldsymbol{w}_l - \text{proj}_{S_l^j}(\boldsymbol{w}_l) + \boldsymbol{w}_l B_l^j Q_l^{j,t}(B_l^j)').\end{aligned} \tag{5.14}$$

Here B_l^j is the bases matrix for subspace S_l^j. Intuitively, the scaled weight projection $\boldsymbol{w}_l B_l^j Q_l^{j,t} (B_l^j)'$ replaces the weight projection $\text{proj}_{S_l^j}(\boldsymbol{w}_l)$ of task j by a scaled version, which transforms the knowledge of task j to the appropriate model of the new task t through the optimization of $Q_l^{j,t}$.

Regime 3 (backward knowledge transfer): We say task $j \in Reg_{l,3}^t$ for layer l if the following holds:

$$\|\text{proj}_{S_l^j}(\nabla \mathcal{L}_t(\boldsymbol{w}_l^{t-1}))\| \geq \epsilon_1 \|\nabla \mathcal{L}_t(\boldsymbol{w}_l^{t-1}))\|,$$
$$\langle \nabla \mathcal{L}_j(\boldsymbol{w}_l^j), \nabla \mathcal{L}_t(\boldsymbol{w}_l^{t-1}) \rangle \geq \epsilon_2 \|\nabla \mathcal{L}_j(\boldsymbol{w}_l^j)\| \|\nabla \mathcal{L}_t(\boldsymbol{w}_l^{t-1})\|.$$

With the sufficient projection and the positive correlation, updating the model along with $\nabla \mathcal{L}_t(\boldsymbol{w}_l)$ could possibly lead to a better model for continual learning and also improve the learning performance of the old task j ((c) in Fig. 5.1). To avoid overly-optimistic modification on the learnt model of task j, we further regularize the projection of the model change on the subspace S_l^j, given that the model projection is indeed frozen for task j to address forgetting with orthogonal projection, i.e., $\text{proj}_{S_l^j}(\boldsymbol{w}_l^{t-1}) = \text{proj}_{S_l^j}(\boldsymbol{w}_l^j)$. This gives the following model update:

$$\boldsymbol{w}_l \longleftarrow \boldsymbol{w}_l - \alpha \nabla [\mathcal{L}_t(\boldsymbol{w}_l) + \lambda \|\text{proj}_{S_l^j}(\boldsymbol{w}_l - \boldsymbol{w}_l^{t-1})\|].$$

Note that the gradient projection on the subspaces of old tasks in Regime 1 and 2 is removed from the gradient in the model update above. Besides, we also learn a scaling matrix $Q_l^{j,t}$ for better forward knowledge transfer from the old task $j \in Reg_{l,3}^t$ to the new task t as in Eq. 5.14.

However, continuous model update with task t gradient $\nabla \mathcal{L}_t(\boldsymbol{w}_l)$ will eventually lead to the task specific model for task t, which usually differs from the model of task j in Regime 3 (Fig. 5.2). To address this problem, note that the second part of Theorem 5.2 characterizes the condition under which updating the model with $\nabla \mathcal{L}_t(\boldsymbol{w}_l)$ will result in backward knowledge transfer. Thus motivated, for a model update iterate $\boldsymbol{w}_{l,k}$ at k-th iteration when learning the new task t, we evaluate the following condition for task $j \in Reg_{l,3}^t$:

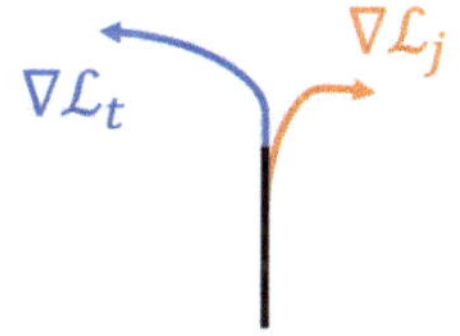

Fig. 5.2 A simple illustration of the case where updating the model with $\nabla \mathcal{L}_t$ benefits the old task $j \in Reg_{l,3}^t$ at the beginning but eventually conflicts with $\nabla \mathcal{L}_j$

$$\langle \nabla \mathcal{L}_j(\boldsymbol{w}_l^j), \nabla \mathcal{L}_t(\boldsymbol{w}_{l,k})\rangle \geq \epsilon_2 \|\nabla \mathcal{L}_j(\boldsymbol{w}_l^j)\| \|\nabla \mathcal{L}_t(\boldsymbol{w}_{l,k})\| \tag{5.15}$$

and degenerate task j to Regime 2, i.e., remove the gradient projection on the subspace S_l^j from the task t gradient and stop modifying the model for task j, if the condition (5.15) does not hold.

Bases extraction: After learning the model $\boldsymbol{w}^t$ for task t, we construct the input subspace for each layer l by extracting bases from its representation $\boldsymbol{x}_{l,i}^t$ based on singular value decomposition (SVD) [218]. More specifically, given a batch of n samples and the learnt model $\boldsymbol{w}^t$, the representation matrix for layer l is denoted as $R_l^t = [\boldsymbol{x}_{l,1}^t, \boldsymbol{x}_{l,2}^t, \ldots, \boldsymbol{x}_{l,n}^t]$. Since the bases of the old tasks may include important bases for the new task t, we determine the bases for task t from the union of the bases of the old tasks and the newly generated bases. Towards this end, we first concatenate the bases B_l^j for $j \in [0, t-1]$ together in a matrix O_l^t and remove the common bases. Then SVD is applied on the residual representation matrix $\tilde{R}_l^t = R_l^t - R_l^t O_l^t (O_l^t)'$, i.e., $\tilde{R}_l^t = U_l^t \Sigma_l^t (V_l^t)'$, where U_l^t is the left singular matrix. We construct B_l^t by selecting the most important bases from the pool of bases in O_l^t and U_l^t depending on their eigenvalues, which yields a low rank matrix approximation of R_l^t.

To conclude, the optimization problem for learning the new task t can be summarized as follows:

$$\begin{aligned}
\min_{\boldsymbol{w}, \{Q_l^{j,t}\}_{l, j \in Reg_{l,2}^t \bigcup Reg_{l,3}^t}} \quad & \mathcal{L}_t(\{\tilde{\boldsymbol{w}}_l\}_l) + \lambda \sum_l \sum_{j \in Reg_{l,3}^t} \|\mathrm{proj}_{S_l^j}(\boldsymbol{w}_l - \boldsymbol{w}_l^{t-1})\|, \\
s.t. \quad & \tilde{\boldsymbol{w}}_l = \boldsymbol{w}_l + \sum_{j \in Reg_{l,2}^t \bigcup Reg_{l,3}^t} [\boldsymbol{w}_l B_l^j Q_l^{j,t} (B_l^j)' - \mathrm{proj}_{S_l^j}(\boldsymbol{w}_l)], \\
& \nabla \mathcal{L}_t(\boldsymbol{w}_l) = \nabla \mathcal{L}_t(\boldsymbol{w}_l) - \sum_{j \in Reg_{l,1}^t \bigcup Reg_{l,2}^t} \mathrm{proj}_{S_l^j}(\nabla \mathcal{L}_t(\boldsymbol{w}_l)).
\end{aligned} \tag{5.16}$$

The key idea is that we conservatively update the model for old tasks in Regime 3 while using orthogonal projection to preserve the knowledge of other old tasks; in the meanwhile, we leverage the scaled weight projection to reuse the model knowledge of old tasks in both Regime 2 and 3 to facilitate forward knowledge transfer. It is worth to note that the task correlation is determined before learning the new task t, as both the strong projection condition and the positive correlation condition only depend on the initial model gradient for the new task. And this can be achieved by a simple forward-backward pass through the initial model with a batch of new task data. The overview of CUBER can be found in Algorithm 1.

Algorithm 1 Continual learning with backward knowledge transfer (CUBER)

Input: task sequence $\mathbb{T} = \{t\}_{t=1}^{T}$;
Learn the first task using vanilla stochastic gradient descent;
Extract the bases $\{B_l^1\}$ based on SVD using the learnt model $\boldsymbol{w}^1$;
for each task t **do**
  Calculate gradient $\nabla \mathcal{L}_t(\boldsymbol{w}_{t-1})$;
  Evaluate the strong projection and the positive correlation conditions for layer-wise task correlation detection to determine $Reg_{l,1}^t$, $Reg_{l,2}^t$ and $Reg_{l,3}^t$;
  for k=1, 2, ... **do**
    Update the model and scaling matrices by solving Eq. 5.16;
    for task $j < t$ and $j \in Reg_{l,3}^t$ **do**
      if $\langle \nabla \mathcal{L}_j(\boldsymbol{w}_l^j), \nabla \mathcal{L}_t(\boldsymbol{w}_{l,k}) \rangle < \epsilon_2 \|\nabla \mathcal{L}_j(\boldsymbol{w}_l^j)\| \|\nabla \mathcal{L}_t(\boldsymbol{w}_{l,k})\|$ **then**
        Degenerate task j to $Reg_{l,2}^t$;
      end if
    end for
  end for
  Store the gradient $\nabla \mathcal{L}_t(\boldsymbol{w}^t)$ in the task memory;
  Extract the bases $\{B_l^t\}$ based on SVD using the learnt model $\boldsymbol{w}^t$;
end for

5.4 Experiments

Datasets. We evaluate the performance of CUBER on four standard CL benchmarks. (1) Permuted MNIST: a variant of the MNIST dataset [189] where random permutations are applied to the input pixels. Following Lopez-Paz and Ranzato [227] and Saha et al. [283], we divide the dataset into 10 tasks with different permutations and each task includes 10 classes. (2) Split CIFAR-100: we divide the CIFAR-100 dataset [178] into 10 different tasks, where each task is a 10-way multi-class classification problem. (3) 5-Datasets: we consider a sequence of 5 datasets, i.e., CIFAR-10, MNIST, SVHN [241], not-MNIST [44], Fashion MNIST [362], and the classification problem on each dataset is a task. (4) Split MiniImageNet: we divide the MiniImageNet dataset [341] into 20 tasks, where each task includes 5 classes.

Baselines. In this work, we compare CUBER with the following baselines on the benchmarks mentioned above. (1) EWC [171]: a regularization-based method that leverages Fisher Information matrix for weights importance evaluation; (2) HAT [294]: learns a hard attention mask to preserve the knowledge of old task; (3) Orthogonal Weight Modulation (OWM) [390]: learns a projector matrix to project the gradient of the new task to the orthogonal direction of input subspace of old tasks; (4) Gradient Projection Memory (GPM) [283]: stores the bases of the input subspace of old tasks and then updates the model with the gradient projection orthogonal to the subspace spanned by these bases; (5) TRGP [218]: proposes a scaled weight projection to facilitate the forward knowledge transfer from related old tasks to the new task while updating the model based on orthogonal gradient projection, which demonstrates the state-of-the-art performance for a fixed capacity network; (6) Averaged GEM (A-GEM) [49]: constrains the new task learning with the gradient calculated using the stored data of old tasks; (7) Experience Replay with Reservior sample (ER_Res) [50]: uses a small episodic memory to store old task samples for addressing forgetting; (8) Multitask: jointly learns all tasks once with a single network using all datasets, which usually serves as a performance upper bound in CL [283].

Network and training details. For a given dataset, we study all CL methods using the same network architecture. More specifically, for Permuted MNIST, we consider a 3-layer fully-connected network including 2 hidden layers with 100 units. And we train the network for 5 epochs with a batch size of 10 for every task. For Split CIFAR-100, we use a version of 5-layer AlexNet by following Saha et al. [283] and Lin et al. [218]. When learning each task, we train the network for a maximum of 200 epochs with early termination based on the validation loss, and use a batch size of 64. For 5-Datasets, we use a reduced ResNet-18 [227] and follow the same training strategy as in Split CIFAR-100. For Split MiniImageNet, a reduced ResNet-18 is also used, and we train the network for a maximum of 100 epoches with early termination. The batch size is 64. Similar to Lin et al. [218], we select at most two tasks to be in Regime 2 and 3 for each layer with the largest gradient projection norm, to reduce the performance sensitivity on the choice of ϵ_1. In the experiments, we set $\epsilon_1 = 0.5$.

To evaluate the learning performance, we consider the following two metrics, i.e., accuracy (ACC) which measures the final accuracy averaged over all tasks, and backward transfer (BWT) which measures the average accuracy change of each task after learning new tasks:

$$ACC = \frac{1}{T}\sum_{i=1}^{T} A_{T,i}, \quad BWT = \frac{1}{T-1}\sum_{i=1}^{T-1}(A_{T,i} - A_{i,i})$$

where $A_{i,j}$ represents the testing accuracy of task j after learning task i.

5.4.1 Main Results

As shown in Table 5.1, CUBER demonstrates the best performance of BWT on all datasets compared to the baselines. In particular, *positive BWT can be obtained by CUBER on Split CIFAR-100 and Split MiniImageNet, which has not been achieved by previous works in a fixed capacity network without data-replay*. For 5-Dataset, since the tasks therein are less related to each other as in GPM [283], we do not expect much knowledge transfer across tasks (but knowledge transfer still exists in terms of the layer-wise features), and there is no forgetting in CUBER. Achieving zero-forgetting is very difficult for Permuted MNIST, because all the tasks share one output layer and there is no task identifier during testing (domain-incremental) [283]. However, even in this case CUBER still achieves the best BWT, i.e., nearly non-forgetting, among all methods. Clearly, the strong performance

Table 5.1 The ACC and BWT with the standard deviation values over 5 different runs on different datasets. Here for Split CIFAR-100, Split MiniImageNet and 5-Dataset we use a multi-head network, while a single-head network is used for Permuted MNIST. Moreover, $\epsilon_2 = 0.0$

Method	Multi-head						Domain-incremental	
	Split CIFAR-100		Split MiniImageNet		5-Dataset		Permuted MNIST	
	ACC (%)	BWT (%)	ACC (%)	BWT (%)	ACC (%)	BWT (%)	ACC (%)	BWT (%)
Multitask	79.58 ± 0.54	–	69.46 ± 0.62	–	91.54 ± 0.28	–	96.70 ± 0.02	–
OWM	50.94 ± 0.60	−30 ± 1	–	–	–	–	90.71 ± 0.11	−1 ± 0
EWC	68.80 ± 0.88	−2 ± 1	52.01 ± 2.53	−12 ± 3	88.64 ± 0.26	−4 ± 1	89.97 ± 0.57	−4 ± 1
HAT	72.06 ± 0.50	0 ± 0	59.78 ± 0.57	−3 ± 0	91.32 ± 0.18	−1 ± 0	–	–
A-GEM	63.98 ± 1.22	−15 ± 2	57.24 ± 0.72	−12 ± 1	84.04 ± 0.33	−12 ± 1	83.56 ± 0.16	−14 ± 1
ER_Res	71.73 ± 0.63	−6 ± 1	58.94 ± 0.85	−7 ± 1	88.31 ± 0.22	−4 ± 0	87.24 ± 0.53	−11 ± 1
GPM	72.48 ± 0.40	−0.9 ± 0	60.41 ± 0.61	−0.7 ± 0.4	91.22 ± 0.20	−1 ± 0	93.91 ± 0.16	−3 ± 0
TRGP	74.46 ± 0.32	−0.9 ± 0.01	61.78 ± 0.60	−0.5 ± 0.6	**93.56** ± 0.10	−0.04 ± 0.01	96.34 ± 0.11	−0.8 ± 0.1
CUBER (ours)	**75.54** ± 0.22	**0.13** ± 0.08	**62.67** ± 0.74	**0.23** ± 0.15	**93.48** ± 0.10	**0.00** ± 0.02	**97.25** ± 0.00	**−0.02** ± 0.00

on BWT indicates that CUBER can effectively facilitate the backward knowledge transfer by wisely modifying the learnt model of old tasks.

Benefiting from the superior performance on the backward knowledge transfer, CUBER also achieves the best or comparable performance on the averaged accuracy. More specifically, CUBER improves around 1% in ACC over the best prior results on Split CIFAR-100, Split MiniImageNet and Permuted MNIST, while showing the comparable performance with the state-of-the-art method TRGP even on 5-datasets where tasks are less correlated. Moreover, CUBER performs better than Multitask on both 5-Dataset and Permuted MNIST, *which implies the importance of studying knowledge transfer in CL:* Both TRGP and CUBER outperform Multitask on 5-Dataset because of the scaled weight projection to facilitate forward knowledge transfer, whereas by facilitating backward knowledge CUBER becomes the only method that outperforms Multitask on Permuted MNIST.

5.4.2 Ablation Studies

Backward knowledge transfer. The value of backward knowledge transfer indicates the average accuracy improvement for each old task after learning all tasks, which implies that the new task learning provides additional useful information for learning features of similar old tasks. Intuitively, this value should depend on the task similarity in CL. To better understand the advantage of CUBER in terms of backward knowledge transfer, we further consider a special setup which includes a sequence of similar and dissimilar tasks. Specifically, different with Split-CIFAR100 where no tasks have overlapping classes, we split the first 50 classes in CIFAR100 into 7 tasks (OL-CIFAR100): Task 0–6 contain classes 0–9, 5–14, 10–19, 20–29, 25–34, 30–39, 40–49, respectively. We compare the performance of CUBER with TRGP in this setup. As shown in Table 5.2, CUBER clearly outperforms TRGP in both ACC and BWT. In particular, CUBER has a positive BWT of 0.28%, where TRGP suffers from forgetting. We further analyze the task selections in Table 5.3, where "selected old task" refers to the task that has the largest number of layers in Regime 3 of the new task. For example, according to the setup of OL-CIFAR100, the selected old task for Task 4 should be Task 3 with a high probability as they have overlapping classes. And we denote a new metric, namely BWT-S, to evaluate the backward knowledge transfer of the selected old task after learning the new task, i.e., BWT-S=$A_{t,j} - A_{t-1,j}$ where j is the "selected old task" of the new task t. As shown in Table 5.3, CUBER correctly identifies the correlated tasks for most new tasks except Task 5. This is a reasonable result because the task correlation detection is based on the initial model gradient which is noisy in general, and hence only serves as an estimation of underlying true correlation. However, such an estimated task correlation characterization is indeed sufficient to effectively facilitate backward knowledge transfer, as corroborated by the positive BWT-S and the superior performance of CUBER.

Forward knowledge transfer. We also evaluate the forward knowledge transfer (FWT) in CUBER, compared to the best two baseline methods GPM and TRGP. Here the FWT measures the gap between $A_{i,i}$ and the accuracy of learning task i only from scratch. For simplicity, we use the FWT of GPM as a baseline and evaluate the improvements of TRGP and CUBER over GPM. It can be seen from Table 5.4 that CUBER performs even better than TRGP although CUBER follows the same strategy, i.e., the scaled weight projection, to facilitate the forward knowledge transfer, and achieves the best FWT among the three methods in most cases. The reason behind is that the characterization of Regime 3 in CUBER not only allows the modification of the learnt model of the old tasks to prompt the backward

Table 5.2 The comparison of ACC and BWT between CUBER and TRGP in OL-CIFAR100

Method	ACC(%)	BWT(%)
CUBER	74.94	0.28
TRGP	74.33	−0.18

Table 5.3 The comparison of BWT-S between CUBER and TRGP in OL-CIFAR100. The selected old task (*) in (b) represents the old task with the largest number of layers in Regime 3 of the new task

	Task 1	Task 2	Task 4	Task 5	Average
Selected old task*	Task 0	Task 1	Task 3	Task 0	–
BWT-S (CUBER)	0.50	0.71	0.03	0.07	0.33
BWT-S (TRGP)	−0.20	0.10	−0.60	−0.20	−0.23

Table 5.4 Comparison of FWT among GPM, TRGP and CUBER. The value for GPM is zero because we treat GPM as the baseline and consider the relative FWT improvement over GPM

Method	Split CIFAR-100	Split MiniImageNet	5-Dataset	Permuted MNIST
GPM	0	0	0	0
TRGP	2.01	2.36	1.98	0.18
CUBER	2.79	3.13	1.96	0.8

Table 5.5 The impact of ϵ_2 on the performance in Split CIFAR-100

$\epsilon_2 = 0.0$		$\epsilon_2 = 0.2$		$\epsilon_2 = 0.5$	
ACC (%)	BWT (%)	ACC (%)	BWT (%)	ACC (%)	BWT (%)
75.54	0.22	75.73	0.03	75.55	0.01

knowledge transfer, but also relaxes the constraint on the gradient update for the new task, i.e., the model can be now updated in the subspace of the selected old tasks for learning the new task. This gradient constraint relaxation consequently leads to better model learning of the new task.

Impact of ϵ_2. It is clear that the selection of layer-wise Regime 3 depends on the value of the threshold ϵ_2. To show the impact of ϵ_2, we evaluate the learning performance under different values of ϵ_2 in Split CIFAR-100. As shown in Table 5.5, the performance on ACC is comparable for all three cases and the BWT decreases as the value of ϵ_2 increases. Intuitively, ϵ_2 characterizes the conservatism in selecting tasks to Regime 3 and modifying the model of selected tasks. Specifically, with a larger ϵ_2, we just consider the backward knowledge transfer to the old tasks that are strongly correlated with the new task, and only slightly modify the learnt model of these selected tasks, because the condition Eq. 5.15 can be quickly violated with the model update and CUBER will stop changing the learnt model of the selected tasks.

5.4.3 More Experimental Results

We also study the accuracy evolution curves for each task during the continual learning procedure in Fig. 5.3. To clearly demonstrate the accuracy evolution behaviors, we selectively study task 2, task 4, task 6 in PMNIST [189], CIFAR-100 Split [178], MiniImageNet [341], and task 1, task 2, task 3 in 5-Dataset, by comparing the performance among CUBER, TRGP and GPM. As shown in Fig. 5.3, CUBER (the red curve) has the most stable performance on

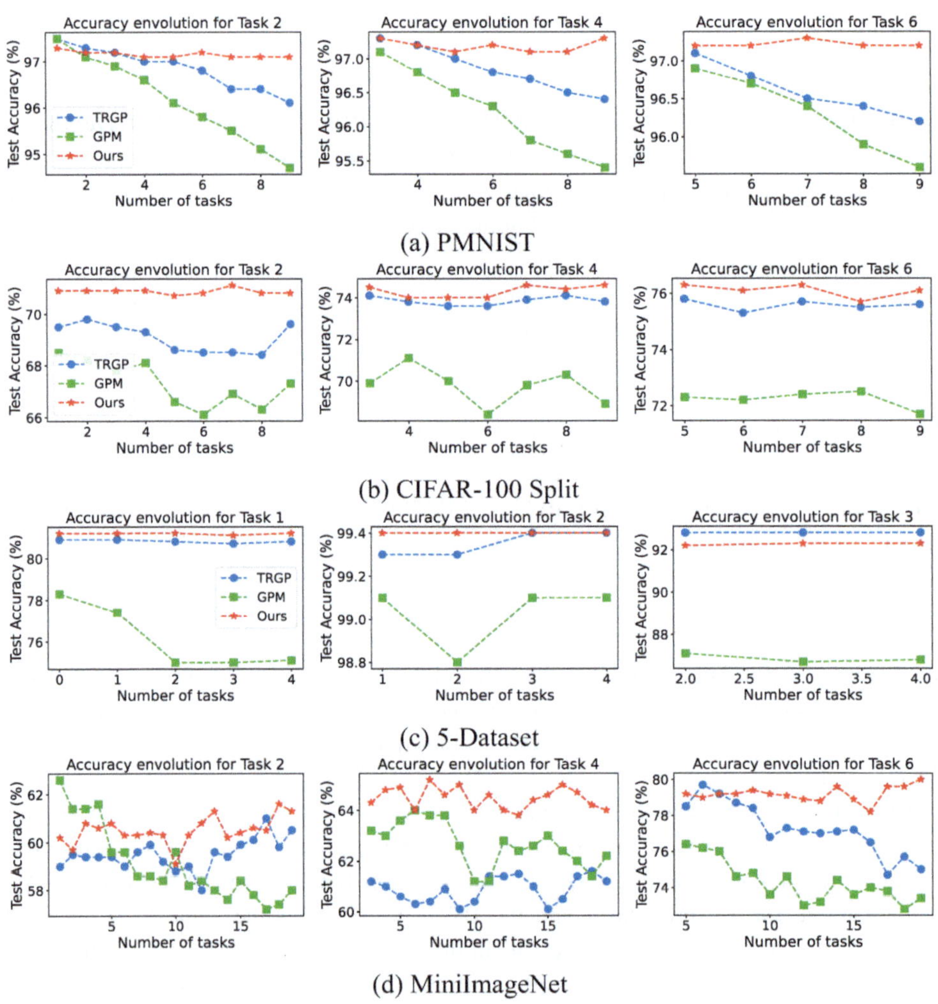

Fig. 5.3 Accuracy evolution for different tasks on PMNIST, CIFAR-100 Split, 5-Datasets and MiniImageNet, respectively

most tasks, as a result of the better backward knowledge transfer facilitated by selectively updating the learnt model of old tasks. All the experiments are conducted by using one Nvidia Quadro RTX 5000 GPU.

5.4.4 More Experimental Details

5.4.4.1 Network Details

3-layer fully-connected network: The network consists of 3 fully-connected layers with 784, 100, 100 units respectively. We use ReLU activation layer after the first two layers.

5-layer AlexNet: Following Saha et al. [283] and Lin et al. [218], the AlexNet used for experiments on Split CIFAR-100 consists of 3 convolutional layers and 2 fully-connected layers, where batch normalization is added in each layer expect the classifier layer. The convolutional layers have 64, 128 and 256 filters with 4×4, 3×3 and 2×2 kernel sizes, respectively, and each fully-connected layer contains 2048 units. We use ReLU activation function and 2×2 max-pooling after the convolutional layers, and dropout of 0.2 for the first two layers and 0.5 for other layers.

A reduced ResNet-18: Following Lopez-Paz and Ranzato [227] and Saha et al. [283], we use a reduced ResNet-18 for experiments on both 5-Datasets and miniImageNet. It includes 17 convolutional layers with 3 short-cut connections. The convolutional layers in 4 stages have 180, 360, 720, 1440 filters, respectively. We adapt the 2×2 average pooling layer before the classifier. In addition, we use convolution with stride 2 in the first layer for miniImageNet, since it has larger input resolution (i.e., 84×84) than 5-dataset (i.e., 32×32).

5.4.4.2 List of Hyperparameters

In what follows, we list the hyperparameters for all the methods considered in this work. For TRGP, we use the hyperparameters provided in Lin et al. [218]. For other baseline methods, we use the hyperparameters for GPM as provided in Saha et al. [283], and provide the hyperparameters for the rest following Saha et al. [283] for being consistent with the corresponding reported results herein.

As shown in the Table 5.6, we use 'lr' to represent the initial learning rate, and 'cifar', 'mini', '5d' and 'pm' to represent 'Split CIFAR-100', 'Split MiniImageNet', '5-Dataset' and 'Permuted MNIST', respectively.

Optimizer/learning rate: For all experiments, we adapt the SGD optimizer by modifying the gradient based on the optimization problem (3). This is consistent with the orthogonal-projection based methods (e.g., Farajtabar et al. [100], Saha et al. [283] and Lin et al. [218]) where the gradient direction used in the plain SGD optimizer is modified to minimize the interference to the old tasks. The learning rates for different datasets are shown in Table 5.6. Here for PMNIST, we use a fixed learning rate, while for other datasets the learning rate decays during the training process.

Table 5.6 List of hyperparameters for CUBER and the related baseline methods

Methods	Hyperparameters
OWM	lr: 0.01 (cifar), 0.3 (pm)
EWC	lr: 0.05 (cifar), 0.03 (mini, 5d, pm)
	Regularization coefficient (λ): 5000 (cifar, mini, 5d), 1000 (pm)
HAT	lr: 0.05 (cifar), 0.03 (mini), 0.1 (5d)
	$s_{\max}$: 400 (cifar, mini, 5d)
	c: 0.75 (cifar, mini, 5d)
A-GEM	lr: 0.05 (cifar), 0.1 (mini, 5d, pm)
	Memory size (number of samples): 2000 (cifar), 500 (mini), 3000 (5d), 1000 (pm)
ER_Res	lr: 0.05 (cifar), 0.1 (mini, 5d, pm)
	Memory size (number of samples): 2000 (cifar), 500 (mini), 3000 (5d), 1000 (pm)
GPM	lr: 0.01 (cifar, pm), 0.1 (mini, 5d)
	Number of samples for base extraction (n): 125 (cifar), 100 (mini, 5d), 300 (pm)
	ϵ_{th}: 0.97, increase by 0.003 with t (cifar); 0.985, increase by 0.003 with t (mini)
	ϵ_{th}: 0.965 (5d); 0.95 for the first layer, otherwise 0.99 (pm)
TRGP	lr: 0.01 (cifar, pm), 0.1 (mini, 5d)
	Number of samples for base extraction (n): 125 (cifar), 100 (mini, 5d), 300 (pm)
	ϵ_{th}: same with GPM
	ϵ^l: 0.5 (cifar, mini, 5d, pm)
CUBER	lr: 0.01 (cifar, pm), 0.1 (mini, 5d)
	Number of samples for base extraction (n): 125 (cifar), 100 (mini, 5d), 300 (pm)
	ϵ_{th}: same with GPM
	ϵ_1: 0.5 (cifar, mini, 5d, pm); ϵ_2: 0 (cifar, mini, 5d, pm); λ: 1 (cifar, mini, 5d, pm)

Early stopping: For all experiments, the minimum learning rate is set to 1e-5, the learning rate decay factor is set to 2, and the number of holds before decaying the learning rate is 6. After each model update, we evaluate the validation loss for the current task using its validation dataset, and count the times whenever the validation loss increases. When the counter is greater than 6, we decay the learning rate by 2 and reset the counter to 0. The training process will be stopped when the learning rate decays to the minimum learning rate.

5.4.5 Memory and Time Cost

Storing gradient: After learning each task, most elements in the gradient matrix for each layer can be close to zero. When evaluating the conditions for regime 3, we flatten the gradient matrix into a vector for each layer, e.g., a gradient matrix in $\mathbb{R}^{m\times n}$ to a gradient

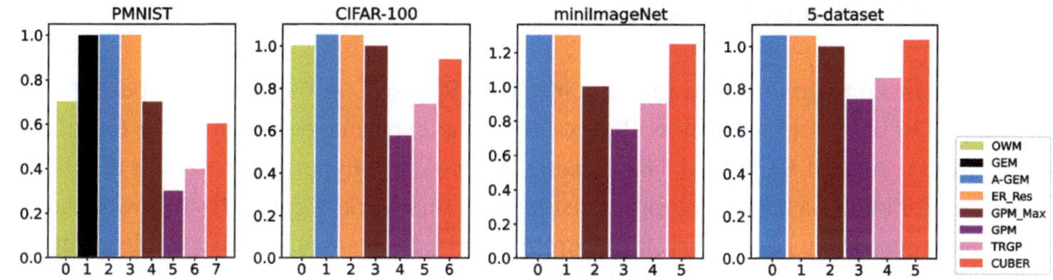

Fig. 5.4 Comparison of memory utilization on PMNIST, CIFAR-100 Split, MiniImageNet and 5-Datasets, respectively

Table 5.7 Training time comparison on CIFAR-100 Split, 5-Datasets and MiniImageNet. Here the training time is normalized with respect to the value of GPM. Please refer [283] for more specific time

Dataset	Methods							
	OWM	EWC	HAT	A-GEM	ER_Res	GPM	TRGP	Ours (CUBER)
CIFAR-100	2.41	1.76	1.62	3.48	1.49	1	1.65	1.86
5-Datasets	–	1.52	1.47	2.41	1.40	1	1.21	1.55
MiniImageNet	–	1.22	0.91	1.79	0.82	1	1.34	1.61

vector in $\mathbb{R}^{mn}$. The inner product between gradient matrices hence becomes the inner product between gradient vectors, where the near-zero elements have little impact on the value of the product. Therefore, to save the computation cost and memory, we prune the layer-wise average gradient vector after learning each task with a certain sparsity ratio, and only store a gradient vector with non-zero elements and the corresponding indices in the original gradient vector.

Memory and training time: As shown in Fig. 5.4 and Table 5.7, we compare the memory utilization and training time between CUBER and the related baseline methods in terms of the normalized value with respect to the value of GPM, following the same strategy as in Lin et al. [218]. The values of OWM, EWC, HAT, A-GEM, ER_Res are from the reported results in GPM [283]. The value of TRGP is from the reported results in Lin et al. [218]. It can be seen that CUBER indeed has comparable complexity with the baseline methods. While SVD is used to extract bases in CUBER, the training time of CUBER is still less than some baseline methods due to its relatively simple and fast model update. And because we only store part of the gradient for each old task, the required memory utilization does not increase a lot.

5.4.6 Baseline Implementations

To ensure fair comparisons, we implement CUBER based on the released code of GPM [283] and TRGP [218], and follow the exactly same experimental setups. We reproduced the reported results of GPM and TRGP using the provided hyperparameters in the papers. Therefore, for other baseline methods considered in this work, we directly follow these two papers and present the reported results therein in Sect. 5. Besides, we also implemented all the baseline methods based on their released official code, and show the reproduced results in Table 5.8.

Table 5.8 The ACC and BWT with the standard deviation values over 5 different runs on different datasets. Here for Split CIFAR-100, Split MiniImageNet and 5-Dataset we use a multi-head network, while we consider domain-incremental setup for Permuted MNIST. Moreover, $\epsilon_2 = 0.0$

Method	Multi-head						Domain-incremental	
	Split CIFAR-100		Split MiniImageNet		5-Dataset		Permuted MNIST	
	ACC (%)	BWT (%)	ACC (%)	BWT (%)	ACC (%)	BWT (%)	ACC (%)	BWT (%)
Multitask	79.58 ± 0.54	–	69.46 ± 0.62	–	91.54 ± 0.28	–	96.70 ± 0.02	–
OWM	50.44 ± 0.72	−30 ± 1	–	–	–	–	89.63 ± 0.21	−1 ± 0
EWC	68.30 ± 0.65	−2 ± 1	50.78 ± 2.98	−12 ± 4	87.67 ± 0.37	−3 ± 1	89.05 ± 0.57	−4 ± 2
HAT	73.21 ± 0.76	0 ± 0	60.23 ± 0.57	−4 ± 1	91.82 ± 0.34	−1 ± 0	–	–
A-GEM	64.16 ± 1.41	−14 ± 3	56.88 ± 0.87	−13 ± 2	82.48 ± 0.56	−13 ± 1	83.05 ± 0.12	−14 ± 2
ER_Res	72.24 ± 0.57	−7 ± 2	58.04 ± 0.73	−7 ± 1	88.54 ± 0.36	−4 ± 1	88.72 ± 0.56	−10 ± 1
GPM	72.48 ± 0.40	−0.9 ± 0	60.41 ± 0.61	−0.7 ± 0.4	91.22 ± 0.20	−1 ± 0	93.91 ± 0.16	−3 ± 0
TRGP	74.46 ± 0.32	−0.9 ± 0.01	61.78 ± 0.60	−0.5 ± 0.6	**93.56** ± 0.10	−0.04 ± 0.01	96.34 ± 0.11	−0.8 ± 0.1
CUBER (ours)	**75.54** ± 0.22	**0.13** ± 0.08	**62.67** ± 0.74	**0.23** ± 0.15	**93.48** ± 0.10	**0.00** ± 0.02	**97.25** ± 0.00	**−0.02** ± 0.00

5.5 Conclusion

In this chapter, we study the problem of backward knowledge transfer in CL. Different from most existing methods that generally freeze the learnt model of the old tasks so as to mitigate catastrophic forgetting, this study seeks to carefully modify the learnt model to facilitate backward knowledge transfer from the new task to the old tasks. To this end, we first introduce notions of strong projection and positive correlation to characterize the task correlation, and show that when the task gradients are sufficiently aligned in the old task subspace, appropriate model change for the old tasks could be beneficial for CL and result in better backward knowledge transfer. Based on the theoretical analysis, we next propose CUBER to carefully learn the model for the new task, which would carefully update the learnt model of the old tasks that are positively correlated with the new task. As shown in the experimental results, CUBER can successfully improve the backward knowledge transfer on the standard CL benchmarks in contrast to related baselines.

Cloud-Edge Collaboration via Pretrained and Federated Continual Learning

6

Clearly, edge-only learning at a single edge device can only work for small-scale AI applications and may not be capable of performing complex AI tasks that require more computation resources and data samples. For example, generative AI becomes an extremely hot topic in the past three years due to the recent breakthroughs in deep learning and large generative models, such as GPT series [43] and Sora [42]. Training and deploying a large generative model from scratch on a single edge device is not feasible due to the large amount of data and computation required therein. Facilitating collaborations between cloud and edge provides a promising solution to this problem, by leveraging the historical data in the cloud through pretrained models or the related knowledge in other edge devices shared through the cloud in a federated manner. In this chapter, we will first introduce our study on edge CL of generative models through cloud-edge collaboration, and then give a brief overview of the recent developments in the literature for CL with pretrained models and federated CL.

6.1 Introduction

Learning tasks across different edge nodes often share model similarity. For instance, different robots may perform similar coordination behaviors according to the environment changes. With this sight, we advocate that the pre-trained generative models from other edge nodes are utilized to speed up the learning at a given edge node, and seek to answer the following critical questions: (1) *"What is the right abstraction of knowledge from multiple pre-trained models for continual learning?"* and (2) *"How can an edge server leverage this knowledge for continual learning of a generative model?"*

H. Wang et al., *Continual and Reinforcement Learning for Edge AI*,
Synthesis Lectures on Learning, Networks, and Algorithms,
https://doi.org/10.1007/978-3-031-84363-1_6

The key to answering the first question lies in efficient model fusion of multiple pre-trained generative models. A common approach is the *ensemble method* [40, 291] where the outputs of different models are aggregated to improve the prediction performance. However, this requires the edge server to maintain all the pre-trained models and run each of them, which would outweigh the resources available at edge servers. Another way for model fusion is direct *weight averaging* [198, 309]. Because the weights in neural networks are highly redundant and no one-to-one correspondence exists between the weights of two different neural networks, this method is known to yield poor performance even if the networks represent the same function of the input. As for the second question, Transfer Learning is a promising learning paradigm where an edge node incorporates the knowledge from the cloud or another node with its local training samples. Wang et al. [347, 351] and Yonetani et al. [379]. Recent work on Transferring Generative Adversarial Networks (GANs) [351] proposed several transfer configurations to leverage pre-trained GANs to accelerate the learning process. However, since the transferred GAN is used only as initialization, transferring GANs suffers from catastrophic forgetting. Specifically, catastrophic forgetting may occur when a pre-trained learning model is further trained using another dataset. Over-training the model with new data causes model to be over-tuned to new data and forget the features learned from previous data [163]. To tackle these challenges, in this chapter we aim to develop a framework which explicitly optimizes the continual learning of generative models for the edge, based on the adaptive coalescence of pre-trained generative models from other edge nodes, using optimal transport theory tailored towards GANs.

To mitigate the mode collapse problem due to the vanishing gradients, multiple GAN configurations have been proposed based on the Wasserstein-p metric W_p, including Wasserstein-1 distance [19] and Wasserstein-2 distance [201, 221]. Despite Wasserstein-2 GANs are analytically tractable, the corresponding implementation often requires regularization and is often outperformed by the Wasserstein-1 GAN (W1GAN). With this insight, we will focus on the W1GAN (WGAN refers to W1GAN throughout).

6.1.1 Basic Setting

Specifically, we consider a setting where a target edge node, denoted Node 0, aims to learn a generative model for representing its underlying distribution. Training a WGAN is intimately related to finding a distribution minimizing the Wasserstein distance from the underlying distribution μ_0 [20]. In practice, an edge node has a limited number of samples with empirical distribution $\hat{\mu}_0$, which is distant from μ_0. A naive approach is to train a WGAN based on the limited local samples only, which can be captured via the optimization problem given by $\min_{\nu \in \mathcal{P}} W_1(\nu, \hat{\mu}_0)$, with $W_1(\cdot, \cdot)$ being the Wasserstein-1 distance between two distributions and $\mathcal{P}$ being the distribution space. The best possible outcome of solving this optimization problem can generate a distribution very close to $\hat{\mu}_0$, which however could still be far away from the true distribution μ_0. Clearly, training a WGAN simply based on the limited local samples at an edge node would not be able to obtain a generative model representing the true distribution μ_0.

As alluded to earlier, learning tasks across different edge nodes may share similarity, e.g., limited local samples at Node 0 should be similar to the data samples from other edge nodes. To facilitate the continual learning at Node 0, pre-trained generative models from other related edge nodes can be leveraged via knowledge transfer. Without loss of generality, we assume that there are a set $\mathcal{K}$ of K edge nodes with pre-trained generative models. Since one of the most appealing benefits of WGANs is the ability to continuously estimate the Wasserstein distance during training [19], we assume that the knowledge transfer from Node k to Node 0 is in the form of a Wasserstein ball with radius η_k centered around its pre-trained generative model μ_k at Node k, for $k = 1, \ldots, K$. Intuitively, radius η_k represents the relevance (hence utility) of the knowledge transfer, and the smaller it is, the more informative the corresponding Wasserstein ball is. Building on this knowledge transfer model, we treat the continual learning problem at Node 0 as the coalescence of K generative models and empirical distribution $\hat{\mu}_0$, and cast it as the constrained optimization problem:

$$\min_{\nu \in \mathcal{P}} W_1(\nu, \hat{\mu}_0), \text{ s.t. } W_1(\nu, \mu_k) \leq \eta_k, \forall k \in \mathcal{K}. \tag{6.1}$$

Observe that the constraints in problem (6.1) dictate that the optimal coalesced generative model, denoted by ν^*, lies within the intersection of K Wasserstein balls (centered around $\{\mu_k\}$), exploiting the knowledge transfer systematically. It is worth noting that the optimization problem (6.1) can be extended to other distance functionals, e.g., Jensen-Shannon divergence.

In this chapter, we propose a systematic framework to enable continual learning of generative models via adaptive coalescence of pre-trained generative models from other edge nodes and local samples at Node 0. In particular, by treating the knowledge transferred from each node as a Wasserstein ball centered around its local pre-trained generative model, we cast the problem as a constrained optimization problem which optimizes the continual learning of generative models. Moreover, by applying Lagrangian relaxation to (6.1), we reduce the optimization problem to finding a Wasserstein-1 barycenter of $K + 1$ probability measures, among which K of them are pre-trained generative models and the last one is the empirical distribution (not a generative model though) corresponding to local data samples at Node 0. We propose a *barycentric fast adaptation approach* to efficiently solve the barycenter problem, where the barycenter ν_K^* for the K pre-trained generative models is found recursively offline in edge servers, and then the barycenter between the empirical distribution $\hat{\mu}_0$ of Node 0 and ν_K^* is solved via fast adaptation at Node 0 (See Fig. 6.1). A salient feature in this barycentric approach is that generative replay, enabled by pre-trained GANs, is used to annihilate catastrophic forgetting. Besides, it is known that the Wasserstein-1 barycenter is notoriously difficult to analyze, partly because of the existence of infinitely many minimizers of the Monge Problem. Appealing to optimal transport theory, we use dis-

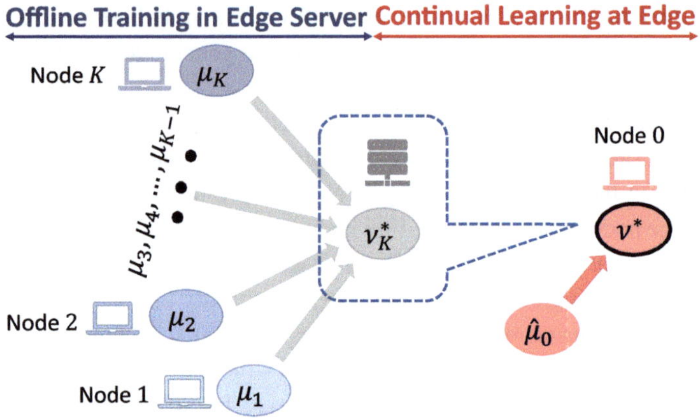

Fig. 6.1 Continual learning of generative models based on coalescence of pre-trained generative models $\{\mu_k, k = 1, \ldots, K\}$ and local dataset at Node 0 (denoted by $\hat{\mu}_0$)

placement interpolation as the theoretic foundation to devise recursive algorithms for finding adaptive barycenters, which ensures the resulting barycenters lie in the baryregion. From the implementation perspective, we also introduce a "recursive" WGAN configuration, where a 2-discriminator WGAN is used per recursive step to find adaptive barycenters sequentially. Then the resulting barycenter in offline training is treated as the meta-model initialization and fast adaptation is carried out to find the generative model using the local samples at Node 0. A weight ternarization method, based on joint optimization of weights and threshold for quantization, is developed to compress the generative model and enable efficient edge learning. Extensive experiments corroborate the efficacy of the proposed framework for fast edge learning of generative models. Further, our experimental studies corroborate that W_1-based recursive WGAN configuration performs better than the W_2^2-based one.

6.1.2 Use Cases

To further illustrate the motivation behind the problem (6.1), we provide two applications in what follows.

(*i*) Privacy is vital in predictive health applications of edge learning. A representative example is a pet disease prediction task, in which a user uploads photos featuring her/his pet to edge server. The user desires to learn whether her/his pet has a disease, but does not want to leak any private information related to her/his location or lifestyle to the edge server. A WGAN model that is trained to contain the essential information associated with pet's health information can generate unique photos to prevent the leakage of private information. However, a private WGAN model cannot be trained on user's phone/laptop, because it would require a lot of data samples as well as processing power. Instead, the user can leverage a barycenter WGAN model from the edge server, which is trained on healthy and unhealthy pet images offline, to train user's personalized WGAN model so as to represent the pet's health condition without revealing any private information. The training process can be done on user's device as barycentric fast adaptation requires less computational power.

(*ii*) Data storage at the edge is gaining popularity due to low delay access to data and ease of local data management. Even though edge servers are more capable than user edge devices/nodes, they still have much less memory and computational power possessed by cloud data centers. To overcome these challenges, barycenter WGAN models can be used to represent datasets for applications requiring less precision. To exemplify, a single barycenter WGAN model could represent the information in hundreds of cat image databases instead of storing the same information in hundreds of WGAN models with minimum accuracy loss at the edge server. This accuracy loss can later be recovered quickly if barycentric fast adaptation is applied to the barycenter WGAN model with desired database.

6.2 Related Work

Optimal transport theory has recently been studied for deep learning applications (see, e.g., Brenier [41], Ambrosio et al. [13] and Villani et al. [340]). Agueh and Carlier [8] has developed an analytical solution to the Wasserstein barycenter problem. Aiming to numerically solve the Wasserstein barycenter problem, Cuturi[70], Cuturi and Doucet [71] and Cuturi and Peyré [72] proposed smoothing through entropy regularization for the discrete setting, based on linear programming. Srivastava et al. [317] employed posterior sampling algorithms in studying Wasserstein barycenters, and Anderes et al. [14] characterized Wasserstein barycenters for the discrete setting (cf. Staib et al. [318], Ye et al. [376] and Singh and Jaggi [306]). Most recent studies leveraged optimal transport theory in domain adaptation [243], cross-domain alignment [57], reinforcement learning [56] and developed further approximations to Wasserstein distance, e.g., Sinkhorn divergence [207]. Nguyen et al. [243] proposed the use of optimal transport theory to quantify the distance between embedded distributions of source and target data in the joint space. By minimizing the distance, TIDOT can mitigate both data and label shifts between different domains. Chen et al. [57] modeled cross-domain alignment problem as a graph matching problem and leveraged Wasserstein and Gromov-Wasserstein distances for node and edge matching, respectively. In Chen et al. [56], authors employ optimal transport theory to solve the difficulties associated with the application of reinforcement learning in language generation. Li et al. [207] studied the Sinkhorn divergence and expanded its applicability to the problems with complex data with nonlinear structure. In particular, authors extended the analytical results in Euclidean space to the reproducing kernel Hilbert space.

GANs [116] have recently emerged as a powerful deep learning tool for obtaining generative models. Arjovsky et al. [19] has introduced Wasserstein metric in GANs, which can help mitigate the vanishing gradient issue to avoid mode collapse. Using optimal transport theory, recent advances of Wasserstein GANs have shed light on understanding generative models. Recent works [176, 201, 221] proposed multiple transport theory based GAN configurations using quadratic Wasserstein-2 cost. In contrast, for the Wasserstein-1 GAN, the discriminator may utilize one of infinitely many transport maps from underlying empirical data distribution to the generative model [13, 340], and it remains open to decipher the relation between the model training and the transport maps. Along a different line, a variety of techniques have been proposed for more robust training of GANs [91, 266, 305, 379].

Pushing the AI frontier to the network edge for achieving edge intelligence has recently emerged as the marriage of AI and edge computing [406]. There are significant challenges since AI model training generally requires tremendous resources that greatly outweigh the capability of resource-limited edge nodes. To address this, various approaches have been proposed in the literature, including model compression [295, 346, 373], knowledge transfer learning [110, 248, 345], hardware acceleration [337, 349], collaboration-based methods [213, 399]. In Shafiee et al. [295], authors proposed an efficient deep neural network of size smaller than 2.4 MB to solve for classification problems. However, the proposed neural network is designed to only work on small number of classes. Osia et al. [248] designed a hybrid architecture where local devices and cloud system collaborate on running a deep neural network that has previously been fine-tuned on the cloud. To achieve privacy, the proposed technique extracts the minimum necessary information from user to transfer to cloud. Venkataramani et al. [337] proposed an energy efficient powerful server architecture to train deep neural networks, which focused on improving core utilization, memory bandwidth and reducing synchronization overheads to achieve its efficiency. Gao et al. [110] proposed an adaptive neural network architecture, in which network architecture changes based on the new task. Continual learning with efficient architecture search technique learns which neurons to leverage while training for the new task. Different from these existing studies, *this work focuses on continual learning of generative models at the edge node*. Rather than learning the new model from scratch, continual learning aims to design algorithms leveraging knowledge transfer from pre-trained models to the new learning task [326], assuming that the training data of previous tasks are unavailable for the newly coming task. Generative replay is gaining more attention where synthetic samples corresponding to earlier tasks are obtained with a generative model and replayed in model training for the new task to mitigate forgetting [277, 280, 319]. In this work, by learning generative models via the adaptive coalescence of pre-trained generative models from other nodes, the proposed "recursive" WGAN configuration facilitates fast edge learning in a continual manner, which can be viewed as an innovative integration of a few key ideas in continual learning, including the replay method [249, 279, 301, 360] which generates pseudo-samples using generative models, and the regularization-based methods [83, 171, 197, 293] which sets the regularization for the model learning based on the learned knowledge from previous tasks, in continual learning [76].

6.3 Adaptive Coalescence of Wasserstein-1 Generative Models

In what follows, we first recast problem (6.1) as a variant of the Wasserstein barycenter problem. Then, we propose a two-stage recursive algorithm, characterize the geometric properties of geodesic curves therein and use displacement interpolation as the foundation to devise recursive algorithms for finding adaptive barycenters.

6.3.1 A Wasserstein-1 Barycenter Formulation via Lagrangian Relaxation

Observe that the Lagrangian for (6.1) is given as follows:

$$\mathcal{L}(\{\lambda_k\}, \nu) = W_1(\nu, \hat{\mu}_0) + \sum_{k=1}^{K} \lambda_k W_1(\nu, \mu_k) - \sum_{k=1}^{K} \lambda_k \eta_k, \tag{6.2}$$

where $\{\lambda_k \geq 0\}_{1:K}$ are the Lagrange multipliers. Based on Volpi et al. [342], problem (6.1) can be solved by using the following Lagrangian relaxation with $\lambda_k = \frac{1}{\eta_k}, \forall k \in \mathcal{K}$, and $\lambda_0 = 1$:

$$\min_{\nu \in \mathcal{P}} \sum_{k=1}^{K} \frac{1}{\eta_k} W_1(\nu, \mu_k) + W_1(\nu, \hat{\mu}_0). \tag{6.3}$$

It is shown in Sinha et al. [307] that the selection $\lambda_k = \frac{1}{\eta_k}, \forall k \in \mathcal{K}$, ensures the same levels of robustness for (6.3) and (6.1). Intuitively, such a selection of $\{\lambda_k\}_{0:K}$ strikes a right balance, in the sense that larger weights are assigned to the knowledge transfer models (based on the pre-trained generative models $\{\mu_k\}$) from the nodes with higher relevance, captured by smaller Wasserstein-1 ball radii. For given $\{\lambda_k \geq 0\}$, (6.3) turns out to be a Wasserstein-1 barycenter problem (cf. Agueh and Carlier [8] and Srivastava et al. [317]), with the new complication that $\hat{\mu}_0$ is an empirical distribution corresponding to local samples at Node 0. Since $\hat{\mu}_0$ is not a generative model per se, its coalescence with other K general models is challenging.

6.3.2 A Two-Stage Adaptive Coalescence Approach for Wasserstein-1 Barycenter Problem

Based on (6.3), we take a two-stage approach to enable efficient learning of the generative model at edge Node 0. The primary objective of Stage I is to find the barycenter for K pre-trained generative models $\{\mu_1, \ldots, \mu_K\}$. Clearly, the ensemble method would not work well due to required memory and computational resources. With this insight, we develop a recursive algorithm for adaptive coalescence of pre-trained generative models. In Stage

II, the resulting barycenter solution in Stage I is treated as the model initialization, and is further trained using the local samples at Node 0. We propose that the offline training in Stage I is performed in the edge server, and the fast adaptation in Stage II is carried out at the edge server (in the same spirit as the model update of Google EDGE TPU), as outlined below:

Stage I: Find the barycenter of K pre-trained generative models across K edge nodes offline. Mathematically, this entails the solution of the following problem:

$$\min_{\nu \in \mathcal{P}} \sum_{k=1}^{K} \frac{1}{\eta_k} W_1(\nu, \mu_k). \tag{6.4}$$

To reduce computational complexity, we propose the following recursive algorithm: Take μ_1 as an initial point, i.e., $\nu_1^* = \mu_1$, and let ν_{k-1}^* denote the barycenter of $\{\mu_i\}_{1:k-1}$ obtained at iteration $k-1$ for $k = 2, \ldots, K$. Then, at each iteration k, a new barycenter ν_k^* is solved between the barycenter ν_{k-1}^* and the pre-trained generative model μ_k.

Stage II: Fast adaptation to find the barycenter between ν_K^* and the local dataset at Node 0. Given the solution ν_K^* obtained in Stage I, we subsequently solve the following problem: $\min_{\nu \in \mathcal{P}} W_1(\nu, \hat{\mu}_0) + W_1(\nu, \nu_K^*)$. By taking ν_K^* as the model initialization, fast adaptation based on local samples is used to learn the generative model at Node 0.

6.3.3 From Displacement Interpolation to Adaptive Barycenters

As noted above, in practical implementation, the W1-GAN often outperforms Wasserstein-p GANs ($p > 1$). However, the Wasserstein-1 barycenter is notoriously difficult to analyze due to the non-uniqueness of the minimizer to the Monge Problem [340]. Appealing to optimal transport theory, we next characterize the performance of the proposed two-stage recursive algorithm for finding the Wasserstein-1 barycenter of pre-trained generative models $\{\mu_k, k = 1, \ldots, K\}$ and the local dataset at Node 0, by examining the existence of the barycenter and characterizing its geometric properties based on geodesic curves.

The seminal work [230] has established the existence of geodesic curves between any two distribution functions σ_0 and σ_1 in the p-Wasserstein space, $\mathcal{P}_p$, for $p \geq 1$. It is shown in Villani [340] that there are infinitely many minimal geodesic curves between σ_0 and σ_1, when $p = 1$. This is best illustrated in N dimensional Cartesian space, where the minimal geodesic curves between $\varsigma_0 \in \mathbb{R}^N$ and $\varsigma_1 \in \mathbb{R}^N$ can be parametrized as follows: $\varsigma_t = \varsigma_0 + s(t)(\varsigma_1 - \varsigma_0)$, where $s(t)$ is an arbitrary function of t, indicating that there are infinitely many minimal geodesic curves between ς_0 and ς_1. This is in stark contrast to the case $p > 1$ where there is a unique geodesic between ς_0 and ς_1. In a similar fashion, there exists infinitely many transport maps, T_0^1, from σ_0 to σ_1 when $p = 1$. For convenience, let $C(\sigma_0, \sigma_1)$ denote an appropriate transport cost function quantifying the minimum cost to move a unit mass from σ_0 to σ_1. It had been shown in Villani [340] that when $p = 1$, two interpolated distribution functions on two distinct minimal curves may have a non-zero distance, i.e., $C(\hat{T}_0^1\#\sigma_0, \tilde{T}_0^1\#\sigma_0) \geq 0$, where # denotes the push-forward operator, thus yielding multiple minimizers to (6.4). For convenience, define $\mathcal{F} := \hat{\mu}_0 \cup \{\mu_k\}_{1:K}$.

Definition 6.1 (*Baryregion*) Let $g_t(\mu_k, \mu_\ell)_{0\leq t\leq 1}$ be any minimal geodesic curve between any pair $\mu_k, \mu_\ell \in \mathcal{F}$, and define the union $\mathcal{R} := \bigcup_{k=1}^{K} \bigcup_{\ell=k+1}^{K+1} g_t(\mu_k, \mu_\ell)_{0\leq t\leq 1}$. The baryregion $\mathcal{B}_{\mathcal{R}}$ is given by $\mathcal{B}_{\mathcal{R}} = \bigcup_{\sigma\in\mathcal{R}} \bigcup_{\varpi\in\mathcal{R},\varpi\neq\sigma} g_t(\sigma, \varpi)_{0\leq t\leq 1}$.

Intuitively, $\mathcal{B}_{\mathcal{R}}$ encapsulates all possible interpolations through distinct geodesics between any two distributions in $\mathcal{F}$. Since each geodesic has finite length, $\mathcal{B}_{\mathcal{R}}$ defines a bounded set in $\mathcal{P}_1$. Next we restate in Lemma 6.1 the renowned *Displacement Interpolation* result [230], which sets the foundation for each recursive step in finding a barycenter in our proposed two-stage algorithm. In particular, Lemma 6.1 leads to the fact that the barycenter ν^* resides in $\mathcal{B}_{\mathcal{R}}$.

Lemma 6.1 (Displacement Interpolation, Villani [339]) *Let $C(\sigma_0, \sigma_1)$ denote the minimum transport cost between σ_0 and σ_1, and suppose $C(\sigma_0, \sigma_1)$ is finite for $\sigma_0, \sigma_1 \in \mathcal{P}(X)$. Assume that $C(\sigma_s, \sigma_t)$, the minimum transport cost between σ_s and σ_t for any $0 \leq s \leq t \leq 1$, is continuous. Then, the following holds true for any given continuous path $g_t(\sigma_0, \sigma_1)_{0\leq t\leq 1}$:*

$$C(\sigma_{t_1}, \sigma_{t_2}) + C(\sigma_{t_2}, \sigma_{t_3}) = C(\sigma_{t_1}, \sigma_{t_3}),\ 0 < t_1 \leq t_2 \leq t_3 \leq 1.$$

In the adaptive coalescence algorithm, the kth recursion defines a baryregion, $\mathcal{B}_{\{\nu_{k_1}^*, \mu_k\}}$, consisting of geodesics between the barycenter ν_{k-1}^* found in $(k-1)$th recursion and generative model μ_k. Clearly, $\mathcal{B}_{\{\nu_k^*, \mu_k\}} \subset \mathcal{B}_{\mathcal{R}}$. Viewing each recursive step in the above two-stage algorithm as adaptive displacement interpolation, we have the following main result on the geodesics and the geometric properties regarding ν^* and $\{\nu_k^*\}_{1:K}$.

Proposition 6.1 (Displacement Interpolation for Adaptive Barycenters) *The adaptive barycenter, ν_k^*, obtained at the output of kth recursive step in Stage I, is a displacement interpolation between ν_{k-1}^* and μ_k and resides inside $\mathcal{B}_{\mathcal{R}}$. Further, the final barycenter ν^* resulting from Stage II of the recursive algorithm resides inside $\mathcal{B}_{\mathcal{R}}$.*

Remark 6.1 It is worth pointing out that different orders for adaptive coalescence may lead to different final barycentric W1GAN models, although the resulting ν^* resides in $\mathcal{B}_{\mathcal{R}}$ always. Had the quadratic Wasserstein cost W_2^2 been used, the final barycenter ν^* would be unique in $\mathcal{B}_{\mathcal{R}}$. However, the corresponding implementation using W_2^2 poses significant challenges [176, 221] and is often outperformed by W1GAN, which we will elaborate further in Sect. 6.4.

6.4 Recursive WGAN Configuration for Continual Learning

Based on the above theoretic results on adaptive coalescence via Wasserstein-1 barycenters, we next turn attention to the implementation of computing adaptive barycenters. Notably, assuming the knowledge of accurate empirical distribution models on discrete support, Cuturi and Doucet [71] introduces a powerful linear program (LP) to compute Wasserstein-p barycenters, but the computational complexity of this approach is excessive. In light of this, we propose a WGAN-based configuration for finding the Wasserstein-1 barycenter, which in turn enables fast learning of generative models based on the coalescence of pre-trained models. Specifically, (6.3) can be rewritten as:

$$\min_{G} \max_{\{\varphi_k\}_{0:K}} \mathbb{E}_{x\sim\hat{\mu}_0}[\varphi_0(x)] - \mathbb{E}_{z\sim\vartheta}[\varphi_0(G(z))] \\ + \sum_{k=1}^{K} \frac{1}{\eta_k} \left\{ \mathbb{E}_{x\sim\mu_k}[\varphi_k(x)] - \mathbb{E}_{z\sim\vartheta}[\varphi_k(G(z))] \right\}, \tag{6.5}$$

where G represents the generator and $\{\varphi_k\}_{0:K}$ are $1-$Lipschitz functions for discriminator models, respectively. Observe that the optimal generator DNN G^* facilitates the barycenter distribution ν^* at its output. We note that the multi-discriminator WGAN configuration has recently been developed [91, 133, 242], by using a common latent space to train multiple discriminators so as to improve stability. In stark contrast, in this work distinct generative models from multiple nodes are exploited to train different discriminators, aiming to learn distinct transport plans among generative models.

A naive approach is to implement the above multi-discriminator WGAN in a one-shot manner where the generator and $K+1$ discriminators are trained simultaneously, which however would require overwhelming computation power and memory. To enable efficient training, we use the proposed two-stage algorithm and develop a *recursive* WGAN configuration to sequentially compute (1) the barycenter ν_K^* for the offline training in the edge servers, as shown in Fig. 6.2; and (2) the barycenter ν^* for the fast adaptation at the target edge node, as shown in Fig. 6.3. The analytical relation between *one-shot* and *recursive* barycenters has been studied for Wasserstein-2 distance, and sufficient conditions for their equivalence is presented in Boissard et al. [37], which, would not suffice for Wasserstein-1 distance, because of the existence of multiple Wasserstein-1 barycenters. Proposition 6.1 shows that any barycenter solution to recursive algorithm resides inside a baryregion, which can be viewed as the counterpart for the *one-shot* solution. We have also developed the bound on the gap between *one-shot* and *recursive* algorithms. We next highlight a few important advantages of the "recursive" WGAN configuration for the barycentric fast adaptation algorithm.

Algorithm 3 Offline Training to Solve the Barycenter of K Pre-trained Generative Models

Inputs: K pre-trained generator-discriminator pairs $\{(G_k, D_k)\}_{1:K}$ of corresponding source nodes $k \in \mathcal{K}$, noise prior $\vartheta(z)$, the batch size m, learning rate α

Set $\mathcal{G}_1^* \leftarrow G_1$, $\tilde{\psi}_1^* \leftarrow$ rand() or $\tilde{\psi}_1^* \leftarrow D_1$; **//Barycenter initialization**
for iteration $k = 2, ..., K$ **do**
    Set $\mathcal{G}_k \leftarrow \mathcal{G}_{k-1}^*$, $\tilde{\psi}_k \leftarrow$ rand(), $\psi_k \leftarrow$ rand() (or $\tilde{\psi}_k \leftarrow \psi \in \{\tilde{\psi}_{k-1}^*, \psi_{k-1}^*\}$, $\psi_k \leftarrow D_k$) and choose $\lambda_{\tilde{\psi}_k}$, λ_{ψ_k};
    while generator $\mathcal{G}_k$ has not converged **do**
        Sample batches of prior samples $\{z^{(i)}\}_{i=1}^m$, $\{z_{\tilde{\psi}_k}^{(i)}\}_{i=1}^m$, $\{z_{\psi_k}^{(i)}\}_{i=1}^m$ independently from prior $\vartheta(z)$;
        Generate synthetic data batches $\{x_{\tilde{\psi}_k}^{(i)}\}_{i=1}^m \sim \nu_{k-1}^*$ and $\{x_{\psi_k}^{(i)}\}_{i=1}^m \sim \mu_k$ by passing $\{z_{\tilde{\psi}_k}^{(i)}\}_{i=1}^m$ and $\{z_{\psi_k}^{(i)}\}_{i=1}^m$ through $\mathcal{G}_{k-1}^*$ and G_k, respectively;
        Compute gradients $g_{\tilde{\psi}_k}$ and g_{ψ_k}: $\left\{ g_\omega \leftarrow \lambda_\omega \nabla_\omega \frac{1}{m} \sum_{i=1}^m \left[\omega(x_\omega^{(i)}) - \omega(\mathcal{G}_k(z^{(i)})) \right] \right\}_{\omega=\tilde{\psi}_k, \psi_k}$;
        Update both discriminators ψ_k and $\tilde{\psi}_k$: $\{\omega \leftarrow \omega + \alpha \cdot \text{Adam}(\omega, g_\omega)\}_{\omega=\psi_k, \tilde{\psi}_k}$;
        Compute gradient $g_{\mathcal{G}_k} \leftarrow -\nabla_{\mathcal{G}_k} \frac{1}{m} \sum_{i=1}^m \left[\lambda_{\psi_k} \psi_k(\mathcal{G}_k(z^{(i)})) + \lambda_{\tilde{\psi}_k} \tilde{\psi}_k(\mathcal{G}_k(z^{(i)})) \right]$;
        Update generator $\mathcal{G}_k$: $\mathcal{G}_k \leftarrow \mathcal{G}_k - \alpha \cdot \text{Adam}(\mathcal{G}_k, g_{\mathcal{G}_k})$;
    end while
    Assign $\mathcal{G}_k^* \leftarrow \mathcal{G}_k$
end for
return generator $\mathcal{G}_K^*$ for barycenter ν_K^*, discriminators $\tilde{\psi}_K^*$, ψ_K^*.

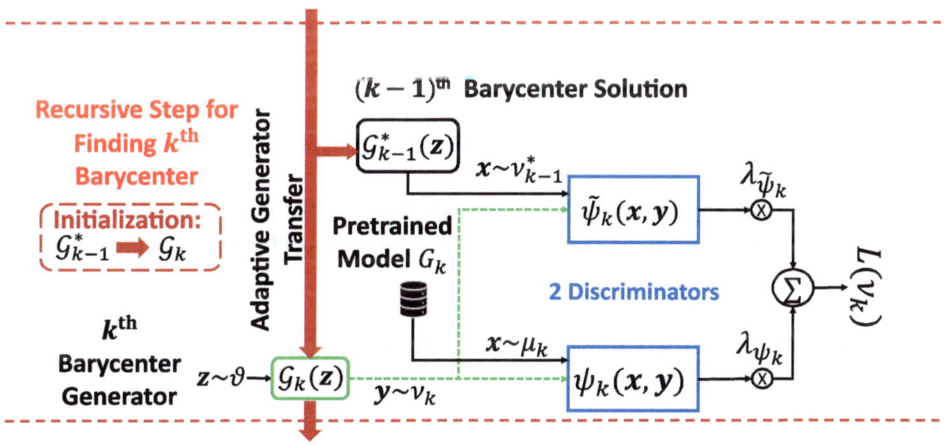

Fig. 6.2 A 2-discriminator WGAN for efficient learning of the kth barycenter generator in offline training, where x denotes the synthetic data generated from pretrained models

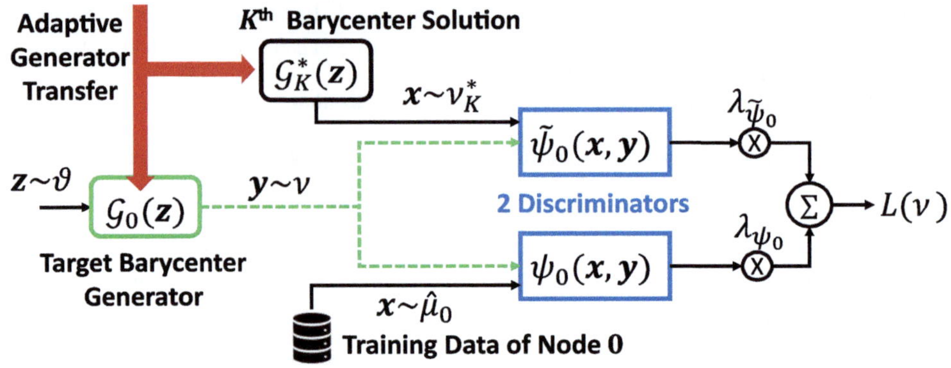

Fig. 6.3 Fast adaptation for learning a generative model at Node 0

6.4.1 A 2-Discriminator WGAN Implementation per Recursive Step to Enable Efficient Training

At each recursive step k, we aim to find the barycenter ν_k^* between pre-trained model μ_k and the barycenter ν_{k-1}^* from last round, which is achieved by training a 2-discriminator WGAN as follows:

$$\min_{\mathcal{G}_k} \max_{\psi_k, \tilde{\psi}_k} \lambda_{\psi_k} \left\{ \mathbb{E}_{x \sim \mu_k}[\psi_k(x)] - \mathbb{E}_{z \sim \vartheta}[\psi_k(\mathcal{G}_k(z))] \right\} \\ + \lambda_{\tilde{\psi}_k} \left\{ \mathbb{E}_{x \sim \nu_{k-1}^*}[\tilde{\psi}_k(x)] - \mathbb{E}_{z \sim \vartheta}[\tilde{\psi}_k(\mathcal{G}_k(z))] \right\}, \tag{6.6}$$

where ψ and $\tilde{\psi}$ denote the corresponding discriminators for pre-trained model G_k and barycenter model $\mathcal{G}_{k-1}^*$ from the previous recursive step, respectively *(See Algorithm 3)*.

6.4.2 Model Initialization in Each Recursive Step

For the initialization of the generator $\mathcal{G}_k$, we use the trained generator $\mathcal{G}_{k-1}^*$ in last step. $\mathcal{G}_{k-1}^*$ corresponds to the barycenter ν_{k-1}^*, and using it as the initialization the displacement interpolation would move along the geodesic curve from ν_{k-1}^* to μ_k [201]. Training GANs with such initializations would accelerate the convergence compared with training from scratch [351]. Finally, ν_K^* is adopted as initialization to enable fast adaptation at the target node. As the barycenter ν_K^* solved via offline training, a new barycenter ν^* between local data (represented by $\hat{\mu}_0$) and ν_K^*, can be obtained by solving the problem:

$$\min_{\mathcal{G}_0} \max_{\psi_0, \tilde{\psi}_0} \lambda_{\psi_0} \{ \mathbb{E}_{x \sim \hat{\mu}_0}[\psi_0(x)] - \mathbb{E}_{z \sim \vartheta}[\psi_0(\mathcal{G}_0(z))] \} \\ + \lambda_{\tilde{\psi}_0} \{ \mathbb{E}_{x \sim \nu_K^*}[\tilde{\psi}_0(x)] - \mathbb{E}_{z \sim \vartheta}[\tilde{\psi}_0(\mathcal{G}_0(z))] \}, \tag{6.7}$$

Algorithm 4 Fast Adaptation Algorithm to Solve for Learning the Generative Model at Node 0.

Inputs: Final generator $\mathcal{G}_K^*$, final discriminators $\tilde{\psi}_K^*$, ψ_K^*, noise prior $\vartheta(z)$, the batch size m, learning rate α

1: Set $\mathcal{G}_0^* \leftarrow \mathcal{G}_K^*$, $\tilde{\psi}_0^* \leftarrow \text{rand}()$ or $\tilde{\psi}_0^* \leftarrow \psi \in \{\tilde{\psi}_K^*, \psi_K^*\}$;
2: **while** generator $\mathcal{G}_0$ has not converged **do**
3: Sample batches of prior samples $\{z^{(i)}\}_{i=1}^m$ and $\{z_{\tilde{\psi}_0}^{(i)}\}_{i=1}^m$ independently from prior $\vartheta(z)$;
4: Get real data batch $\{x_{\psi_0}^{(i)}\}_{i=1}^m \sim \hat{\mu}_0$ and generate synthetic data batch $\{x_{\tilde{\psi}_0}^{(i)}\}_{i=1}^m \sim \nu_K^*$ by passing $\{z_{\tilde{\psi}_0}^{(i)}\}_{i=1}^m$ through $\mathcal{G}_K^*$;
5: Compute gradients $g_{\tilde{\psi}_0}$ and g_{ψ_0}: $\left\{g_\omega \leftarrow \lambda_\omega \nabla_\omega \frac{1}{m}\sum_{i=1}^m \left[\omega(x_\omega^{(i)}) - \omega(\mathcal{G}_0(z^{(i)}))\right]\right\}_{\omega=\tilde{\psi}_0,\psi_0}$;
6: Update both discriminators ψ_0 and $\tilde{\psi}_0$: $\{\omega \leftarrow \omega + \alpha \cdot \text{Adam}(\omega, g_\omega)\}_{\omega=\psi_0,\tilde{\psi}_0}$;
7: Compute gradient $g_{\mathcal{G}_0} \leftarrow -\nabla_{\mathcal{G}_0} \frac{1}{m}\sum_{i=1}^m \left[\lambda_{\psi_0}\psi_0(\mathcal{G}_0(z^{(i)})) + \lambda_{\tilde{\psi}_0}\tilde{\psi}_0(\mathcal{G}_0(z^{(i)}))\right]$;
8: Update generator $\mathcal{G}_0$: $\mathcal{G}_0 \leftarrow \mathcal{G}_0 - \alpha \cdot \text{Adam}(\mathcal{G}_0, g_{\mathcal{G}_0})$;
9: **end while**
10: Assign $\mathcal{G}_0^* \leftarrow \mathcal{G}_0$
11: **return** Generator $\mathcal{G}_0^*$ for barycenter ν_0^*.

i.e., by training a 2-discriminator WGAN, and fine-tuning the generator $\mathcal{G}_0$ from $\mathcal{G}_K^*$ would be notably *faster and more accurate* than learning the generative model from local data only *(See Algorithm 4)*.

6.4.3 Fast Adaptation for Training Ternary WGAN at Node 0

As outlined in Algorithm 2, fast adaptation is used to find the barycenter between ν_K^* and the local dataset at Node 0. To further enhance edge learning, we adopt the weight ternarization method to compress the WGAN model during training. The weight ternarization method not only replaces computationally-expensive multiplication operations with efficient addition/subtraction operations, but also enables the sparsity in model parameters [131]. Specifically, the ternarization process is formulated as:

$$w_l' = S_l \cdot Tern\left(w_l, \Delta_l^{\pm}\right) = S_l \cdot \begin{cases} +1, & w_l > \Delta_l^+ \\ 0, & \Delta_l^- \le w_l \le \Delta_l^+ \\ -1, & w_l < \Delta_l^- \end{cases} \tag{6.8}$$

where $\{w_l\}$ are the full precision weights for lth layer, $\{w_l'\}$ are the weights after ternarization, $\{S_l\}$ is the layer-wise weight scaling coefficient and $\Delta_l^{\pm}$ are the layer-wise thresholds. Since the fixed weight thresholds may lead to accuracy degradation, S_l is approximated as a differentiable closed-form function of $\Delta_l^{\pm}$ so that both weights and thresholds can be optimized simultaneously through back-propagation [142]. Let the generator and the dis-

Algorithm 5 Fast Adaptive Learning of the Ternary Generative Model for Edge Node 0

Inputs: Training dataset $\mathcal{S}_0$, generator $\mathcal{G}_K^*$ for the barycenter ν_K^*, offline barycenter discriminators ψ_K^*, $\tilde{\psi}_K^*$, noise prior $\vartheta(z)$, the batch size m, learning rate α, the number of layers $L_{\mathcal{G}} = L_\psi = L_{\tilde{\psi}} = L$;

1: Set $\mathcal{G}_0 \leftarrow \mathcal{G}_K^*$, $\tilde{\psi}_0 \leftarrow$ rand() and $\psi_0 \leftarrow$ rand() (or $\tilde{\psi}_0 \leftarrow \tilde{\psi}_K^*$ and $\psi_0 \leftarrow \psi_K^*$); **//Initialization**
2: **while** generator $\mathcal{G}_0$ has not converged **do**
3:   **for** $l := 1$ to L **do** **//Weight ternarization**
4:     $\left\{\mathbf{w}'_{l_\omega} \leftarrow Tern(\mathbf{w}_{l_\omega}, S_{l_\omega}, \Delta^{\pm}_{l_\omega})\right\}_{\omega=\tilde{\psi}_k,\psi_k}$;
5:     $\mathbf{w}'_{l_\mathcal{G}} \leftarrow Tern(\mathbf{w}_{l_\mathcal{G}}, S_{l_\mathcal{G}}, \Delta^{\pm}_{l_\mathcal{G}})$;
6:   **end for**
7:   Sample batches of prior samples $\{z^{(i)}\}_{i=1}^m$ from prior $\vartheta(z)$;
8:   Sample batches of training samples $\{x_0^i\}_{i=1}^m$ from local dataset $\mathcal{S}_0$;
9:   **for** $l := L$ to 1 **do** **//Update thresholds**
10:     Compute gradients $\left\{g_{\Delta^{\pm}_{l_\omega}}\right\}_{\omega=\tilde{\psi}_k,\psi_k} : \left\{g_{\Delta^{\pm}_{l_\omega}} \leftarrow \nabla_{\Delta^{\pm}_{l_\omega}} \frac{1}{m}\sum_{i=1}^m \left[\omega_0(x_0^{(i)}) - \omega_0(\mathcal{G}_0(z^{(i)})\right]\right\}_{\omega=\tilde{\psi}_k,\psi_k}$;
11:     Update $\left\{\Delta^{\pm}_{l_\omega}\right\}_{\omega=\tilde{\psi}_k,\psi_k} : \left\{\Delta^{\pm}_{l_\omega} \leftarrow \Delta^{\pm}_{l_\omega} + \alpha \cdot g_{\Delta^{\pm}_{l_\omega}}\right\}_{\omega=\tilde{\psi}_k,\psi_k}$;
12:     Compute gradient $g_{\Delta^{\pm}_{l_\mathcal{G}}} : g_{\Delta^{\pm}_{l_\mathcal{G}}} \leftarrow -\nabla_{\Delta^{\pm}_{l_\mathcal{G}}} \frac{1}{m}\sum_{i=1}^m \left[\psi_0(\mathcal{G}_0(z^{(i)})) + \tilde{\psi}_0(\mathcal{G}_0(z^{(i)}))\right]$;
13:     Update $\Delta^{\pm}_{l_\mathcal{G}} : \Delta^{\pm}_{l_\mathcal{G}} \leftarrow \Delta^{\pm}_{l_\mathcal{G}} - \alpha \cdot g_{\Delta^{\pm}_{l_\mathcal{G}}}$;
14:   **end for**
15:   Repeat steps 3-5 using updated thresholds;
16:   **for** $l := L$ to 1 **do** **//Update the full-precision weights**
17:     Compute gradients $\left\{g_{\mathbf{w}_{l_\omega}}\right\}_{\omega=\tilde{\psi}_k,\psi_k} : \left\{g_{\mathbf{w}_{l_\omega}} \leftarrow \nabla_{\mathbf{w}_{l_\omega}} \frac{1}{m}\sum_{i=1}^m \left[\omega_0(x_0^{(i)}) - \omega_0(\mathcal{G}_0(z^{(i)})\right]\right\}_{\omega=\tilde{\psi}_k,\psi_k}$;
18:     Update $\{\mathbf{w}_{l_\omega}\}_{\omega=\tilde{\psi}_k,\psi_k} : \left\{\mathbf{w}_{l_\omega} \leftarrow \mathbf{w}_{l_\omega} + \alpha \cdot \text{Adam}(\mathbf{w}_{l_\omega}, g_{\mathbf{w}_{l_\omega}})\right\}_{\omega=\tilde{\psi}_k,\psi_k}$;
19:     Compute gradient $g_{\mathbf{w}_{l_\mathcal{G}}} : g_{\mathbf{w}_{l_\mathcal{G}}} \leftarrow -\nabla_{\mathbf{w}_{l_\mathcal{G}}} \frac{1}{m}\sum_{i=1}^m \left[\psi_0(\mathcal{G}_0(z^{(i)})) + \tilde{\psi}_0(\mathcal{G}_0(z^{(i)}))\right]$;
20:     Update $\mathbf{w}_{l_\mathcal{G}} : \mathbf{w}_{l_\mathcal{G}} \leftarrow \mathbf{w}_{l_\mathcal{G}} - \alpha \cdot \text{Adam}(\mathbf{w}_{l_\mathcal{G}}, g_{\mathbf{w}_{l_\mathcal{G}}})$;
21:   **end for**
22:   Repeat step 3-5 using updated full-precision weights;
23: **end while**
24: **return** Ternary generator $\mathcal{G}_0$.

criminators of WGAN at Node 0 be denoted by $\mathcal{G}_0$, $\tilde{\psi}_0$ and ψ_0, which can be parametrized by the ternarized weights $\{\mathbf{w}'_{l_\mathcal{G}}\}_{l_\mathcal{G}=1}^{L_\mathcal{G}}$, $\{\mathbf{w}'_{l_{\tilde{\psi}}}\}_{l_{\tilde{\psi}}=1}^{L_{\tilde{\psi}}}$ and $\{\mathbf{w}'_{l_\psi}\}_{l_\psi=1}^{L_\psi}$, respectively. The barycenter ν^* at Node 0, captured by $\mathcal{G}_0^*$, can be obtained by training the ternary WGAN via iterative updates of both weights and thresholds, which takes three steps in each iteration: (1) calculating the scaling coefficients and the ternary weights for $\mathcal{G}_0$, $\tilde{\psi}_0$ and ψ_0, (2) calculating the loss function using the ternary weights via forward-propagation and 3) updating the full precision weights and the thresholds via back-propagation *(See Algorithm 5)*.

6.4.4 Implementation Challenges in W_2^2-Based GAN

The practical success of W1GAN can be largely attributed to the elegant structural relation between Kantorovich potentials, i.e., $\varphi = -\psi$. Unfortunately, when the quadratic cost W_2^2 is used, Kantorovich potentials translate to $\varphi = \psi^*$, where * denotes the convex conjugate. Note that W_2 is a metric but W_2^2 is not. When implementing the W_2^2-based GAN, both Kantorovich potentials are estimated by 2 distinct DNNs that must satisfy the convex conjugate constraint, which is practically challenging. Very recent studies [176, 201, 221] attempt to enforce the convex conjugate constraint between these DNNs through approximations or regularization under certain assumptions, but it remains not well understood. In this chapter, we have carried out experimental studies to compare the performance of W_1 and W_2^2 based recursive WGAN configurations, and our findings corroborate that W1GAN performs better.

6.5 Experiments

6.5.1 Datasets, Models and Evaluation

We extensively examine the performance of learning a generative model, using the barycentric fast adaptation algorithm, on a variety of widely adapted datasets in the GAN literature, including CIFAR10, CIFAR100, LSUN and MNIST [79, 177, 384]. In experiments, we used various DCGAN-based architectures [268] depending on the dataset as different datasets vary in image size, feature diversity and in sample size, e.g., image samples in MNIST have less diversity compared to the rest of the datasets, while LSUN contains the largest number of samples with larger image sizes. Further, we used the weight ternarization method [142] to jointly optimize weights and quantizers of the generative model at the target edge node, reducing the memory burden of generative models in memory-limited edge devices.

The Frechet-Inception Distance (FID) score [143] is widely adopted for evaluating the performance of GAN models in the literature [64, 119, 351], since it provides a quantitative assessment of the similarity of a dataset to another reference dataset. Therefore, we use the FID score to evaluate the performance evolution of the two-stage adaptive coalescence algorithm and all baseline algorithms during training. We here emphasize that a smaller FID score of a GAN indicates that it has better performance. Note that to avoid one-sided scores and make a fair comparison, other evaluation metrics, in addition to the FID score, are also leveraged to quantify the performance of all algorithms. In all experiments, we use the entire dataset as the reference dataset.

To demonstrate the improvements by using the proposed framework based on *barycentric fast adaptation*, we conduct extensive experiments and compare performance with 3 distinct baselines: (1) *Transferring GANs* [351]: a pre-trained GAN model is used as initialization at Node 0 for training a new WGAN model by using local data samples. (2) *Ensemble method*: The model initialization, obtained by using pre-trained GANs at other edge nodes, is further trained using both local data from Node 0 and synthetic data samples. (3) *Edge-only*: only local dataset at Node 0 is used in WGAN training. Due to the lack of sample diversity at the target edge node, the WGAN model trained using local data only is not expected to attain good performance. In stark contrast, the WGAN model trained using the proposed two-stage adaptive coalescence algorithm, inherits the diversity from the pre-trained models at other edge nodes, and results in better performance compared to its counterparts. Needless to say, if the entire dataset were available at Node 0, the best performance would be achieved.

6.5.2 Experiment Setup

We consider the following two scenarios: (1) *The overlapping case*: the classes of the data samples at other edge nodes and at Node 0 overlap; (2) *The non-overlapping case*: the classes of the data samples at other edge nodes and at Node 0 are mutually exclusive. In overlapping experiments, the corresponding dataset is equally split into 2 sub-datasets and sub-datasets are used to pre-train 2 WGAN models independently. Subsequently, Algorithms 3 and 4 are used consecutively to find the barycenter and the final WGAN model at Node 0, respectively. The few data samples to be used in the fast adaptation stage are randomly selected from all classes in the dataset. For the *transferring GANs* method, the first pre-trained model is further trained on the second sub-dataset using transfer learning to compute a fused model. In the final stage, the few data samples from all classes in the dataset are used to train a WGAN model at Node 0 by leveraging the fused model via transfer learning again. In the Ensemble method, the pre-trained models are used to generate a synthetic dataset. The synthetic dataset is combined with the few data samples from all the classes in the dataset and the combined dataset is leveraged to train a final WGAN model at Node 0. Lastly, the *edge-only* method only leverages the few data samples from all the classes in the dataset to train a WGAN model at Node 0.

In non-overlapping experiments, randomly drawn samples from the first 40% of the classes in the dataset are allocated into the first node and randomly drawn samples from the second 40% of the classes are allocated into the second node. The same steps as in the overlapping case are followed by using these two sub-datasets until the final stage. In the final stage, a few data samples are randomly selected from the remaining 20% of the classes and are placed in Node 0.

6.5.3 Continual Learning Against Catastrophic Forgetting

We investigate the convergence and the generated image quality of various training scenarios on CIFAR100 and MNIST datasets. As illustrated in Figs. 6.4 and 6.5, *barycentric fast adaptation* clearly outperforms all baselines. *Transferring GANs* suffers from catastrophic forgetting, because the continual learning is performed over local data samples at Node 0

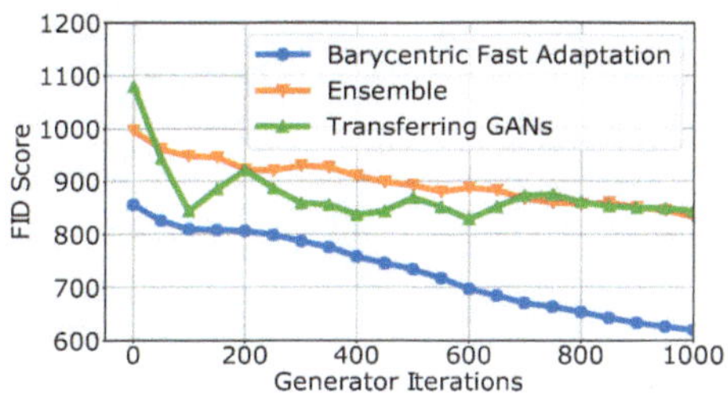

(a) MNIST: Non-overlapping case.

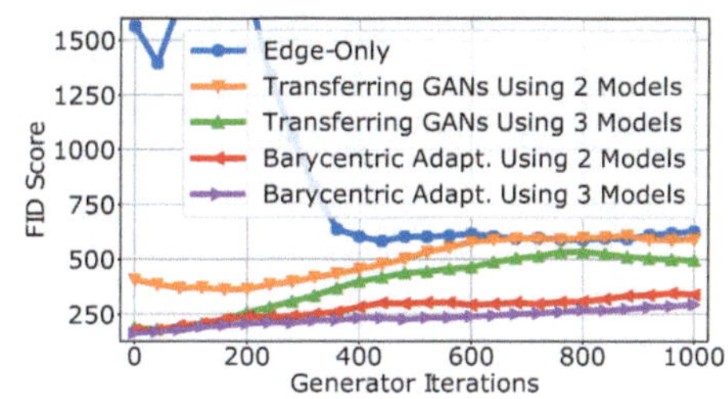

(b) MNIST: Overlapping case.

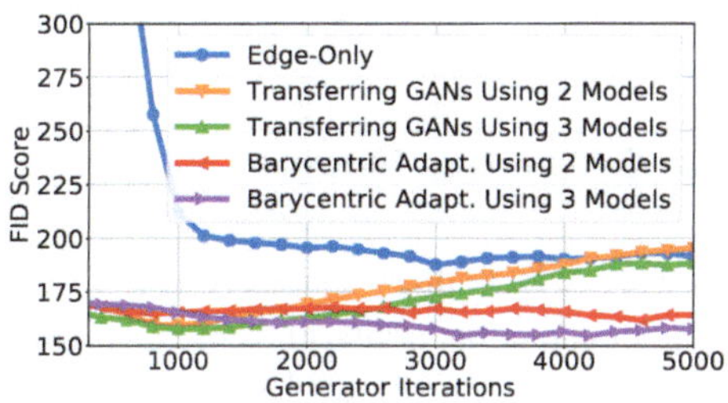

(c) CIFAR10: Overlapping case.

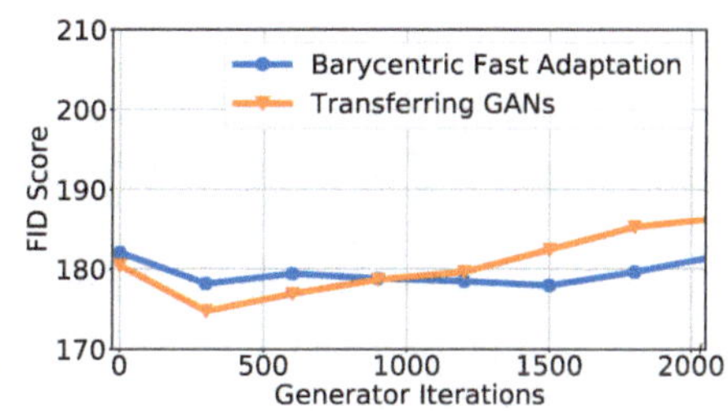

(d) CIFAR100: Overlapping case.

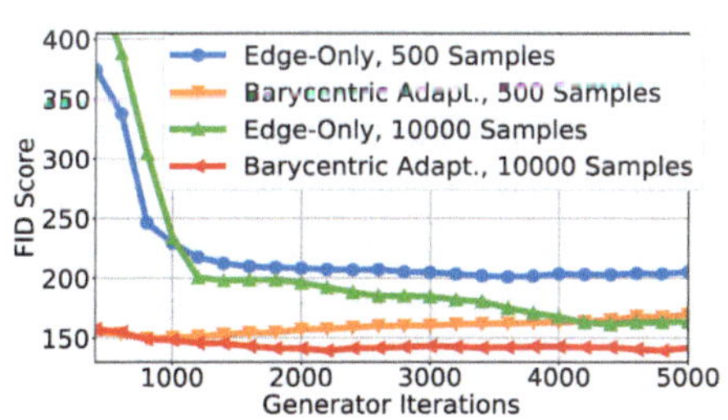

(e) CIFAR10: Overlapping case.

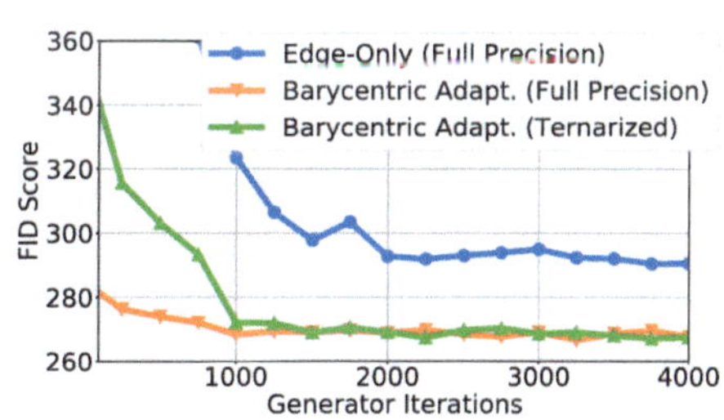

(f) LSUN: Overlapping case.

Fig. 6.4 Comparison of convergence of *barycentric fast adaptation* with various baselines

only. On the contrary, the *barycentric fast adaptation* and the ensemble method leverage generative replay, which mitigates the negative effects of catastrophic forgetting. Further, observe that the ensemble method suffers because of the limited data samples at Node 0, which are significantly outnumbered by synthetic data samples from pre-trained GANs, and this imbalance degrades the applicability of the ensemble method for continual learning. On the other hand, the *barycentric fast adaptation* can obtain the barycenter between the local data samples at Node 0 and the barycenter model trained offline, and hence can effectively leverage the abundance of data samples from edge nodes and the accuracy of local data samples at Node 0 for better continual learning.

Fig. 6.5 Image samples from 4 different approaches for CIFAR10: From top row to bottom row; the images generated by *edge-only* (at 90000 iterations), ternary *barycentric fast adaptation* (at 5000 iterations), *transferring GANs* (at 1000 iterations) and *barycentric fast adaptation* (at 1000 iterations) algorithms. Last row illustrates some real images

6.5.4 Impact of Number of Pre-trained Generative Models

To quantify the impact of cumulative model knowledge from pre-trained generative models on the learning performance at the target node, we consider the scenario where 10 classes in CIFAR10/MNIST are split into 3 subsets, e.g., the first pre-train model has classes {0, 1, 2}, the second pre-trained model has classes {2, 3, 4} and the third pre-trained model has the remaining classes. One barycenter model is trained offline by using the first two pre-trained models and the second barycenter model is trained using all 3 pre-trained models, respectively, based on which we evaluate the performance of *barycentric fast adaptation* with 1000 data samples at the target node. Figure 6.4b and c showcase that the more model knowledge is accumulated in the barycenter computed offline, the higher image quality is achieved at Node 0. As expected, more model knowledge can help new edge nodes in training higher-quality generative models. In both figures, the *barycentric fast adaptation* outperforms *transferring GANs*.

6.5.5 Impact of the Number of Data Samples at Node 0

Figure 6.4e further illustrates the convergence across different number of data samples at the target node on CIFAR10 dataset. As expected, the FID score gap between *barycentric fast adaptation* and *edge-only* method decreases as the number of data samples at the target node increases, simply because the empirical distribution becomes more 'accurate'. In particular, the significant gap of FID scores between *edge-only* and the *barycentric fast adaptation* approaches in the initial stages indicates that the barycenter found via offline training and adopted as the model initialization for fast adaptation, is indeed close to the underlying model at the target node, hence enabling faster and more accurate edge learning than *edge-only*.

6.5.6 Impact of Wasserstein Ball Radii

The Wasserstein ball radius η_k for pre-trained model k represents the relevance (hence utility) of the knowledge transfer which is intimately related to the capability to generalize beyond the pre-trained generative models, and the smaller it is, the more informative the corresponding Wasserstein ball is. Hence, larger weights $\lambda_k = 1/\eta_k$ would be assigned to the nodes with higher relevance. We note that the weights are determined by the constraints and thus are fixed. Since we introduce the recursive WGAN configuration, the order of coalescence (each corresponding to a geodesic curve) may impact the final barycentric WGAN model, and hence the performance of *barycentric fast adaptation*. To this end, we compute the coalescence of models of nodes with higher relevance at latter recursions to ensure that the final barycentric model is closer to the models of nodes with higher relevance.

6.5.7 Ternary WGAN-Based Barycentric Fast Adaptation

With the model initialization in the form of a full-precision barycenter model computed in offline training, we next train a ternary WGAN with 2 discriminators for the target node to compress the generative model further. In particular, we use the same split of classes as the experiment illustrated in Fig. 6.4e, and compare the image quality obtained by ternary WGAN-based fast adaptation against both full precision counterpart and *edge-only*. It can be seen from the FID scores (Fig. 6.4f), the ternary WGAN-based *barycentric fast adaptation* results in negligible performance degradation compared to its full precision counterpart, and is still much better compared to the *edge-only* approach.

6.5.8 Performance Evaluation Using Inception Score

In addition to the FID score, we also use Inception Score (IS), another widely used metric, to signify the robustness of the performance evaluation. Each of the 3 different numerical experiments is repeated 5 times, and the performance evaluation using FID and Inception scores is illustrated in Fig. 6.6. Clearly, both FID and IS evaluations corroborate the superior performance of the *barycentric fast adaptation* algorithm, as well as the small deviation from the mean performance. The worst and best case performances of the *barycentric fast adaptation* and its counterparts are illustrated in Table 6.1. *Best-Mean, Worst-Mean, Best* and *Worst* denote the best mean performance, the worst mean performance, the best performance in all 5 runs and the worst performance in all 5 runs for the corresponding metrics, respectively. Table 6.1 further showcases the superior performance of *barycentric fast adaptation* in comparison to its counterparts, particularly when the number of available samples at Node 0 is limited.

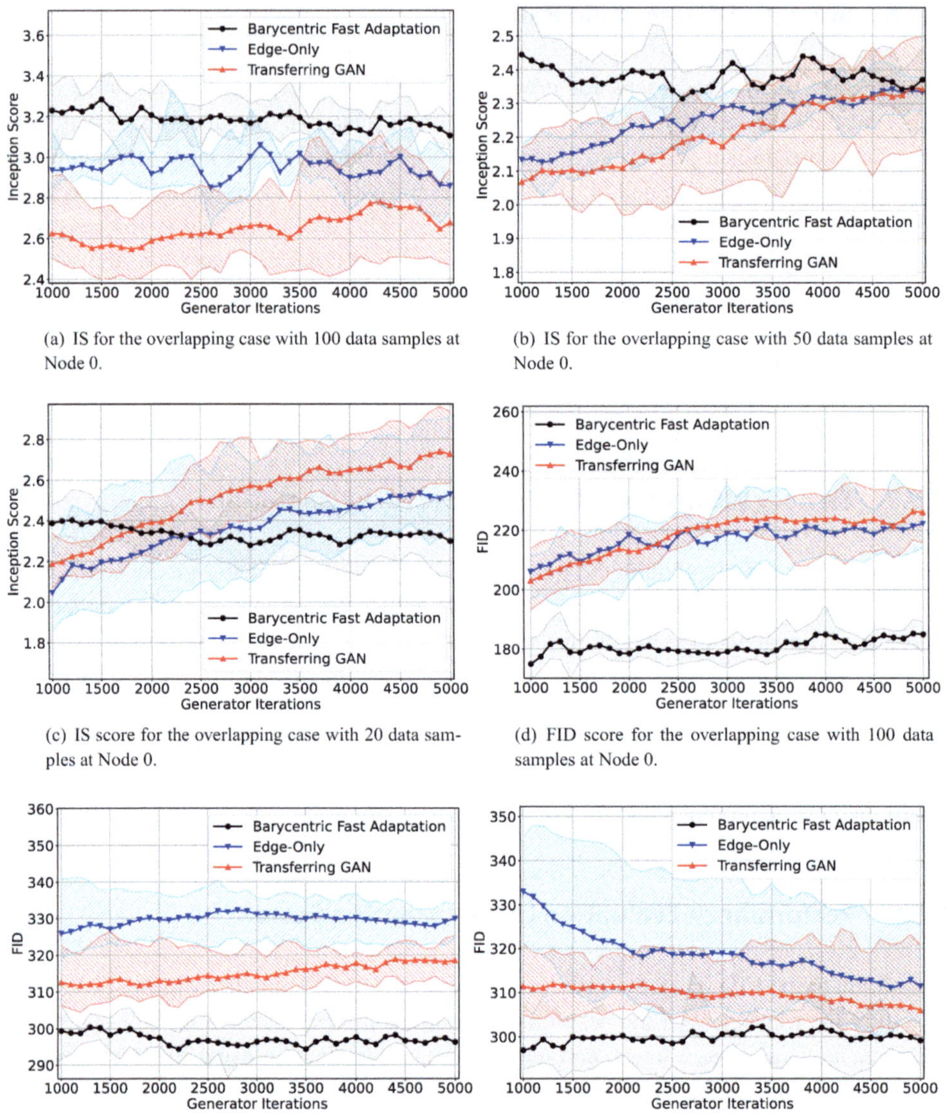

(a) IS for the overlapping case with 100 data samples at Node 0.

(b) IS for the overlapping case with 50 data samples at Node 0.

(c) IS score for the overlapping case with 20 data samples at Node 0.

(d) FID score for the overlapping case with 100 data samples at Node 0.

(e) FID score for the overlapping case with 50 data samples at Node 0.

(f) FID score for the overlapping case with 20 data samples at Node 0.

Fig. 6.6 Performance evaluation comparisons of various WGAN model training techniques using FID and Inception scores tested on CIFAR10. FID and Inception scores conclude similar performance results. Note that higher Inception score and lower FID score indicate better performance

Table 6.1 Performance comparisons of different WGAN model training algorithms

	Metric\experiment	Inception score				Frechet-inception distance			
		Best-Mean	Worst-Mean	Best	Worst	Best-Mean	Worst-Mean	Best	Worst
Figure 6.6a, d	Fast Adaptation	3.28 ± 0.13	3.11 ± 0.08	3.42	2.95	175 ± 7	185 ± 4	169	195
	Edge-Only	3.06 ± 0.11	2.85 ± 0.18	3.33	2.67	206 ± 7	222 ± 9	194	239
	Transferring GANs	2.78 ± 0.32	2.55 ± 0.15	3.11	2.40	203 ± 11	226 ± 8	193	236
Figure 6.6b, e	Fast Adaptation	2.45 ± 0.13	2.31 ± 0.13	2.58	2.12	294 ± 5	300 ± 4	287	306
	Edge-Only	2.34 ± 0.08	2.13 ± 0.09	2.42	2.03	326 ± 11	332 ± 7	319	341
	Transferring GANs	2.35 ± 0.16	2.07 ± 0.07	2.50	1.97	312 ± 8	319 ± 4	305	326
Figure 6.6c, f	Fast Adaptation	2.40 ± 0.12	2.28 ± 0.16	2.55	2.12	297 ± 7	302 ± 4	291	314
	Edge-Only	2.53 ± 0.27	2.05 ± 0.19	2.91	1.36	311 ± 13	333 ± 12	300	348
	Transferring GANs	2.74 ± 0.19	2.19 ± 0.13	2.96	2.07	306 ± 10	312 ± 7	300	325

An important observation herein is that both FID and IS quantify the quality and the class diversity of the generated images. Specifically, the FID score leverages another large dataset (reference dataset) (the whole dataset in our experiments) to *relatively* compute the quality and class diversity of the generated images, whereas IS does not utilize a reference dataset, i.e., IS is *absolute*. A significant implication of this difference is that the FID score of a generated dataset can be different with respect to different reference datasets, while IS will be constant. Therefore, IS cannot quantify the effects of generator overfitting as well as the FID score does. In Fig. 6.6c and f, only 20 data samples are used to train the WGAN generator models, and hence the final WGAN models are prone to extreme overfitting. The WGAN models might generate the same images even for different values of $\mathbf{z}$, i.e., the image diversity within every class might be very low. In accordance with the generator overfitting phenomenon, we observe that the FID scores for all 3 methods are stationary after 2000 iterations in Fig. 6.6f, whereas the IS curves continue to improve for *edge-only* and *transferring GANs* in Fig. 6.6c. This indicates generator overfitting occurs in WGAN model trained with the *edge-only* and *transferring GANs* methods, whereas both the IS and FID scores for the *barycentric fast adaptation* method are stationary, indicating no generator overfitting.

6.5.9 Continual Learning Performance Across Dissimilar Data Samples

This experiment explores the performance of the *barycentric fast adaptation* when offline barycenter is trained from pre-trained WGAN models that are learned from 2 distinct datasets, e.g., CIFAR10 samples are placed in Node 1 and CIFAR100 samples are placed in Node 2. Specifically, samples from all 10 classes in CIFAR10 are placed in Node 1 and the samples from the random 20 classes of CIFAR100 are placed in Node 2. In the *overlapping* scenario, Node 0 contained few samples from the same classes placed in Node 1 and Node 2. In the *non-overlapping* scenario, Node 0 had few samples from 20 classes of CIFAR100 that are different from the classes in Node 2. We run experiments for both *overlapping* and *non-overlapping* scenarios for various number of samples at Node 0.

Figure 6.7 demonstrates that *barycentric fast adaptation* outperforms other baselines for any sample size. The performance gap between *barycentric fast adaptation* and its counterparts increases significantly as fewer data samples are available at Node 0, because *barycentric fast adaptation* is better at transferring the previous knowledge from edge network and mitigating catastrophic forgetting. Clearly, the performance gap between *barycentric fast adaptation* and its counterparts is much higher in this experiment, because less model similarity of pre-trained WGAN models amplifies the effects of catastrophic forgetting.

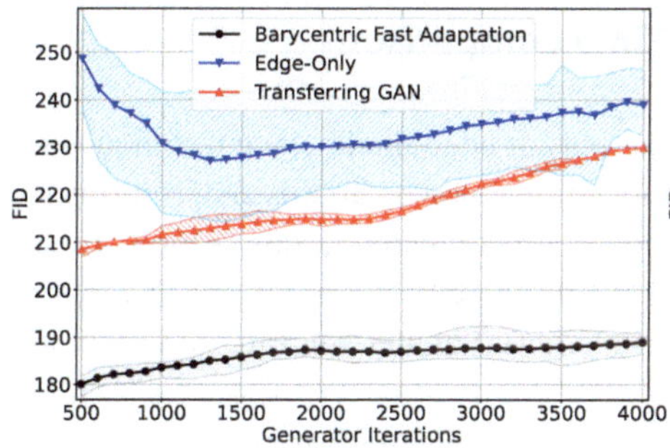

(a) Non-overlapping case: 100 samples at Node 0.

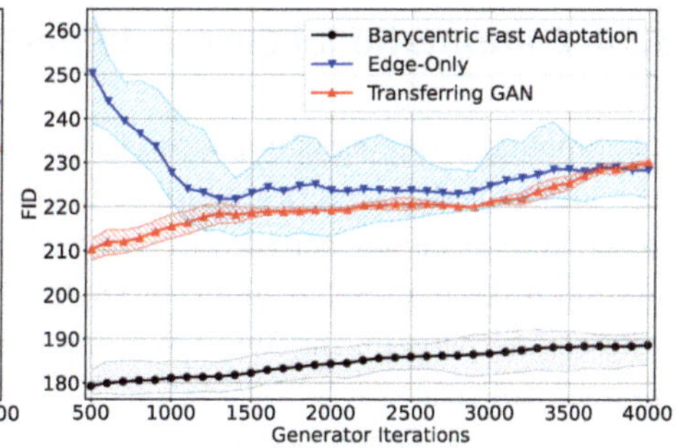

(b) Overlapping case: 150 samples at Node 0.

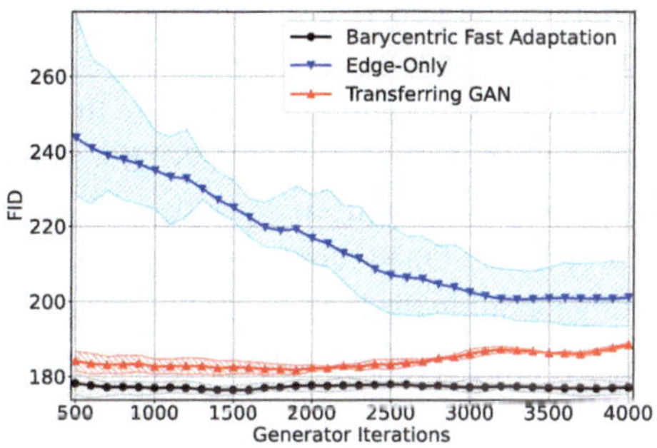

(c) Non-overlapping case: 400 samples at Node 0.

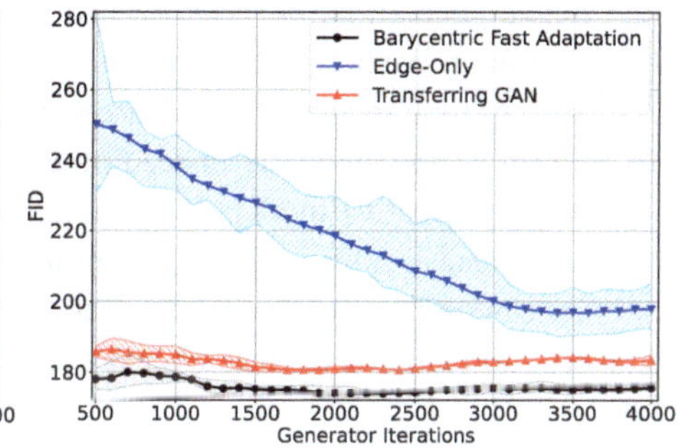

(d) Overlapping case: 600 samples at Node 0.

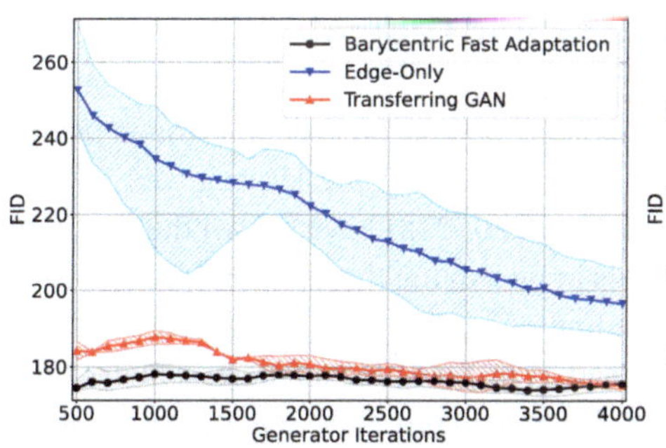

(e) Non-overlapping case: 2000 samples at Node 0.

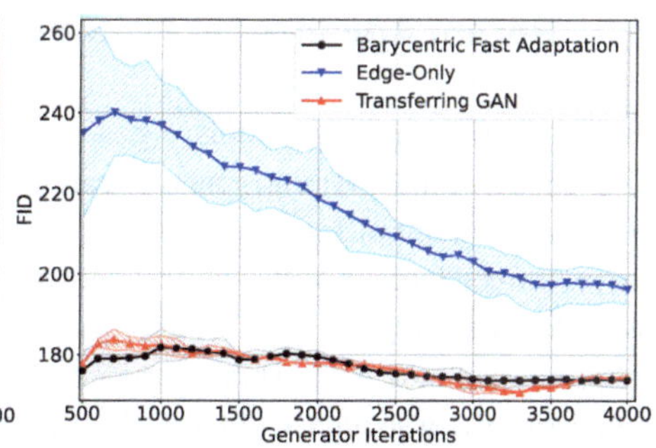

(f) Overlapping case: 3000 samples at Node 0.

Fig. 6.7 Performance evaluation comparisons of various WGAN model training techniques using FID score tested on the mixture of datasets CIFAR10 and CIFAR100. Note that lower FID score indicate better performance

6.5.10 Computational Cost and Run-Time Comparison

Total computational cost is important for the success of edge intelligence techniques since edge nodes and edge servers are limited in computational power. To this end, run-times and corresponding accuracy across different techniques are compared in this section. Table 6.2 shows run-times required to attain certain level of FID score for *barycentric fast adaptation* and its counterparts during WGAN training at Node 0. Smaller run-time indicates lower computational cost.

Results clearly indicate a significant improvement over *edge-only* in computational complexity. While *barycentric fast adaptation* perform similar to *transferring GANs* in lower quality image generation, *barycentric fast adaptation* can generate higher quality images with less training. We here notice that each training iteration for *edge-only* and *transferring GANs* take less time in average compared to *barycentric fast adaptation*, because the latter needs to train 2 discriminators instead of 1. In particular, *transferring GANs* fails to train good enough generative models for generating the highest quality images because of the occurrence of catastrophic forgetting as training progresses. On the contrary, *barycentric fast adaptation* can train generative models capable of generating the highest quality images in a reasonably short amount of time since it overcomes catastrophic forgetting as demonstrated in Table 6.2.

6.6 Recent Advances in Pretrained CL and Federated CL

Pretrained CL Pretrained models, e.g., large language models, have demonstrated phenomenal capabilities in learning general features, benefiting from the large model sizes and tremendous amount of training data. Recent studies [250, 273] have shown that leveraging these pretrained large models has great potentials in improving the performance and robustness of downstream CL tasks, which inspired a line of studies on investigating CL starting from pretrained models. [93] built upon pretrained Transformers and incrementally trained adapters to solve a sequence of CL tasks. L2P [354] treated prompts as learnable parameters and optimized the prompts to a pretrained model, so as to explicitly manage general and task-specific knowledge for CL. CODA-Prompt [310] constructed a decomposed prompt with a weighted sum of learnable prompt components, where the weighting is calculated based on the attention mechanism. Pretrained models are also used to generate class prototypes [260], which would be directly used for prediction at inference based on nearest neighbors. SLCA [395] identified a progressive overfitting problem in pretrained CL and proposed to align the classification layers based on modeling the class-wise distributions.

Federated CL (FCL) The study of FCL was initiated in [382] which proposed an architecture-based algorithm by dividing the local model into a global model and a task-specific model. The global model was updated jointly across all clients to capture common knowledge as in standard FL, whereas the task-specific model was updated locally to retain

Table 6.2 Computational complexity comparison of barycentric fast adaption against its counterparts

	Database	MNIST			CIFAR10				CIFAR10 and CIFAR100			
	Method\Target FID	600	400	200	200	175	165	160	200	190	180	175
Iterations	Fast adaptation	1	1	20	1	1	3000	4500	1	1	1	2200
	Transferring GANs	1	100	DNA	1	1	1000	DNA	1	1	1600	DNA
	Edge-Only	360	DNA	DNA	1200	DNA	DNA	DNA	2500	3400	DNA	DNA
Seconds	Fast adaptation	1.0	1.0	20.4	1.5	1.5	4440.0	6660.0	1.5	1.5	1.5	3300.0
	Transferring GANs	0.5	53.5	DNA	0.8	0.8	755.0	DNA	0.8	0.8	1208.0	DNA
	Edge-Only	192.6	DNA	DNA	906.0	DNA	DNA	DNA	1887.5	2567.0	DNA	DNA

task-specific information for mitigating forgetting. Following this work, there have been several studies that further explored FCL algorithm design based on this parameter isolation strategy [145, 236, 400]. Given the superior performance of rehearsal-based approaches in standard continual learning to alleviate forgetting, some recent studies [88, 89, 228, 265, 355, 409] have also investigated rehearsal-based algorithm design for FCL, such as locally storing a small number of samples for old tasks and the corresponding probability outputs [355], constructing a global buffer to share data among clients with differential privacy [409], storing gradients of the local model in the previous round [228], learning generative models at each client to generate synthetic data for old tasks [265].

6.7 Conclusion

In this chapter, we propose a systematic framework for continual learning of generative models via adaptive coalescence of pre-trained models from other edge nodes. Particularly, we cast the continual learning problem as a constrained optimization problem that can be reduced to a Wasserstein-1 barycenter problem. Appealing to optimal transport theory, we characterize the geometric properties of geodesic curves therein and use displacement interpolation as the foundation to devise recursive algorithms for finding adaptive barycenters. Next, we take a two-stage approach to efficiently solve the barycenter problem, where the barycenter of the pre-trained models is first computed offline in the edge servers via a "recursive" WGAN configuration based on displacement interpolation. Then, the resulting barycenter is treated as the meta-model initialization and fast adaptation is used to find the generative model using the local samples at the target edge node. A weight ternarization method, based on joint optimization of weights and threshold for quantization, is developed to compress the edge generative model further. Extensive experimental studies corroborate the efficacy of the proposed framework. Finally, we present a brief summary of recent advances in pretrained CL and federated CL towards achieving continual edge-AI.

7 Cloud-Edge Collaboration for Continual Reinforcement Learning

Cloud-Edge Collaboration for Continual Reinforcement Learning aims to leverage cloud and edge resources to create more efficient and adaptive reinforcement learning systems. One of the key challenges in deploying cloud-trained models to edge devices is ensuring that the models can effectively adapt to real-world, time-varying conditions while minimizing issues like overestimation bias. Recall that in Chap. 3, in order to reduce the suboptimality gap in Warm-Start Reinforcement Learning, it is essential to reduce the bias in the approximation error, which can be achieved by using techniques such as ensemble learning and planning.

In this chapter, we introduce Adaptive Ensemble Q-learning (AdaEQ) as a promising solution to enhance cloud-edge collaboration for Continual Reinforcement Learning. AdaEQ addresses the overestimation issue in Q-learning by using an ensemble of Q-function approximators to estimate action values. The method dynamically adjusts the ensemble size to maintain a near-zero estimation bias, compensating for the time-varying approximation errors that occur during the learning process. By incorporating Model Identification Adaptive Control (MIAC) into this framework, AdaEQ provides a mechanism for flexible ensemble size adaptation based on real-time feedback. This adaptability makes it especially suited for edge environments, where continual learning and model fine-tuning are critical. Thus, Cloud-Edge Collaboration and AdaEQ can work hand in hand, ensuring that cloud-trained models can be effectively optimized and deployed on edge devices in dynamic environments.

H. Wang et al., *Continual and Reinforcement Learning for Edge AI*,
Synthesis Lectures on Learning, Networks, and Algorithms,
https://doi.org/10.1007/978-3-031-84363-1_7

7.1 Introduction

Thanks to recent advances in function approximation methods using deep neural networks [212], Q-learning [356] has been widely used to solve reinforcement learning (RL) problems in a variety of applications, e.g., robotic control [149, 235], path planning [174, 239] and production scheduling [229, 352]. Despite the great success, it is well recognized that Q-learning may suffer from the notorious overestimation bias [109, 328, 333, 334, 398], which would significantly impede the learning efficiency. Recent work [108, 127] indicates that this problem also persists in the actor-critic setting. To address this issue, the ensemble method [7, 85, 181, 259] has emerged as a promising solution in which multiple Q-function approximators are used to get better estimation of the action values. Needless to say, the ensemble size, i.e., the number of Q-function approximators used in the target, has intrinsic impact on Q-learning. Notably, it is shown in Chen et al. [58] and Lan et al. [186] that while a large ensemble size could completely remove the overestimation bias, it may go to the other extreme and result in underestimation bias and unstable training, which is clearly not desirable. Therefore, instead of simply increasing the ensemble size to mitigate the overestimation issue, a fundamental question to ask is:"*Is it possible to determine the right ensemble size on the fly so as to minimize the estimation bias*?"

Some existing ensemble methods [17, 186, 194] adopt a trial-and-error strategy to search for the ensemble size, which would be time-consuming and require a lot of human engineering for different RL tasks. The approximation error of the Q-function during the learning process plays a nontrivial role in the selection of the ensemble size, since it directly impacts the Q-target estimation accuracy. This however remains not well understood. In particular, the fact that the approximation error is time-varying, due to the iterative nature of Q-learning [32, 393], gives rise to the question that whether a *fixed* ensemble size should be used in the learning process. To answer this question, we show in Sect. 7.2.2 that using a fixed ensemble size is likely to lead to either overestimation or underestimation bias, and the bias may shift between overestimation and underestimation because of the time-varying approximation error, calling for an adaptive ensemble size so as to drive the bias close to zero based on the underlying learning dynamics (Fig. 7.1).

Thus motivated, in this chapter we study effective ensemble size adaptation to minimize the estimation bias that hinges heavily on the time-varying approximation errors during the

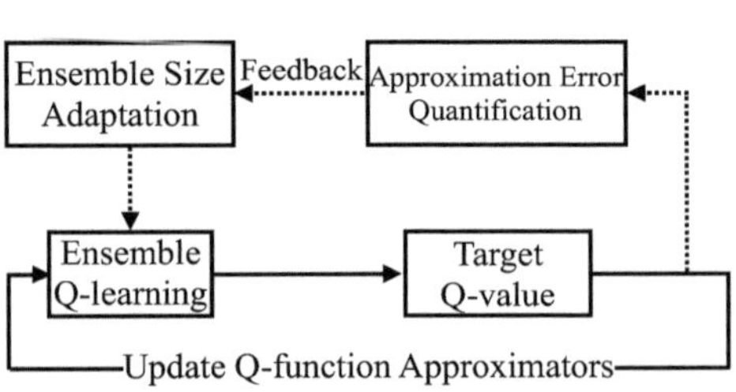

Fig. 7.1 A sketch of the adaptive ensemble Q-learning (AdaEQ)

learning process. To this end, we first characterize the relationship among the ensemble size, the function approximation errors, and the estimation bias, by deriving an upper bound and a lower bound on the estimation bias. Our findings reveal that the ensemble size should be selected adaptively in a way to cope with the impact of the time-varying approximation errors. Building upon the theoretic results, we cast the estimation bias minimization as an adaptive control problem where the approximation error during the learning process is treated as the control object, and the ensemble size is adapted based on the feedback of the control output, i.e., the value of the approximation error from the last iteration. The key idea in this approach is inspired from the classic Model Identification Adaptive Control (MIAC) framework [23, 246], where at each step the current system identification of the control object is fed back to adjust the controller, and consequently a new control signal is devised following the updated control law.

One main contribution of this chapter lies in the development of AdaEQ, a generalized ensemble method for the ensemble size adaptation, aiming to minimize the estimation bias during the learning process. Specifically, the approximation error in each iteration is quantified by comparing the difference between the Q-estimates and the Monte Carlo return using the current learned policy over a testing trajectory [186, 328]. Inspired by MIAC, the approximation error serves as the feedback to adapt the ensemble size. Besides, we introduce a 'tolerance' parameter in the adaptation mechanism to balance the control tendency towards positive or negative bias during the learning process. In this way, AdaEQ can encompass other existing ensemble methods as special cases, including Maxmin [186], by properly setting this hyperparameter. A salient feature of the feedback-adaptation mechanism is that it can be used effectively in conjunction with both standard Q-learning [234] and actor-critic methods [127, 321]. Experimental results on the continuous-control MuJoCo benchmark [329] show that AdaEQ is robust to the initial ensemble size in different environments, and achieves higher average return, thanks to keeping the estimation bias close to zero, when compared to the state-of-the-art ensemble methods such as REDQ [58] and Average-DQN [17].

Related Work. Bias-corrected Q-learning [192] introduces the bias correction term to reduce the overestimation bias. Double Q-learning is proposed in Hasselt [135] and Van Hasselt et al. [333] to address the overestimation issue in vanilla Q-learning, by leveraging two independent Q-function approximators to estimate the maximum Q-function value in the target. S-DQN and S-DDQN use the softmax operator instead of the max operator to further reduce the overestimation bias [316]. Self-correcting Q-learning aims to balance the underestimation in double Q-learning and overestimation in classic Q learning by introducing a new self-correcting estimator [407]. Weighted Q-learning proposes a new estimator based on the weighted average of the sample means, and conducts the empirical analysis in the discrete action space [81]. Weighted Double Q-learning [398] uses the Q-approximator together with the double Q-approximator to balance the overestimation and underestimation bias. Nevertheless, acquiring independent approximators is often intractable for large-scale tasks. To resolve this issue, the Twin-Delayed Deep Deterministic policy gradient algorithm

(TD3) [108] and Soft Actor-Critic (SAC) [127] have been devised to take the minimum over two approximators in the target network. Along a different avenue, the ensemble-based methods generalize double Q-learning to correct the overestimation bias by increasing the number of Q-function approximators. Particularly, Average-DQN [17] takes the average of multiple approximators in the target to reduce the overestimation error, and Random Ensemble Mixture (REM) [7] estimates the target value using the random convex combination of the approximators. It is worth noting that both Average-DQN and REM cannot completely eliminate the overestimation bias. Most recently, Maxmin Q-learning [186] defines a proxy Q-function by choosing the minimum Q-value for each action among all approximators. Similar to Maxmin, Random Ensembled Q-learning (REDQ) [58] formulates the proxy Q-function by choosing only a subset of the ensemble. Nevertheless, both Maxmin and REDQ use a fixed ensemble size. In this study, we introduce an adaptation mechanism for the ensemble size to drive the estimation bias to be close to zero, thereby mitigating the possible overestimation and underestimation issues.

7.2 Impact of Ensemble Size on Estimation Bias

7.2.1 Ensemble Q-Learning

As is standard, we consider a Markov decision process (MDP) defined by the tuple $\langle \mathcal{S}, \mathcal{A}, P, r, \gamma \rangle$, where $\mathcal{S}$ and $\mathcal{A}$ denote the state space and the action space, respectively. $P(s'|s,a) : \mathcal{S} \times \mathcal{A} \times \mathcal{S} \rightarrow [0,1]$ denotes the probability transition function from current state s to the next state s' by taking action $a \in \mathcal{A}$, and $r(s,a) : \mathcal{S} \times \mathcal{A} \rightarrow \mathbb{R}$ is the corresponding reward. $\gamma \in (0,1]$ is the discount factor. At each step t, the agent observes the state s_t, takes an action a_t following a policy $\pi : \mathcal{S} \rightarrow \mathcal{A}$, receives the reward r_t, and evolves to a new state s_{t+1}. The objective is to find an optimal policy π^* to maximize the discounted return $R = \sum_{t=0}^{\infty} \gamma^t r_t$.

By definition, Q-function is the expected return when choosing action a in state s and following with the policy π: $Q^\pi = \mathbf{E}[\sum_{t=0}^{\infty} \gamma^t r_t(s_t, a_t)|s_0 = s, a_0 = a]$. Q-learning is an off-policy value-based method that aims at learning the optimal Q-function $Q^* : \mathcal{S} \times \mathcal{A} \rightarrow \mathbb{R}$, where the optimal Q-function is a fixed point of the Bellman optimality equation [29]:

$$\mathcal{T} Q^*(s,a) = r(s,a) + \gamma \mathbf{E}_{s' \sim P(s'|s,a)} \left[\max_{a' \in \mathcal{A}} Q^*(s', a') \right]. \tag{7.1}$$

Given a transition sample (s, a, r, s'), the Bellman operator can be employed to update the Q-function as follows:

$$Q(s,a) \leftarrow (1-\alpha) Q(s,a) + \alpha y, \quad y := r + \gamma \max_{a' \in \mathcal{A}} Q(s', a'). \tag{7.2}$$

where α is the step size and y is the target. Under some conditions, Q-learning can converge to the optimal fixed-point solution asymptotically [330]. In deep Q-learning, the Q-function is approximated by a neural network, and it has been shown [333] that the approximation error, amplified by the max operator in the target, results in the overestimation phenomena. One promising approach to address this issue is the ensemble Q-learning method, which is the main subject of this study.

The Ensemble Method. Specifically, the ensemble method maintains N separate approximators $Q^1, Q^2, \ldots, Q^N$ of the Q-function, based on which a subset of these approximators is used to devise a proxy Q-function. For example, in Average-DQN [17], the proxy Q-function is obtained by computing the average value over all N approximators to reduce the overestimation bias:

$$Q^{\text{ave}}(\cdot) = \tfrac{1}{N}\textstyle\sum_{i=1}^{N} Q^i(\cdot).$$

However, the average operation cannot completely eliminate the overestimation bias, since the average of the overestimation bias is still positive. To tackle this challenge, Maxmin [186] and REDQ [58] take the 'min' operation over a subset $\mathcal{M}$ (size M) of the ensemble:

$$Q^{\text{proxy}}(\cdot) = \min_{i \in \mathcal{M}} Q^i(\cdot). \tag{7.3}$$

The target value in the ensemble-based Q-learning is then computed as $y = r + \max_{a' \in \mathcal{A}} Q^{\text{proxy}}$. It is worth noting that in the existing studies, the in-target ensemble size M, pre-determined for a given environment, remain fixed in the learning process.

7.2.2 An Illustrative Example

It is known that the determination of the optimal ensemble size is highly nontrivial, and a poor choice of the ensemble size would degrade the performance of ensemble Q-learning significantly [186]. As mentioned earlier, it is unclear a priori if a fixed ensemble size should be used in the learning process. In what follows, we use an example to illustrate the potential pitfalls in the ensemble methods by examining the sensitivity of the estimation bias to the ensemble size [58, 186].

Along the same line as in Van Hasselt et al. [333], we consider an example with a real-valued continuous state space. In this example, there are two discrete actions available at each state and the optimal action values depend only on the state, i.e., in each state both actions result in the same optimal value $Q^*(s, \cdot)$, which is assumed to be $Q^*(s, \cdot) = \sin(s)$. Figure 7.2 demonstrates how the ensemble method is carried out in four stages:

(I) For each Q-function approximator Q^i, $i = 1, 2, \ldots, 5$, we first generate 10 noisy action-value samples independently (green dots in Fig. 7.2a). Let $e^i(s, a)$ denote the *approximation error* of Q^i:

$$Q^i(s, a) = Q^*(s, a) + e^i(s, a), \quad \text{with } e^i(s, a) \sim \mathcal{U}(-\tau_i, \tau_i), \tag{7.4}$$

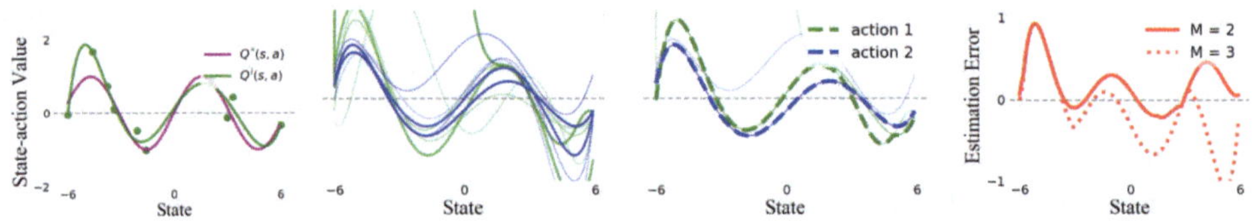

(a) Function approximation. (b) Five function approximators. (c) Ensemble via 'min' operator. (d) Estimation error.

Fig. 7.2 Illustration of estimation bias in the ensemble method. **a** Each approximator is fitted to the noisy values (green dots) at the sampled states independently. **b** Five Q-function approximators are obtained for both actions (green lines and blue lines). **c** Apply the min operator over M ($M = 3$) randomly selected approximators to obtain a proxy approximator for each action. **d** The estimation error is obtained by comparing the underlying true value (purple line in (**a**)) and the target value using the proxy approximator

where $\tau_i \sim \mathcal{U}(0, \tau)$ models the approximation error distribution for the i-th approximator. Note that the assumption on the uniform error distribution is commonly used to indicate that both positive and negative approximation error are possible in Q-function approximators [58, 186, 328].

(II) Next, Fig. 7.2b illustrates the ensemble ($N = 5$) of approximators for two actions, where each approximator is a 6° polynomial that fits the noisy values at sampled states.

(III) Following the same ensemble approach in Chen et al. [58] and Lan et al. [186], we randomly choose M approximators from the ensemble and take the minimum over them to obtain a proxy approximator for each action, resulting in the dashed lines in Fig. 7.2c.

(IV) Finally, the maximum action value of the proxy approximator is used as the target to update the current approximators. To evaluate the target value *estimation error*, Fig. 7.2d depicts the difference between the obtained target value and the underlying true value when using different ensemble size M. As in Van Hasselt et al. [333], we utilize the average estimation error (i.e., estimation bias) to quantify the performance of current approximators. For example, when the ensemble size $M = 2$, the red line is above zero for most states, implying the overestimation tendency in the target. Clearly, Fig. 7.2d indicates that the estimation bias is highly dependent on the ensemble size, and even a change of M can lead the shift from overestimation to underestimation. Since the Q-function approximation error of each approximator changes over time in the training process [32], we next analyze the impact of the ensemble size on the estimation bias under different approximation error distributions. As shown in Fig. 7.3a, with a fixed ensemble size M, the estimation bias may shift between positive and negative and be 'dramatically' large for some error distributions. In light of this observation, departing from using a fixed size, we advocate to adapt the in-target ensemble size, e.g., set $M = 4$ when the noise parameter $\tau > 1.5$ and $M = 3$ otherwise. The estimation bias resulted by this adaptation mechanism is much closer to zero. Besides, Fig. 7.3b characterizes the estimation bias under different action spaces, which is also important considering that different tasks normally have different action spaces and the

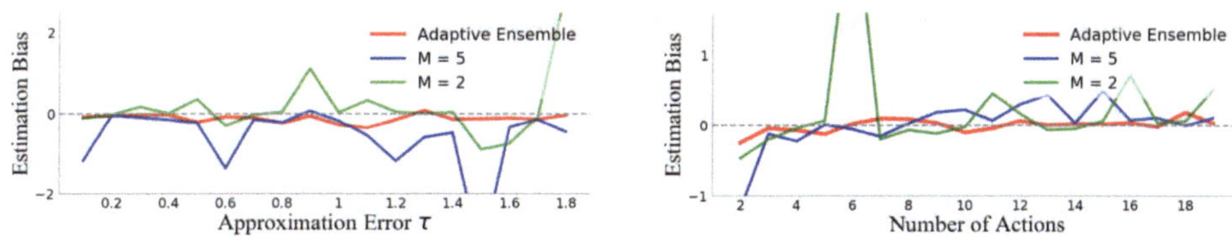

(a) Estimation bias vs. τ.

(b) Estimation bias vs. numbers of actions.

Fig. 7.3 Illustration of overestimation and underestimation phenomena for different ensemble sizes

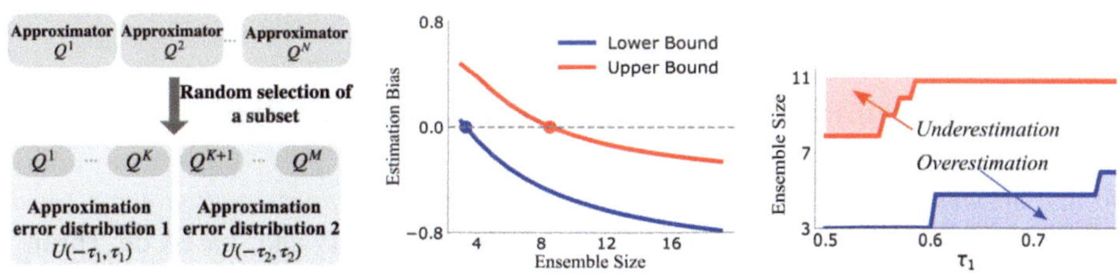

(a) Q-function approximation error distributions.

(b) Lower bound and upper bound on estimation bias.

(c) Impact of approximation error on the estimation bias: overestimation vs. underestimation.

Fig. 7.4 Illustration of upper bounds and lower bounds on estimation bias in Theorem 7.1. **a** The case where the approximation errors of the Q-approximators can be categorized into two uniform distributions. **b** The lower bound and the upper bound corresponding to Eqs. (7.5) and (7.6), for given τ_1, τ_2, A. The blue point represents the 'critical' point where decreasing the ensemble size may lead to overestimation (the lower bound is positive), and the red point denotes the 'critical' point where increasing the ensemble size may lead to underestimation (the upper bound is negative). **c** Due to the time-varying feature of the approximation errors, the blue curve and the red curve depict the 'critical' points for the lower bound and the upper bound, respectively

number of available actions may vary in different states even for the same task. The adaptive ensemble approach is clearly more robust in our setting. In a nutshell, both Fig. 7.3a and b suggest that a fixed ensemble size would not work well to minimize the estimation bias during learning for different tasks. This phenomenon has also been observed in the empirical results [186]. In stark contrast, adaptively changing the ensemble size based on the approximation error indeed can help to reduce the estimation bias in different settings.

Parameter Setting in Numerical Illustration of Fig. 7.4. In Fig. 7.4a, the approximation error parameter $\tau_1 = 0.5$, $\tau_2 = 0.4$ and the number of actions is set as $A = 30$. The number of approximators for which the approximation errors follow the first distribution is $K = 2$. In Fig. 7.4c, we fix $\tau_2 = 0.4$, $A = 30$, $K = 2$ and τ_1 ranges from 0.5 to 0.8. The changes of the 'critical' points (red dot and blue dot) are depicted in Fig. 7.4c.

7.3 Adaptive Ensemble Q-Learning (AdaEQ)

Motivated by the illustrative example above, we next devise a generalized ensemble method with ensemble size adaptation to drive the estimation bias to be close to zero, by taking into consideration the time-varying feature of the approximation error during the learning process. Formally, we consider an ensemble of N Q-function approximators, i.e., $\{Q_i\}_{i=1}^N$, with each approximator initialized independently and randomly. We use the minimum of a subset $\mathcal{M}$ of the N approximators in the Q-learning target as in Eq. (7.3), where the size of subset $|\mathcal{M}| = M \leq N$.

7.3.1 Lower Bound and Upper Bound on Estimation Bias

We first answer the following key question: "*How does the approximation error, together with the ensemble size, impact the estimation bias*?". To this end, based on Thrun and Schwartz [328], we characterize the intrinsic relationship among the ensemble size M, the Q-function approximation error and the estimation bias, and derive an upper bound and a lower bound on the bias in the tabular case. Without loss of generality, we assume that for each state s, there are A available actions.

Let $e^i(s, a) \triangleq Q^i(s, a) - Q^\pi(s, a)$ be the approximation error for the i-th Q-function approximator, where $Q^\pi(s, a)$ is the ground-truth of the Q-value for the current policy π. By using Eq. (7.3) to compute the target Q-value, we define the estimation error in the Bellman equation for transition (s, a, r, s') as Z_M:

$$Z_M \triangleq r + \gamma \max_{a' \in \mathcal{A}} \min_{i \in \mathcal{M}} Q^i(s', a') - \left(r + \max_{a' \in \mathcal{A}} Q^\pi(s', a')\right).$$

Here a positive $\mathbf{E}[Z_M]$ implies overestimation bias while a negative $\mathbf{E}[Z_M]$ implies underestimation bias. Note that we use the subscription M to emphasize that the estimation bias is intimately related to M.

The case with two distributions for Q-function approximation errors. For ease of exposition, we first consider the case when the approximation errors follow one of the two uniform distributions, as illustrated in Fig. 7.4a. Specifically, assume that for $i \in \mathcal{K} \subset \mathcal{M}$ with $|\mathcal{K}| = K$, $e^i(s, a) \sim \mathcal{U}(-\tau_1, \tau_1)$, and for $i \in \mathcal{M} \setminus \mathcal{K}$, $e^i(s, a) \sim \mathcal{U}(-\tau_2, \tau_2)$. Without loss of generality, we assume that $\tau_1 > \tau_2 > 0$. It is worth noting that in Thrun and Schwartz [328], Lan et al. [186] and Chen et al. [58], the approximation error for all approximators is assumed to follow the same uniform distribution, i.e., $\tau_1 = \tau_2$, which is clearly more restrictive than the case here with two error distributions. For instance, when only one approximator is chosen to be updated at each step [186], the approximation error distribution of this approximator would change over time and hence differ from the others. We have the following results on the upper bound and lower bound of the estimation bias $\mathbf{E}[Z_M]$.

Theorem 7.1 *For the case with two distributions for Q-function approximation errors, the estimation bias* $\mathbf{E}[Z_M]$ *satisfies that*

$$\mathbf{E}[Z_M] \geq \gamma\left(\tau_1(1 - f_{AK} - 2f_{AM}) + \tau_2(1 - f_{AM})\right); \tag{7.5}$$

$$\mathbf{E}[Z_M] \leq \gamma\left(\tau_1 + \tau_2(1 - 2f_{A(M-K)} - (1-\beta_K)^A)\right), \tag{7.6}$$

where $\beta_K = (\frac{1}{2} - \frac{\tau_2}{2\tau_1})^K$, $f_{AK} = \frac{1}{K}B(\frac{1}{K}, A+1) = \frac{\Gamma(A+1)\Gamma(1+\frac{1}{K})}{\Gamma(A+\frac{1}{K}+1)} = \frac{A(A-1)\cdots 1}{(A+\frac{1}{K})(A+\frac{1}{K}-1)\cdots(1+\frac{1}{K})}$ *with* $B(\cdot,\cdot)$ *being the Beta function.*

Proof We first restate the following results from order statistics [186].

Lemma 7.1 *Let* $X_1, X_2, \ldots, X_M$ *be M i.i.d random variables from an absolutely continuous distribution with probability density function (PDF)* $f(x)$ *and cumulative distribution function (CDF)* $F(x)$. *Denote* $\mu = \mathbf{E}[X_i]$ *and* $\sigma^2 = Var[X_i] < +\infty$. *Let* $X_{1:M} \leq X_{2:M} \leq \cdots \leq X_{M:M}$ *be the order statistics obtained by reordering these random variables in increasing order of magnitude. Denote the PDF and CDF of* $X_{1:M} = \min_{i\in\mathcal{M}} X_i$ *as* $f_{1:M}(x)$ *and* $F_{1:M}(x)$. *Denote* $f_{M:M}(x)$ *and* $F_{M:M}(x)$ *to be the PDF and CDF of* $X_{M:M} = \max_{i\in\mathcal{M}} X_i$. *Then we have*

(i) $f_{1:M}(x) = Mf(x)(1-F(x))^{M-1}$, $F_{1:M}(x) = 1-(1-F(x))^M$.
(ii) $f_{M:M}(x) = Mf(x)(F(x))^{M-1}$, $F_{M:M}(x) = (F(x))^M$.

Now we prove Theorem 7.1.

Theorem 7.2 *For the case with two distributions for Q-function approximation errors, the estimation bias* $\mathbf{E}[Z_M]$ *satisfies that*

$$\mathbf{E}[Z_M] \geq \gamma\left(\tau_1(1 - f_{AK} - 2f_{AM}) + \tau_2(1 - f_{AM})\right); \tag{7.7}$$

$$\mathbf{E}[Z_M] \leq \gamma\left(\tau_1 + \tau_2(1 - 2f_{A(M-K)} - (1-\beta_K)^A)\right), \tag{7.8}$$

where $\beta_K = (\frac{1}{2} - \frac{\tau_2}{2\tau_1})^K$, $f_{AK} = \frac{1}{K}B(\frac{1}{K}, A+1) = \frac{\Gamma(A+1)\Gamma(1+\frac{1}{K})}{\Gamma(A+\frac{1}{K}+1)} = \frac{A(A-1)\cdots 1}{(A+\frac{1}{K})(A+\frac{1}{K}-1)\cdots(1+\frac{1}{K})}$ *with* $B(\cdot,\cdot)$ *being the Beta function.*

Proof Assume that for $i \in \mathcal{K} \subset \mathcal{M}$, $e^i(s,a) \sim \mathcal{U}(-\tau_1, \tau_1)$ with PDF $f_1(x)$ and CDF $F_2(x)$, and for $i \in \mathcal{M} \setminus \mathcal{K}$, $e^i(s,a) \sim \mathcal{U}(-\tau_2, \tau_2)$ with PDF $f_2(x)$ and CDF $F_2(x)$. Without loss of generality, we assume that $\tau_1 > \tau_2 > 0$.

From Lemma 7.1, it is clear that

$$
\begin{aligned}
F_{1:M}(x) =&\mathbf{P}(X_{1:M} \leq x) = \begin{cases} 1-(1-F_1(x))^K & x < -\tau_2 \\ 1-(1-F_1(x))^K(1-F_2(x))^{M-K} & x \in [-\tau_2, \tau_2] \\ 1 & x > \tau_2 \end{cases}, \\
f_{1:M}(x) =&\frac{dF_{1:M}(x)}{dx},
\end{aligned}
\tag{7.9}
$$

where $F_1(x) = \frac{1}{2} + \frac{x}{2\tau_1}$, $f_1(x) = \frac{1}{2\tau_1}$, $x \in [-\tau_1, \tau_1]$ and $F_2(x) = \frac{1}{2} + \frac{x}{2\tau_2}$, $f_2(x) = \frac{1}{2\tau_2}$, $x \in [-\tau_2, \tau_2]$. Denote $F_{1:K}(x) = 1-(1-F_1(x))^K$ and $f_{1:K}(x) = Kf_1(x)(1-F_1(x))^{K-1}$. Then, the estimation bias can be obtained as

$$
\begin{aligned}
\mathbf{E}[Z_M] = &\gamma\ \mathbf{E}[\max_{a'} \min_{i \in \mathcal{M}} e^i(s,a)] \\
=&\gamma \int_{-\tau_1}^{\tau_1} Axf_{1:M}(x)F_{1:M}(x)^{A-1}dx \\
=&\gamma(\underbrace{\int_{-\tau_1}^{-\tau_2}}_{①} + \underbrace{\int_{-\tau_2}^{\tau_2}}_{②} + \underbrace{\int_{\tau_2}^{\tau_1}}_{③})dx.
\end{aligned}
\tag{7.10}
$$

We first consider term ① in Eq. (7.10), where

$$
\begin{aligned}
① =&\int_{-\tau_1}^{-\tau_2} Axf_{1:K}(x)F_{1:K}(x)^{A-1}dx \\
=&(\int_{-\tau_1}^{\tau_1} - \int_{-\tau_2}^{\tau_2} - \int_{\tau_2}^{\tau_1})Axf_{1:K}(x)F_{1:K}(x)^{A-1}dx.
\end{aligned}
$$

It follows that ① can be bounded above as follows:

$$
\begin{aligned}
① \leq&(\int_{-\tau_1}^{\tau_1} - \int_{-\tau_2}^{\tau_2})Axf_{1:K}(x)F_{1:K}(x)^{A-1}dx \\
=&\int_{-\tau_1}^{\tau_1} AxKf_1(x)(1-F_1(x))^{K-1}\left(1-(1-F_1(x))^K\right)^{A-1}dx \\
&-\int_{-\tau_2}^{\tau_2} AxKf_1(x)(1-F_1(x))^{K-1}\left(1-(1-F_1(x))^K\right)^{A-1}dx \\
=&\tau_1[1-2f_{AK}] - \tau_2\left(1-(\frac{1}{2}-\frac{\tau_2}{2\tau_1})^K\right)^A - \tau_2\left(1-(\frac{1}{2}+\frac{\tau_2}{2\tau_1})^K\right)^A + 2\tau_1\underbrace{\int_{\frac{1}{2}-\frac{\tau_2}{2\tau_1}}^{\frac{1}{2}+\frac{\tau_2}{2\tau_1}}(1-y^K)^A dy}_{\leq f_{AK}} \\
\leq&\tau_1 - \tau_2\left(1-(\frac{1}{2}-\frac{\tau_2}{2\tau_1})^K\right)^A - \tau_2\left(1-(\frac{1}{2}+\frac{\tau_2}{2\tau_1})^K\right)^A,
\end{aligned}
$$

where $f_{AK} = \frac{1}{K}B(\frac{1}{K}, A+1) = \frac{\Gamma(A+1)\Gamma(1+\frac{1}{K})}{\Gamma(A+\frac{1}{K}+1)} = \frac{A(A-1)\cdots 1}{(A+\frac{1}{K})(A+\frac{1}{K}-1)\cdots(1+\frac{1}{K})}$ with $B(\cdot,\cdot)$ being beta function and $y := \frac{1}{2} - \frac{x}{2\tau_1}$.

We can also have the following lower bound on ①:

$$
\begin{aligned}
① =&\tau_1(1-2f_{AK}) - \tau_2\left(1-(\frac{1}{2}-\frac{\tau_2}{2\tau_1})^K\right)^A - \tau_2\left(1-(\frac{1}{2}+\frac{\tau_2}{2\tau_1})^K\right)^A + 2\tau_1\int_{\frac{1}{2}-\frac{\tau_2}{2\tau_1}}^{\frac{1}{2}+\frac{\tau_2}{2\tau_1}}(1-y^K)^A dy \\
&+\tau_2\left(1-(\frac{1}{2}-\frac{\tau_2}{2\tau_1})^K\right)^A + 2\tau_1\int_0^{\frac{1}{2}-\frac{\tau_2}{2\tau_1}}(1-y^K)^A dy \\
=&\tau_1(1-2f_{AK}) - \tau_2\left(1-(\frac{1}{2}+\frac{\tau_2}{2\tau_1})^K\right)^A + 2\tau_1\int_0^{\frac{1}{2}+\frac{\tau_2}{2\tau_1}}(1-y^K)^A dy \\
\geq&\tau_1(1-2f_{AK}) - \tau_2\left(1-(\frac{1}{2}+\frac{\tau_2}{2\tau_1})^K\right)^A + \tau_1 f_{AK} \\
=&\tau_1(1-f_{AK}) - \tau_2\left(1-(\frac{1}{2}+\frac{\tau_2}{2\tau_1})^K\right)^A .
\end{aligned}
$$

For the term ② in Eq. (7.10), it follows that

$$
\begin{aligned}
② =& \int_{-\tau_2}^{\tau_2} Axf_{1:M}(x)F_{1:M}(x)^{A-1}dx \\
=& \int_{-\tau_2}^{\tau_2} xd\left[1-(\frac{1}{2}-\frac{x}{2\tau_1})^K(\frac{1}{2}-\frac{x}{2\tau_2})^{M-K}\right]^A dx \\
=&\tau_2+\tau_2\left(1-(\frac{1}{2}+\frac{\tau_2}{2\tau_1})^K\right)^A - \underbrace{\int_{-\tau_2}^{\tau_2}\left[1-(\frac{1}{2}-\frac{x}{2\tau_1})^K(\frac{1}{2}-\frac{x}{2\tau_2})^{M-K}\right]^A dx}_{(*)},
\end{aligned}
$$

where $(*)$ satisfies that

$$
\begin{aligned}
(*) >& \int_{-\tau_2}^{\tau_2}\left[1-(\frac{1}{2}-\frac{x}{2\tau_2})^{M-K}\right]^A dx = 2\tau_2 f_{A(M-K)}, \\
(*) =& \int_{-\tau_2}^{0} + \int_0^{\tau_2} \\
<& \int_{-\tau_2}^{0}[1-(\frac{1}{2}-\frac{x}{2\tau_2})^M]^A dx + \int_0^{\tau_2}[1-(\frac{1}{2}-\frac{x}{2\tau_1})^M]^A dx \\
<&\tau_2 f_{AM} + 2\tau_1 f_{AM}.
\end{aligned}
$$

From the definition of $F_{1:M}(x)$, we have that term ③ in Eq. (7.10) equals to zero.

Summarizing, we obtain the following upper bound and lower bound on the estimation bias $\mathbf{E}[Z_M]$, for the case with two distributions for Q-function approximation errors:

$$\mathbf{E}[Z_M] \geq \gamma \left(\tau_1(1 - f_{AK} - 2f_{AM}) + \tau_2(1 - 2f_{AM})\right)$$
$$\mathbf{E}[Z_M] \leq \gamma \left(\tau_1 + \tau_2[1 - 2f_{A(M-K)}] - \tau_2(1 - \beta_K)^A\right)$$

where $\beta_K = (\frac{1}{2} - \frac{\tau_2}{2\tau_1})^K$. □

Theorem 7.1 reveals that the estimation bias depends on the ensemble size as well as the approximation error distributions. To get a more concrete sense of Theorem 7.1, we consider an example where $\tau_1 = 0.5$ and $\tau_2 = 0.4$, as depicted in Fig. 7.4b, and characterize the relationship between the estimation bias and the ensemble size M. Notably, the estimation bias turns negative when the ensemble size $M > M_u = 9$ (red point: the value of M where the upper bound is 0) and becomes positive when $M < M_l = 4$ (blue point: the value of M where the lower bound is 0). In Fig. 7.4c, we fix $\tau_2 = 0.4$ and show how those two critical points (M_u and M_l) change along with τ_1. Here the red shaded area indicates underestimation bias when $M > M_u$, and the blue shaded area indicates overestimation bias when $M < M_l$. Clearly, in order to avoid the positive bias (blue shaded area), it is desirable to increase the ensemble size when the approximation error is large, e.g., $\tau_1 > 0.6$. On the other hand, decreasing the ensemble size is more preferred to avoid underestimation (red shaded area) when the approximation error is small, e.g., $\tau_1 < 0.6$.

The general case with heterogeneous distributions for Q-function approximation errors. Next, we consider a general case, in which the approximation errors for different approximators $\{Q^i\}$ are independently but non-identically distributed. Specifically, we assume that the approximation error $e^i(s, a)$ for $Q^i(s, a), i = 1, 2, \ldots, M$, follows the uniform distribution $\mathcal{U}(-\tau_i, \tau_i)$, where $\tau_i > 0$. We use a multitude of tools to devise the upper bound and lower bound on the estimation bias $\mathbf{E}[Z_M]$. As expected, this general case is technically more challenging and the bounds would be not as sharp as in the special case with two distributions.

Theorem 7.3 *For the general case with heterogeneous error distributions, the estimation bias* $\mathbf{E}[Z_M]$ *satisfies that*

$$\mathbf{E}[Z_M] \geq \gamma \left(\tau_{\min} - \tau_{\max}(f_{A(M-1)} + 2f_{AM})\right); \quad (7.11)$$
$$\mathbf{E}[Z_M] \leq \gamma \left(2\tau_{\min} - \tau_{\max}\left(f_{AM} - 2g_{AM}\right)\right), \quad (7.12)$$

where $\tau_{\min} = \min_i \tau_i$ *and* $\tau_{\max} = \max_i \tau_i$. $g_{AM} = \frac{1}{M} I_{0.5}(\frac{1}{M}, A + 1)$ *with* $I_{0.5}(\cdot, \cdot)$ *being the regularized incomplete Beta function.*

Numerical Illustration of Theorem 2.2 For a better understanding of Theorem 7.3, we next provide an example in Fig. 7.9 following the same line as in Sect. 7.3.1. Figure 7.9b shows the case when $\tau_{\max}$ is small, i.e., $\tau_{\max} = 0.1$. Consistent with our analysis in Sect. 7.3.1, increasing the ensemble size $M > 10$ (red point) will lead to underestimation bias (upper bound is negative). It can be seen clearly in Fig. 7.9c that, when $\tau_{\min}$ is small, the underestimation is the major issue when increasing the ensemble. On the other hand, Fig. 7.9d demonstrates the case when $\tau_{\min}$ is large, i.e., $\tau_{\min} = \tau_{\max} = 1$. In this case, decreasing the ensemble size is 'likely' to cause overestimation. Similarly, we can observe from Fig. 7.9e that when increasing $\tau_{\min}$, the ensemble size M should be increased to avoid the overestimation (the blue shaded area). In this example, we set the number of actions as $A = 75$.

Proof Assume that the approximation error $e^i(s, a)$ for each Q approximator follows uniform distribution $\mathcal{U}(-\tau_i, \tau_i)$ with PDF $f_i(x)$ and CDF $F_i(x)$. The PDF and CDF for each approximation error distribution is as follows,

$$f_i(x) = \begin{cases} \frac{1}{2\tau_i}, & x \in [-\tau_i, \tau_i] \\ 0, & \text{otherwise} \end{cases}, \quad F_i(x) = \begin{cases} 0, & x < -\tau_i \\ \frac{x+\tau_i}{2\tau_i}, & x \in [-\tau_i, \tau_i] \\ 1, & x > \tau_i \end{cases}$$

From Lemma 7.1, it can be seen that

$$\begin{aligned} F_{1:M}(x) =& \mathbf{P}(X_{1:M} \leq x) = 1 - \prod_{i=1}^{M}(1 - F_i(x)), \\ f_{1:M}(x) =& \sum_{i=1}^{M} \left(f_i(x) \prod_{j \neq i} \left(1 - F_j(x)\right) \right). \end{aligned}$$

Assume that at state s', there are A actions applicable. Then the estimation bias Z_M is

$$\begin{aligned} \mathbf{E}[Z_M] =& \gamma \mathbf{E}_{\tau_1,\ldots,\tau_M}[\max_{a'} \min_{i \in \mathcal{M}} e^i(s, a)] \\ =& \gamma \left[\int_{-\tau_{\max}}^{\tau_{\max}} A x f_{1:M}(s) F_{1:M}(x)^{A-1} dx \right]. \end{aligned}$$

Considering the integration terms, we conclude that

$$
\begin{aligned}
&\int_{-\tau_{\max}}^{\tau_{\max}} Axf_{1:M}(s)F_{1:M}(x)^{A-1}dx \\
=&\int_{-\tau_{\max}}^{\tau_{\max}} xd(F_{1:M}(x)^{A}) \\
=&x(F_{1:M}(x)^{A})\Big|_{-\tau_{\max}}^{\tau_{\max}} - \int_{-\tau_{\max}}^{\tau_{\max}} F_{1:M}(x)^{A}dx \\
=&\tau_{\max} - \int_{-\tau_{\max}}^{\tau_{\max}} \left(1-\prod_{i=1}^{M}(1-F_i(x))\right)^{A} dx \\
=&\tau_{\max} - \underbrace{\int_{-\tau_{\max}}^{\tau_{\max}} (1-(1-F_1(x))(1-F_2(x))\cdots(1-F_M(x)))^{A}\, dx,}_{①}
\end{aligned}
$$

where $\tau_{\min} = \min_i \tau_i$ and $\tau_{\max} = \max_i \tau_i$. Denote $f_{AM} = \frac{1}{M}B(\frac{1}{M}, A+1) = \frac{\Gamma(A+1)\Gamma(1+\frac{1}{M})}{\Gamma(A+\frac{1}{M}+1)}$ $= \frac{A(A-1)\cdots 1}{(A+\frac{1}{M})(A+\frac{1}{M}-1)\cdots(1+\frac{1}{M})}$ with $B(\cdot,\cdot)$ being beta function.

It first can be seen that ① satisfies that

$$
\begin{aligned}
① \leq &\int_{-\tau_{\min}}^{0} \left(1-(\frac{1}{2}-\frac{1}{2\tau_{\max}}x)^{M}\right)^{A} dx + \int_{0}^{\tau_{\min}} \left(1-(\frac{1}{2}-\frac{1}{2\tau_{\min}}x)^{M}\right)^{A} dx \\
&+\int_{\tau_{\min}}^{\tau_{\max}} dx + \int_{-\tau_{\max}}^{-\tau_{\min}} (1-(\frac{1}{2}-\frac{x}{2\tau_{\max}})^{M-1})^{A}dx \\
=&2\tau_{\max}\int_{\frac{1}{2}}^{\frac{1}{2}+\frac{\tau_{\min}}{2\tau_{\max}}} (1-y^{M})^{A}dy + 2\tau_{\min}\int_{0}^{\frac{1}{2}} (1-y^{M})^{A}dy \\
&+\tau_{\max}-\tau_{\min}+2\tau_{\max}\int_{\frac{1}{2}+\frac{\tau_{\min}}{2\tau_{\max}}}^{1} (1-y^{M-1})^{A}dy \quad (y := \frac{1}{2}-\frac{x}{2\tau_{\max}}) \\
\leq&2\tau_{\max}\int_{0}^{1} (1-y^{M})^{A}dy + \tau_{\max}-\tau_{\min}+\tau_{\max} f_{A(M-1)} \\
\leq&2\tau_{\max} f_{AM} + \tau_{\max}-\tau_{\min}+\tau_{\max} f_{A(M-1)}.
\end{aligned}
$$

And the lower bound on ① can be obtained as

$$
\begin{aligned}
① \geq & \int_{-\tau_{\min}}^{0} \left(1-(\frac{1}{2}-\frac{1}{2\tau_{\min}}x)^M\right)^A dx + \int_{0}^{\tau_{\min}} \left(1-(\frac{1}{2}-\frac{1}{2\tau_{\max}}x)^M\right)^A dx \\
& + \tau_{\max} - \tau_{\min} + \int_{-\tau_{\max}}^{-\tau_{\min}} (1-(\frac{1}{2}-\frac{x}{2\tau_{\max}}))^A dx \\
=& 2\tau_{\min} \int_{\frac{1}{2}}^{1} (1-y^M)^A dy + 2\tau_{\max} \int_{\frac{1}{2}-\frac{1}{2}\frac{\tau_{\min}}{\tau_{\max}}}^{\frac{1}{2}} (1-y^M)^A dy + \tau_{\max} - \tau_{\min} \\
& + \frac{\tau_{\max}(1-\tau_{\min}/\tau_{\max})^{A+1}}{2^A(A+1)} \\
\geq & \tau_{\min} \frac{2}{M} I_{0.5}(\frac{1}{M}, A+1) + 2\tau_{\max} \left(\int_0^1 - \int_{\frac{1}{2}}^1\right) (1-y^M)^A dy + \tau_{\max} - \tau_{\min} \\
& + \frac{\tau_{\max}(1-\tau_{\min}/\tau_{\max})^{A+1}}{2^A(A+1)} \\
\geq & 2\tau_{\min} g_{AM} + 2\tau_{\max} f_{AM} - 2\tau_{\max} g_{AM} + \tau_{\max} - \tau_{\min} + \frac{\tau_{\max}(1-\tau_{\min}/\tau_{\max})^{A+1}}{2^A(A+1)},
\end{aligned}
$$

where $g_{AM} = \frac{1}{M} I_{0.5}(\frac{1}{M}, A+1)$ with $I_{0.5}(\cdot,\cdot)$ being the regularized incomplete Beta function. The last inequality is true due to the following result:

$$
\begin{aligned}
\int_{\frac{1}{2}}^{1} (1-y^M)^A dy =& \frac{1}{M} \int_{(\frac{1}{2})^M}^{1} t^{\frac{1}{M}-1} (1-t)^A dt \\
\geq & \frac{1}{M} \int_{\frac{1}{2}}^{1} t^{\frac{1}{M}-1} (1-t)^A dt \\
=& \frac{1}{M} I_{0.5}(\frac{1}{M}, A+1) \qquad (t := y^M) \\
:=& g_{AM}.
\end{aligned}
$$

Consequently, we have that ① satisfies

$$
\begin{aligned}
① \geq & \tau_{\max} - \tau_{\min} + \frac{\tau_{\max}(1-\tau_{\min}/\tau_{\max})^{A+1}}{2^A(A+1)} + 2\tau_{\min} g_{AM} + 2\tau_{\max} f_{AM} - 2\tau_{\max} g_{AM} \\
\geq & \tau_{\max} - \tau_{\min} + \frac{\tau_{\max} - (A+1)\tau_{\min}}{2^A(A+1)} + \tau_{\max} f_{AM} - 2\tau_{\max} g_{AM} + 2\tau_{\min} g_{AM} \\
\geq & \tau_{\max}(1 + \frac{1}{2^A(A+1)} + f_{AM} - 2g_{AM}) - \tau_{\min}(1+\frac{1}{2^A}) \\
\geq & \tau_{\max}(1 + f_{AM} - 2g_{AM}) - \frac{3}{2}\tau_{\min} \\
\geq & \tau_{\max}(1 + f_{AM} - 2g_{AM}) - 2\tau_{\min},
\end{aligned}
$$

Summarizing, we obtain the following upper bound and lower bound on the estimation bias $\mathbf{E}[Z_M]$, in the general case with heterogeneous distributions for Q-function approximation errors:

$$\begin{aligned}\mathbf{E}[Z_M] \geq& \gamma \left(\tau_{\min} - \tau_{\max}(f_{A(M-1)} + 2f_{AM})\right);\\ \mathbf{E}[Z_M] \leq& \gamma \left(2\tau_{\min} - \tau_{\max}\left(f_{AM} - 2g_{AM}\right)\right).\end{aligned}$$

□

Observe from Theorem 7.3 that the lower bound in (7.11) is positive when $\tau_{\min}(1 - 2f_{AM}) > \tau_{\max} f_{A(M-1)}$, indicating the existence of the overestimation issue. On thew contrary, the upper bound in (7.12) is negative when $2\tau_{\min} < \tau_{\max}\left(1 + f_{AM} - 2g_{AM}\right)$, pointing to the underestimation issue. In general, when $\tau_{\min}$ is large enough, decreasing ensemble size M is likely to cause overestimation, e.g., $\mathbf{E}[Z_M] \geq 0$ when $M < 2$. On the other hand, when $\tau_{\max}$ is small enough, increasing ensemble size M is likely to cause underestimation, e.g., $\mathbf{E}[Z_M] \leq 0$ when M is sufficiently large.

Determination of parameter c**.** As illustrated in Fig. 7.4c, for given approximation error characterization, a threshold c can be chosen such that increasing the ensemble size would help to correct the overestimation bias when $\tau_{\max} > c$, and decreasing the ensemble size is more conductive to mitigate the underestimation bias when $\tau_{\max} < c$. Specifically, parameter c is determined in two steps. *Step 1: To estimate approximation error distribution parameters* $\tau_{\min}$ *and* $\tau_{\max}$ *by running an ensemble based algorithm (e.g., Algorithm* 6*) for a few epochs with a fixed ensemble size.* In particular, a testing trajectory is generated from a random initial state using the current policy to compute the (discounted) MC return Q^{π} and the estimated Q-function value $Q^i, i = 1, 2, \ldots, N$. We next fit a uniform distribution model $\mathcal{U}(-\tau_i, \tau_i)$ of the approximation error $(Q^i - Q^{\pi})$ for each Q-function approximator Q^i. Then, $\tau_{\min}$ and $\tau_{\max}$ can be obtained by choosing the minimum and maximum values among $\tau_i, i = 1, 2, \ldots, N$. *Step 2: To obtain the upper bound and the lower bound in Theorem* 2.2 *by using* $\{\tau_{\min}, \tau_{\max}, A, \gamma\}$. We investigate the relationship between ensemble size M and the estimation bias by studying the bounds and identifying the 'critical' points as illustrated in Fig. 7.4b. Observe that a 'proper' ensemble size should be chosen between the 'critical' points, so as to reduce the overestimation and underestimation bias as much as possible. Since the approximation error is time-varying during the learning process, these two 'critical' points vary along with $\{\tau_{\max}\}$ and $\{\tau_{\min}\}$ (as shown in Fig. 7.4c). Intuitively, it is desirable to drive the system to avoid both the red region (underestimation) and the blue region (overestimation). It can be clearly observed that there is a wide range of choice for parameter c (e.g., [0.5, 0.7] in Fig. 7.4c) for the algorithm to stay in the white region, indicating that even though the pre-determined c above is not optimized, it can still serve the purpose well.

Summarizing, both Theorems 7.1 and 7.3 indicate that the approximation error characterization plays a critical role in controlling the estimation bias. In fact, both the lower bound and the upper bound in Theorem 7.3 depends on $\tau_{\min}$ and $\tau_{\max}$, which are time-varying due to the iterative nature of the learning process, indicating that it is sensible to use an adaptive ensemble size to drive the estimation bias to be close to zero, as much as possible.

7.3.2 Practical Implementation

Based on the theoretic findings above, we next propose AdaEQ that adapts the ensemble size based on the approximation error feedback on the fly, so as to drive the estimation bias close to zero. Particularly, as summarized in Algorithm 6, AdaEQ introduces two important steps at each iteration t, i.e., approximation error characterization (line 3) and ensemble size adaptation (line 4), which can be combined with the framework of either Q-learning or actor-critic methods.

Characterization of the time-varying approximation error. As outlined in Algorithm 6, the first key step is to quantify the time-varying approximation error at each iteration t (for ease of exposition, we omit the subscript t when it is clear from the context). Along the same line as in Van Hasselt et al. [58, 108, 333], we run a testing trajectory of length H, $\mathcal{T} = (s_0, a_0, s_1, a_1, \ldots, s_H, a_H)$, from a random initial state using the current policy π, and compute the discounted Monte Carlo return $Q^{\pi}(s, a)$ and the estimated Q-function value $Q^i(s, a)$, $i = 1, \ldots, N$ for each visited state-action pair (s, a). The empirical standard derivation of $Q^i(s, a) - Q^{\pi}(s, a)$ can be then obtained to quantify the approximation error of each approximator Q^i. Then, we take the average of the empirical standard derivation over all approximators to characterize the approximation error at the current iteration t, i.e.,

$$\tilde{\tau}_t = \frac{1}{N} \sum_{i=1}^{N} \operatorname{std}(Q^i(s, a) - Q^{\pi}(s, a)), \quad (s, a) \in \mathcal{T}. \tag{7.13}$$

Error-feedback based ensemble size adaptation. Based on the theoretic results and Fig. 7.4c, we update the ensemble size M at each iteration t based on the approximation error Eq. (7.13), using the following piecewise function:

$$M_t = \begin{cases} \operatorname{rand}(M_{t-1} + 1, N) & \tilde{\tau}_{t-1} > c, \; M_{t-1} + 1 \leq N \\ \operatorname{rand}(2, M_{t-1} - 1) & \tilde{\tau}_{t-1} < c, \; M_{t-1} - 1 \geq 2 \\ M_{t-1} & \text{otherwise,} \end{cases} \tag{7.14}$$

where $\operatorname{rand}(\cdot, \cdot)$ is a uniform random function and c is a pre-determined parameter to capture the 'tolerance' of the estimation bias during the adaptation process. Recall that parameter c can be determined by using the upper bound and the lower bound in Theorem 7.3 (Theorem 7.1). Particularly, a larger c implies that more tolerance of the underestimation bias is allowed when adapting the ensemble size M_t. A smaller c, on the other hand, admits more tolerance of the overestimation. In this way, AdaEQ can be viewed as a generalization of Maxmin and REDQ with ensemble size adaptation. In particular, when $c = 0$ and $M_t + 1 \leq N$, the adaptation mechanism would increase the ensemble size until it is equal to N. Consequently, AdaEQ degenerates to Maxmin [186] where $M = N$, leading to possible underestimation bias. Meantime, when c is set sufficiently large, the ensemble size M would decrease until reaching the minimal value 2 during the learning process, where the estimation bias would be positive according to Theorem 7.3. In this case, AdaEQ is degenerated to REDQ [58] with ensemble size $M = 2$.

Remark. We use random sampling in Eq. (7.14) for two reasons. Firstly, the characterization of the approximation error in Eq. (7.13) is noisy in nature. In particular, Monte Carlo returns with finite-length testing trajectory may introduce empirical errors when estimating the underlying ground true value of Q^{π}. This noisy estimation is often the case when the policy is not deterministic, or the environment is not deterministic. Thus, we use random sampling to 'capture' the impact of this noisy estimation. Secondly, in general it is infeasible to characterize the exact relationship between estimation bias Z_M and ensemble size M. Without any further prior information except from the bounds we obtained in Theorems 2.1 and 2.2 about the approximation error, the random sampling can be viewed as the 'exploration' in AdaEQ.

Algorithm 6 Adaptive Ensemble Q-learning (AdaEQ)

Empty replay buffer $\mathcal{D}$, step size α, number of the approximators N, initial in-target ensemble size $M_0 \leq N$, initial state s. Initialize N approximators with different training samples.
for Iteration $t = 1, 2, 3, \cdots$ **do**
  Identify approximation error parameter $\tilde{\tau}_t$ using Equation (7.13)
  Update ensemble size M_t according to Equation (7.14)
  Sample a set $\mathcal{M}$ of M_t different indices from $\{1, 2, \cdots, N\}$
  Obtain the proxy approximator $Q^{\text{proxy}}(s, a) \leftarrow \min_{i \in \mathcal{M}} Q^i(s, a), \forall a \in \mathcal{A}$
  Choose action a from current state s using policy derived from Q^{proxy} (e.g., ε-greedy)
  Take action a, observe r and next state s'
  Update replay buffer $\mathcal{D} \leftarrow \mathcal{D} \cup \{s, a, r, s'\}$
  for $i = 1, 2, \cdots, N$ **do**
    Sample a random mini-batch B from $\mathcal{D}$
    Compute the target: $y(s, a, r, s') \leftarrow r + \gamma \max_{a' \in \mathcal{A}} Q^{\text{proxy}}(s', a'), (s, a, r, s') \in B$
    Update Q-function Q^i: $Q^i(s, a) \leftarrow (1 - \alpha)Q^i(s, a) + \alpha y(s, a, r, s'), (s, a, r, s') \in B$
  end for
  $s \leftarrow s'$
end for

7.4 Experimental Results

In this section, we evaluate the effectiveness of AdaEQ by answering the following questions: (1) Can AdaEQ minimize the estimation bias and further improve the performance in comparison to existing ensemble methods? (2) How does AdaEQ perform given different initial ensemble sizes? (3) How does the 'tolerance' parameter c affect the performance?

To make a fair comparison, we follow the setup of Chen et al. [58] and use the same code base to compare the performance of AdaEQ with REDQ [58] and Average-DQN (AVG) [17], on three MuJoCo continuous control tasks: Hopper, Ant and Walker2d. The same hyperparameters are used for all the algorithms. Specifically, we consider $N = 10$ Q-function approximators in total. The ensemble size $M = N = 10$ for AVG, while the initial M for AdaEQ is set as 4. The ensemble size for REDQ is set as $M = 2$, which is the fine-tuned result from Chen et al. [58]. For all the experiments, we set the 'tolerance' parameter c in Eq. (7.14) as 0.3 and the length of the testing trajectories as $H = 500$. The ensemble size is updated according to Eq. (7.14) every 10 epochs in AdaEQ. The discount factor is 0.99. Moreover, we summarize the implementation details as follows.

Hyperparameters and Implementation Details In the empirical implementation, our code for AdaEQ is partly based on REDQ authors' open source code (https://github.com/watchernyu/REDQ) [58] and we use the identical hyperparameter setting, for the sake of fair comparison. For all three methods compared in our experiments, the first 5000 data points are obtained by randomly sampling from the action space without updating the Q-networks. For REDQ, we use the fine-tuned ensemble size $M = 2$ for all the MuJoCo benchmark tests. The results are similar with the reported results in the original paper. The detailed hyperparameter setting is summarized in Table 7.1.

Additional Empirical Results on MuJoCo Benchmark **Training Time Comparison.** In Table 7.2, we compare the average wall-clock training time over three training seeds. All the tasks are trained on the same 2080Ti GPU. It can be seen that the ensemble size adaptation mechanism does not significantly increase the training time for most tasks.

Illustrative Examples for Time-Varying Approximation Errors We present in Fig. 7.5 an example to illustrate the time-varying nature of approximation errors during the training process. Following the setting in Sect. 7.2.2, we use 3 different approximators to approximate the true action-value. Each approximator is a 6-degree polynomial that fits the samples. The initial approximation errors used to generate samples are set to follow uniform distribution with parameter $\tau = 0.3, 0.5, 0.7$, respectively. At each iteration, we first generate new samples from the approximator obtained in the last iteration and then update the approximator using these samples. The mean and standard derivation of the approximation error over different states are depicted in Fig. 7.5a and b. Clearly, the approximation error is time-varying and can change dramatically during the training process.

Performance Comparison with TD3 and SAC **Evaluation of estimation bias.** To investigate the impact of the adaptation mechanism in AdaEQ, we begin by examining how the

Table 7.1 A comparison of hyperparameter settings among AdaEQ, REDQ [58] and AVG [17] implementation

Hyperparameter	AdaEQ (Our Method)	REDQ	AVG
Learning rate	$3 \cdot 10^{-4}$	$3 \cdot 10^{-4}$	$3 \cdot 10^{-4}$
Discount factor (γ)	0.99	0.99	0.99
Optimizer	Adam	Adam	Adam
Target smoothing coefficient (ρ)	$5 \cdot 10^{-3}$	$5 \cdot 10^{-3}$	$5 \cdot 10^{-3}$
Batch size	256	256	256
Replay buffer size	10^6	10^6	10^6
Non-linearity	ReLU	ReLU	ReLU
Number of hidden layers	2	2	2
Number of hidden unites per layer	256	256	256
Number of approximators (N)	10	10	10
Testing trajectory length H	500	500	500
Initial ensemble size (M_0)	4	2	10
Ensemble size adaptation	True	False	False
Adaptation frequency	Every 10 epochs	–	–
'Tolerance' parameter c	0.3	–	–

Table 7.2 A comparison of training time among AdaEQ, REDQ [58] and AVG [17] implementation

Environment	Hopper-v2 125 K	Walker2d 300 K	Ant 300 K	Humanoid 250 K
AdaEQ	62509.38	120598.13	130673.32	148786.28
REDQ	61565.28	118604.28	122954.04	104462.43
AVG	172115.94	151058.98	129526.58	124834.21
AdaEQ/REDQ	1.01×	1.02×	1.06×	1.42×

estimation bias changes in the training process. After each epoch, we run an evaluation episode of length $H = 500$, starting from an initial state sampled from the replay buffer. We calculate the estimation error based on the difference between the Monte Carlo return value and the Q-estimates as in Van Hasselt et al. [333], Chen et al. [58] and Janner et al. [152]. For each experiment, the shaded area represents a standard deviation of the average evaluation over 3 training seeds. As shown in the first row of Fig. 7.6, AdaEQ can reduce the estimation bias to nearly zero in all three benchmark environments, in contrast to REDQ and AVG. The AVG approach tends to result in positive bias in all three environments during the learning procedure, which is consistent with the results obtained in Chen et al. [58]. Notably, it can be clearly observed from Hopper and Walker2d tasks that the estimation bias for AdaEQ is driven to be close to zero, thanks to the dynamic ensemble size adjustment

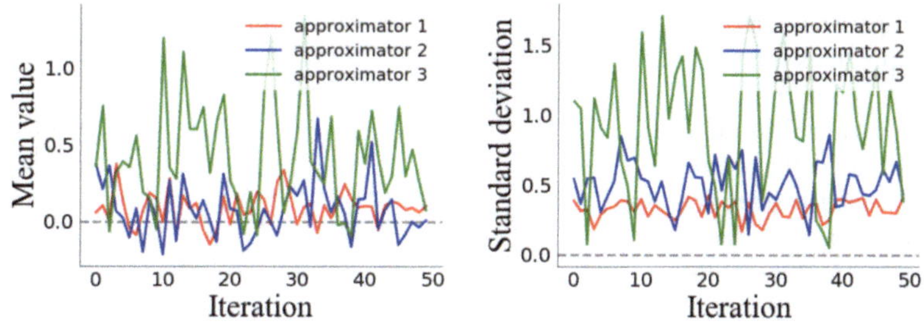

(a) Mean approximation error over states.

(b) Standard deviation of the approximation error over states.

Fig. 7.5 Illustration of the time-varying approximation error during the training process. **a** Mean approximation error over states. **b** Standard deviation of the approximation error over states

Table 7.3 A comparison of average return among AdaEQ (proposed method), REDQ [58], AVG [17], SAC [127] and TD3 [108]

Environment	Hopper-v2 125 K	Walker2d 300 K	Ant 300 K	Humanoid 250 K
AdaEQ	3372	4012	5241	4982
REDQ	3117	3871	5013	4521
AVG	1982	2736	2997	2015
SAC	2404	2556	2485	1523
TD3	982	3624	2048	–

during the learning process. Meantime, in the Ant task, even though the fine-tuned REDQ can mitigate the overestimation bias, it tends to have underestimation bias, whereas AdaEQ is able to keep the bias closer to zero (gray dashed line) even under a 'non-optimal' choice of the initial ensemble size (Table 7.3).

Performance on MuJoCo benchmark. We evaluate the policy return after each epoch by calculating the undiscounted sum of rewards when running the current learnt policy [58, 152]. The second row of Fig. 7.6 demonstrates the average return during the learning process for AdaEQ, AVG and REDQ, respectively. Especially, we choose the fine-tune ensemble size for REDQ [58]. As observed in Fig. 7.6, AdaEQ can efficiently learn a better policy and achieve higher average return in all three challenging MuJoCo tasks, without searching the optimal parameters beforehand for each of them. Meantime, AdaEQ only incurs slightly more computation time than REDQ in most MuJoCo tasks. We also provide the wall-clock training time comparison in Table 5.7

Robustness to the initial ensemble size. Next, we investigate the performance of AdaEQ under different settings of the initial ensemble size in the Hopper-v2 and Ant environment, i.e., $M = (2, 3, 5)$ and $M = (3, 5, 7)$. As shown in Fig. 7.7, AdaEQ consistently outperforms the others in terms of the average performance over different setups, which implies the benefit

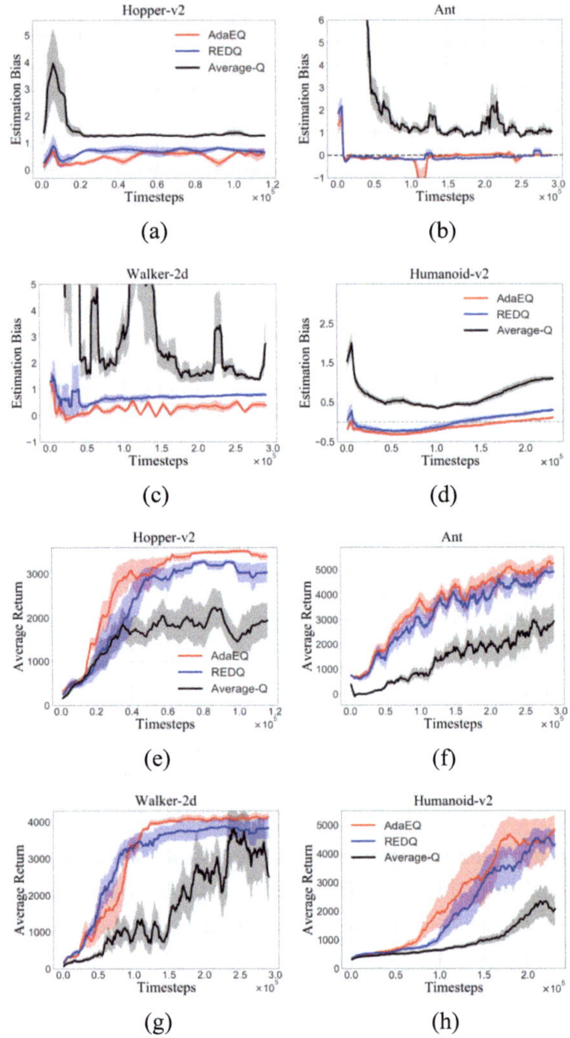

Fig. 7.6 Comparison in terms of the estimation bias (first row) and average return (second row) in three MuJoCo tasks. Solid lines are the mean values and the shaded areas are the standard deviations across three random seeds. We use the undiscounted sum of all the rewards in the testing episode to evaluate the performance of the current policy after each epoch. The estimation error is evaluated by comparing the difference between the Monte Carlo return and the average Q-value for each state-action pair visited during the testing episode. We take the average of those error values as the estimation bias. For AdaEQ, we use the same hyperparameter c for ensemble size adaptation

of adjusting the in-target ensemble size based on the error feedback. It can be seen from the shaded area that the performance of AVG and REDQ, may vary significantly when the ensemble size changes.

Robustness to parameter c in a wide range. As illustrated in Fig. 7.8, we conduct the ablation study by setting $c = 0.001, 0.3, 0.5, 0.7, 1.5$ on the Hopper-v2 and Ant tasks. Clearly, AdaEQ works better for $c \in [0.3, 0.7]$. The experiment results corroborate our analysis in Sect. 7.3.1 that our algorithm is not sensitive to parameter c in a wide range. As mentioned in Sect. 7.3.2, when parameter c is close to zero, AdaEQ degenerates to Maxmin, which is known to suffer from underestimation bias when the ensemble size is large [58].

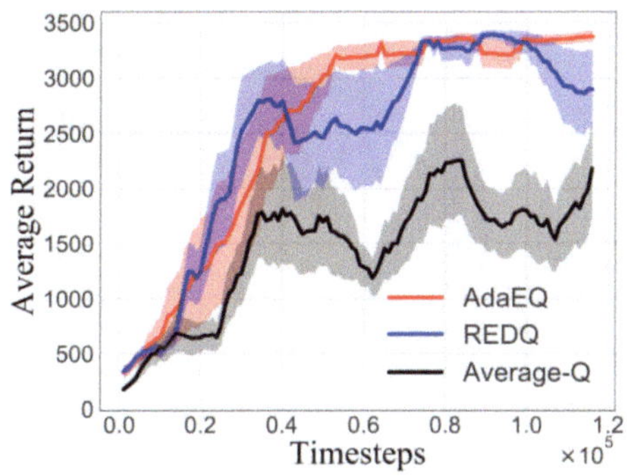

(a) Hopper-v2 task. Average returns over different initial ensemble size $M = 2, 3, 5$.

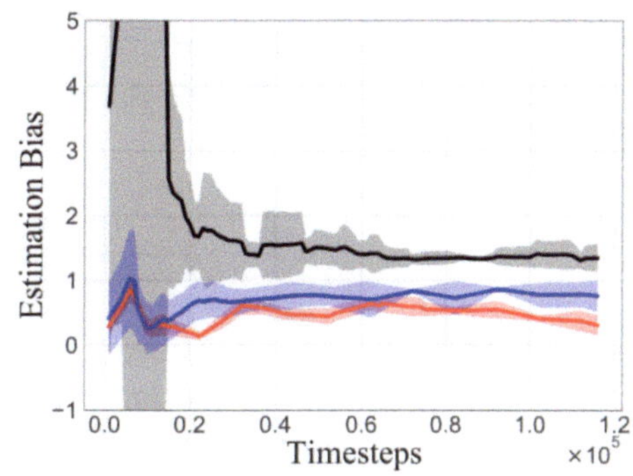

(b) Hopper-v2 task. Estimation bias over different initial ensemble size $M = 2, 3, 5$.

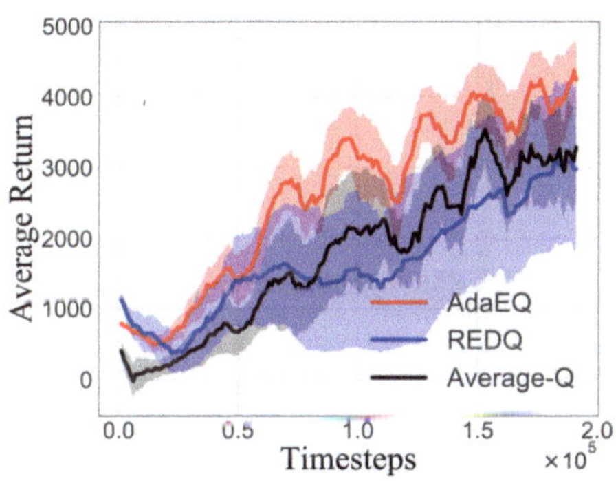

(c) Ant task. Average returns over different initial ensemble size $M = 3, 5, 7$.

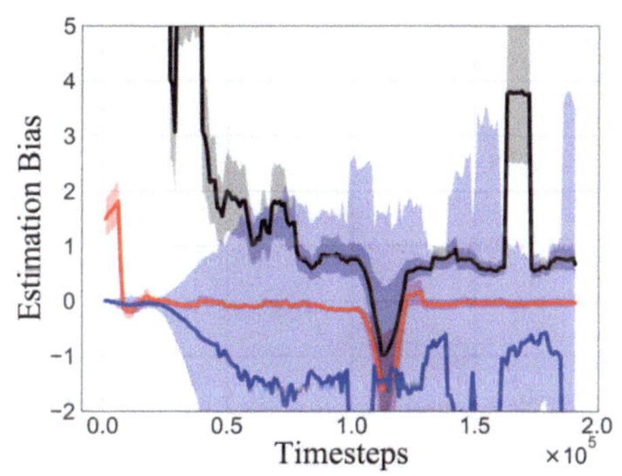

(d) Ant task. Estimation bias over different initial ensemble size $M = 3, 5, 7$.

Fig. 7.7 Impacts of the initial ensemble size M on the performance of AdaEQ in Hopper-v2 and Ant tasks. The solid lines are the mean values and the shaded areas are the standard deviations across three ensemble size settings

Further, as illustrated in Fig. 7.8b, when c is large, e.g., $c = 1.5$, the ensemble size would gradually decrease to the minimum and hence would not be able to throttle the overestimation tendency during the learning process.

Convergence Analysis of AdaEQ Assume that at iteration t, the ensemble size is $M_t \leq N$, where N is the number of approximators in the ensemble method. It is straightforward to verify that the stochastic approximation noise term has the contraction property as stated in Chen et al. [58] and Lan et al. [186]. It follows that AdaEQ converges to the optimal Q-function with probability 1.

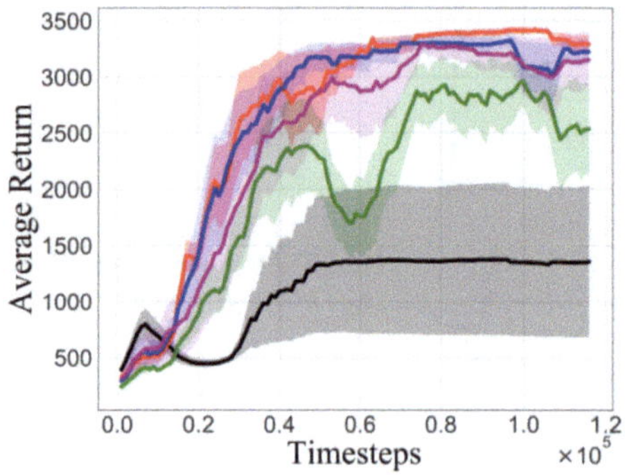

(a) Hopper-v2 task. Average returns over different parameter c in AdaEQ.

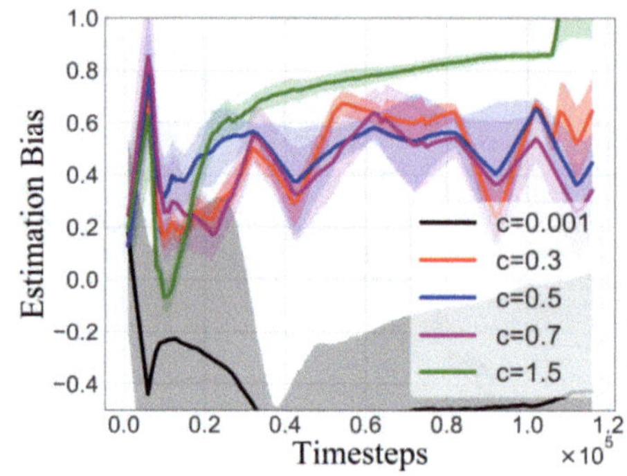

(b) Hopper-v2 task. Estimation bias over different parameter c in AdaEQ.

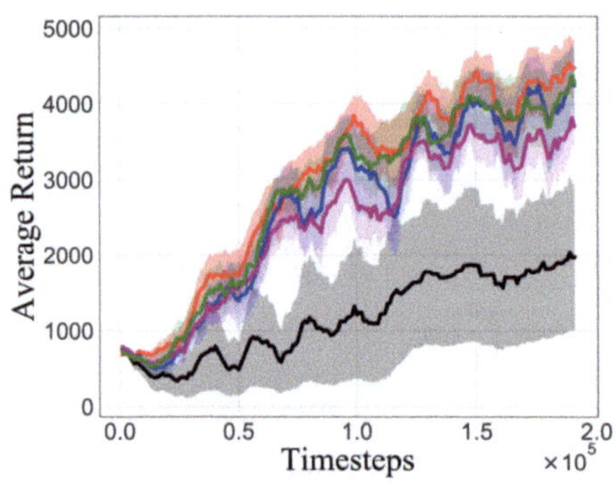

(c) Ant task. Average returns over different parameter c in AdaEQ.

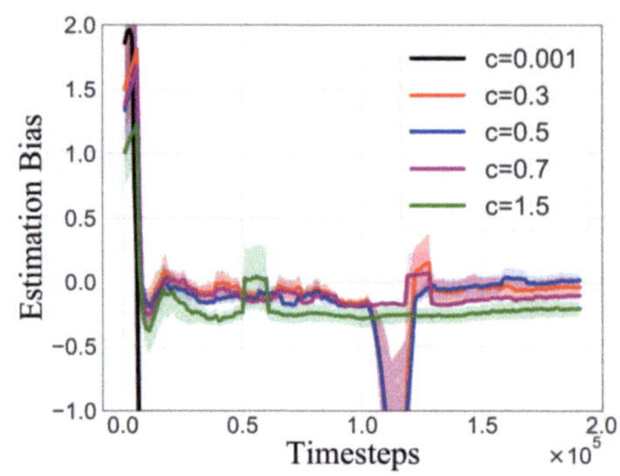

(d) Ant task. Estimation bias over different parameter c in AdaEQ.

Fig. 7.8 Impacts of parameter c on the performance of AdaEQ in Hopper-v2 and Ant tasks. The initial ensemble size is set to $M = 4$. The mean value and the standard deviation are evaluated across three training seeds

7.5 Conclusion

Determining the right ensemble size is highly nontrivial for the ensemble Q-learning to correct the overestimation without introducing significant underestimation bias. In this chapter, we devise AdaEQ, a generalized ensemble Q-learning method for the ensemble size adaptation, aiming to minimize the estimation bias during the learning process. More specifically, by establishing the upper bound and the lower bound of the estimation bias, we first characterize the impact of both the ensemble size and the time-varying approximation error on the estimation bias. Building upon the theoretic results, we treat the estimation bias minimization as an adaptive control problem, and take the approximation error as feedback to adjust the ensemble size adaptively during the learning process. Our experiments show that

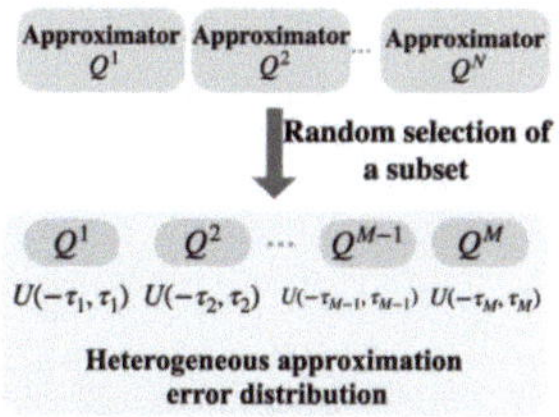

(a) Approximation error distributions settings.

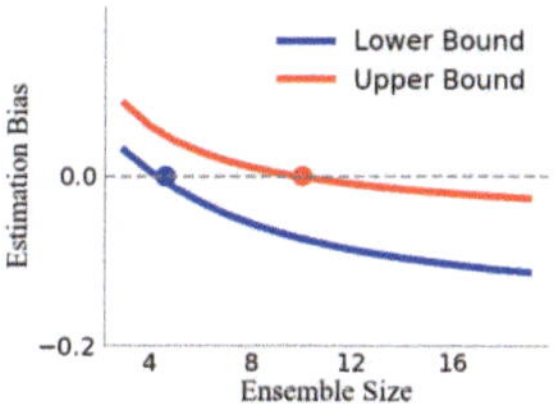

(b) $\tau_{\min} = 0.07, \tau_{\max} = 0.1$.

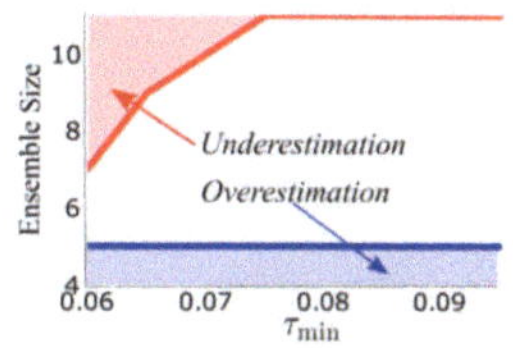

(c) Impact of approximation error on the estimation bias when $\tau_{\max} = 0.1$: overestimation vs. underestimation.

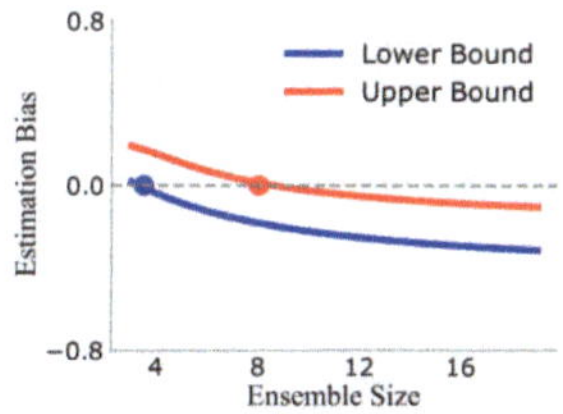

(d) $\tau_{\min} = 1, \tau_{\max} = 1$.

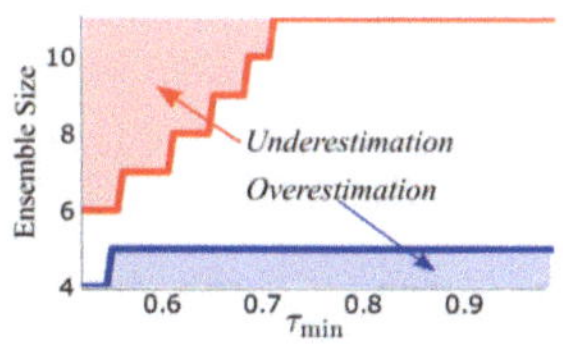

(e) Impact of approximation error on the estimation bias when $\tau_{\max} = 1$: overestimation vs. underestimation.

Fig. 7.9 Illustration of upper bounds and lower bounds on estimation bias in Theorem 7.3. **a** The case where the approximation errors of the Q-approximators are heterogeneous. **b**, **d** The lower bound and the upper bound for given τ_1, τ_2, A. **c**, **e** Due to the time-varying feature of the approximation errors, the blue curve and the red curve depict the 'critical' points for the lower bound and the upper bound, respectively

AdaEQ consistently and effectively outperforms the existing ensemble methods, such as REDQ and AVG in MuJoCo tasks, corroborating the benefit of using AdaEQ to drive the estimation bias close to zero.

There are many important avenues for future work. In terms of the bounds of the estimation bias, our analysis builds upon the standard independent assumption as in previous works. It's worth noting that in practice, the errors are generally correlated [328] and the theoretical analysis for this case remains not well understood. Additionally, in this work, we use a heuristic tolerance parameter in the adaptation mechanism to strike the balance in controlling the positive bias and negative bias. It is of great interest to develop a systematic approach to optimize this tolerance parameter.

Edge-Edge Collaboration via Decentralized Online Meta-Learning

8

Despite the superior fast learning performance of online meta-learning, to learn a good meta-model for within-task fast adaptation, a single edge device alone still has to learn over many tasks, which inevitably encounters the cold-start problem. Observe that in a multi-device edge network, the learning tasks across different devices in the same environment often share some model similarity [312]. For example, different robots may perform similar coordination behaviors according to the environment changes. In fact, one of the most remarkable abilities of human being is to continuously speed up learning of new tasks based on previous experiences from oneself as well as from others. Thus inspired, one may wonder if the cold-start problem for a single device could be mitigated via limited collaboration among multiple edge devices by leveraging the task similarity therein. Here by "limited collaboration" we mean limited communication between neighboring devices only, as the communication cost usually is a bottleneck in wireless communication systems [214]. In this chapter, we seek to answer the following open questions: *(1) Can we accelerate the online meta-learning at a single device on average in a multi-device edge network, with only one communication step among neighbors per learning task? (2) If yes, how much can we improve upon the single-device case?*

8.1 Introduction

We give an affirmative answer to the first question and show that the optimal task-average regret can be achieved at a faster rate for each device (i.e., agent) in the multi-agent network via limited communication, compared to single-agent online meta-learning. More specifically, we propose MAOML, a multi-agent online meta-learning framework, which generalizes the single-agent online meta-learning framework, ARUBA, in Khodak et al.

H. Wang et al., *Continual and Reinforcement Learning for Edge AI*,
Synthesis Lectures on Learning, Networks, and Algorithms,
https://doi.org/10.1007/978-3-031-84363-1_8

[167] to a multi-agent online meta-learning setting. In particular, we cast the multi-agent online meta-learning into an equivalent *two-level nested* OCO problem, where we treat the within-task adaptation as a standard *task-level* OCO problem, and the meta-model update as a distributed *network-level* OCO problem across the multi-agent network. Mathematically, it can be shown that the performance ceiling of the multi-agent online meta-learning, in terms of the task-average regret, heavily depends on the performance of the distributed network-level OCO for the meta-model update. This is intuitive as a good meta-model should be able to capture the most important information across different tasks in the multi-agent network for enabling fast learning of a new task. *Therefore, the problem of accelerating online meta-learning boils down to improving the performance per agent of distributed network-level OCO via limited communication.*

Then, the next key question is "*how much can an agent benefit from distributed OCO through limited communication with its neighbors?*" To this end, consider a multi-agent network with N agents. Intuitively, the more agents there are and the more information exchange, the smaller the average regret would be, and this is of interest particularly in a networked system. It is well known that the optimal regret in single-agent OCO is of order $O(\sqrt{T})$ after T iterations, achievable by either online gradient descent (OGD) or follow-the-regularized-leader (FTRL) [137, 299]. Interestingly, Khodak et al. [77, 159] suggest that an average regret of $O(\sqrt{T/N})$, i.e., a factor of $\sqrt{1/N}$ speedup, can be obtained at each agent for multi-agent stochastic OCO, by performing the synchronizations of local model predictions after each (or multiple) iteration. However, the required synchronization (for the model predictions) where all agents need to communicate until reaching consensus [77, 159], incurs a significant communication burden, requiring $\Theta(QT)$ communication steps with Q being the diameter of the network, and hence inevitably suffers from the latency which degrades the learning performance. *In a nutshell, it remains unclear a priori if distributed OCO algorithms can achieve significant improvement in terms of the average regret per agent, with only one communication step per iteration.*

In this chapter, we first propose a multi-agent online meta-learning framework to address the cold-start problem in single-agent online meta-learning, by leveraging the task similarity, i.e., the tasks follow some unknown distribution as in standard meta-learning [103], across multiple agents via limited communication. Along the line of the ARUBA framework introduced in Khodak et al. [167], we treat the multi-agent online meta-learning as a two-level nested OCO problem, where the within-task adaptation and the meta-model update are formulated as a standard *task-level* OCO problem and a distributed *network-level* OCO problem across the multi-agent network, respectively. Next, we characterize the performance upper bound of multi-agent online meta-learning in terms of the *agent-task-averaged regret*, and show that it heavily depends on how much an agent can benefit from the distributed network-level OCO for updating the meta-models through limited communication with its neighbors, which is unclear a priori. To tackle this challenge, we further consider a distributed online gradient descent algorithm (DOGD-GT) with *gradient tracking* [262, 267]. We show that by carefully tracking of the accumulated gradient consensus error through only limited communication among multiple agents, the average regret per agent can be significantly reduced to $O(\sqrt{T/N})$ compared with the single-agent case, thus revealing a linear speedup of the learning performance. Moreover, building on the agent-level performance speedup benefiting from the multi-agent collaboration via gradient tracking in the distributed network-level OCO, we next propose a multi-agent online meta-learning algorithm called MAOML. It can be shown that each agent in MAOML can achieve a notable performance improvement in terms of the average regret per agent, i.e., approaching the optimal within-task regret at a faster rate of $O(1/\sqrt{NT})$ compared with the rate of $\tilde{O}(1/\sqrt{T})$ in the single-agent online meta-learning ARUBA. To the best of our knowledge, this is the first work to the address the cold-start problem by studying multi-agent online meta-learning under limited communication. At the end, we conduct extensive experiments on various datasets to demonstrate the performance of DOGD-GT and MAOML. The experimental results clearly indicate the improvement of MAOML over the single-agent online meta-learning in terms of the agent-task-averaged performance, corroborating the benefits of utilizing the task similarity across multiple agents through limited communication in both convex and nonconvex setups.

8.2 Related Work

Distributed OCO [368] in a multi-agent network has recently garnered much interest, where each agent first learns the model parameters based on its local data and then communicates its local model information with its neighbors. However, little attention has been paid to understand the impact of the network size on the average regret achievable at each individual agent therein. To reap the potential benefits that an agent can achieve when carrying out distributed OCO, a convex loss function with both adversarial and stochastic components is considered in Zhao et al. [402]. Assuming that the expected gradient is bounded above by G and the stochastic variance is bounded above by σ^2, they have shown that the network expected regret is $O(\sqrt{N^2G^2T + NT\sigma^2})$. In contrast, Dekel et al. [77] studies distributed OCO in a stochastic setup and proposes a distributed mini-batch algorithm, which leads to a network regret of order $O(\sqrt{NT})$, i.e., an agent-average regret of $O(\sqrt{T/N})$, indicating a possible linear speed-up of the average regret per agent. However, there is a hidden cost associated with the needed synchronization among all agents, required at each iteration, which could incur a significant communication burden and learning performance degradation. To reduce the communication cost, a dynamic synchronization strategy is proposed in Kamp et al. [159] by reducing the frequency of synchronization, which however requires a central coordinator and still suffers from learning latency because of the synchronization.

A gradient-tracking based distributed OGD algorithm is considered in Zhang et al. [396] for distributed OCO problem. However, the results therein are different from ours, as outlined next: (1) [396] aims to show that the dynamic regret of distributed OCO has no explicit dependence on the time horizon, as in the centralized case, whereas we focus on characterizing the performance speedup by cleverly exploiting the limited multi-agent collaboration.

(2) The results in Zhang et al. [396] rely on the assumption that the loss function is strongly-convex, which is required even in the centralized case so as to remove the dependence on the time horizon. Since our focus is on the multi-agent speedup, strong-convexity is not a necessity and we consider convex loss functions instead. (3) The dynamic regret defined in Zhang et al. [396] cannot simply generalize to the problem setup in our setting, and a non-trivial analysis of the regret bound is needed to quantify the performance speedup.

8.3 Multi-agent Online Meta-Learning

In this section, we first introduce the multi-agent online meta-learning framework, and cast it into an equivalent two-level nested OCO problem. By characterizing the upper bound of the agent-task-averaged regret, we show that the performance of the distributed network-level OCO for the meta-model update, is the bottleneck for the performance of multi-agent online meta-learning.

8.3.1 Problem Formulation

As is standard in a multi-agent network, we assume that the agents communicate in an undirected and connected communication graph $\mathcal{G} = (\mathcal{V}, \mathcal{E})$, where $\mathcal{V} \triangleq \mathcal{N} = \{1, \ldots, N\}$ is the set of vertices (agents) and $\mathcal{E} \subset \mathcal{V} \times \mathcal{V}$ is the set of edges connecting agents. Agent i and j can communicate with each other if and only if $(i, j) \in \mathcal{E}$. We further denote $\mathcal{N}_i = \{j | j \neq i, (i, j) \in \mathcal{E}\}$ as the set of neighbors of agent i. Each agent can make its decision based on the local information and the information obtained from its neighbors via weighted averaging. To model this 'weighting' process, a consensus weight matrix, $W = [w_{ij}] \in \mathbb{R}^{N \times N}$, is usually introduced with the following properties:

- For any $(i, j) \in \mathcal{E}$, we have $w_{ij} > 0$; otherwise, $w_{ij} = 0$. In particular, $w_{ii} > 0$.
- Matrix W is doubly stochastic, i.e., $\sum_{i'} w_{i'j} = \sum_{j'} w_{ij'} = 1$ for all $i, j \in \mathcal{N}$.

In the multi-agent online meta-learning framework, each agent $n \in \mathcal{N}$ faces with a sequence of online learning tasks $\mathcal{T}_{t,n}$ indexed by $t = 1, \ldots, T$, as illustrated in Fig. 8.1. We assume that all agents are synchronized at the task level, i.e., new tasks arrive at all agents at the same time. For each learning task $\mathcal{T}_{t,n}$, the agent n must sequentially choose $m_{t,n}$ actions $\theta^i_{t,n}$ from some convex compact set Θ and incur loss $l^i_{t,n} : \Theta \to \mathbb{R}$ which is convex and Lipschitz, for $i \in [1, m_{t,n}]$. After learning one task, each agent would share learned model knowledge with its neighbors through one communication step to facilitate the learning of new tasks.

Let $\theta^*_{t,n}$ denote the optimal model parameter for task $\mathcal{T}_{t,n}$, i.e., $\theta^*_{t,n} = \arg\min_{\theta \in \Theta} \sum_{i=1}^{m_{t,n}} l^i_{t,n}(\theta)$. Following the standard assumption in meta-learning [103], we assume that

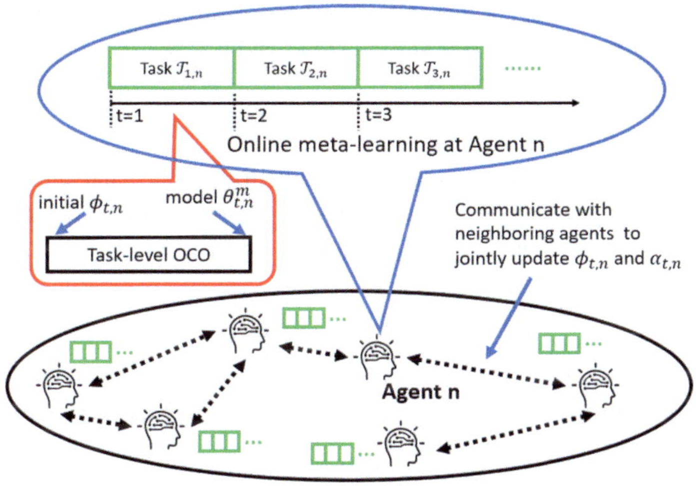

Fig. 8.1 The framework of multi-agent online meta-learning. Each agent in the multi-agent network has a sequence of online learning tasks, and it shares the learned model knowledge ($\phi_{t,n}$ and $\alpha_{t,n}$) with its neighbors to facilitate the learning of new tasks at time t

all the optimal model parameters $\theta^*_{t,n}$ for any $t \in [1, T]$ and $n \in \mathcal{N}$ follow some unknown distribution $\mathcal{P}_{\mathcal{T}}$, so as to capture the task similarity across the network. In multi-agent online meta-learning, the agents aim to obtain good learning performance for each individual task. In the same spirit with [167], we study the **agent-task-averaged regret** (ATAR) after each agent encounters T tasks:

$$\mathbf{R}_a = \frac{1}{NT}\sum_{n=1}^{N}\sum_{t=1}^{T}\left(\sum_{i=1}^{m_{t,n}} l^i_{t,n}(\theta^i_{t,n}) - \sum_{i=1}^{m_{t,n}} l^i_{t,n}(\theta^*_{t,n})\right).$$

A low ATAR ensures that the individual task regret of an algorithm is small on average over the network, compared to that of the optimal within-task parameter. To this end, every agent in the network can collaboratively learn, through limited communications with neighboring agents, the meta-models, i.e., *a model initialization* $\phi_{t,n}$ *and a task-dedicated learning rate* $\alpha_{t,n}$ by utilizing other agents' information, such that good within-task performance can be achieved with $\theta^i_{t,n}$ adapted from $\phi_{t,n}$ during the online meta-learning.

8.3.2 Two-Level Nested OCO

Based on the ARUBA framework [167], we treat the multi-agent online meta-learning as a *two-level nested OCO* problem, and develop a theoretical framework for understanding the performance of multi-agent meta-learning through the lens of distributed OCO. For simplicity, we assume $m_{t,n} = m$, for any $t \in [1, T]$ and $n \in \mathcal{N}$.

8.3.2.1 Task-Level OCO

For the task $\mathcal{T}_{t,n}$ at the agent n, given the model initialization $\phi_{t,n}$ and within-task learning rate $\alpha_{t,n}$ learned jointly based on the previous tasks, the agent seeks to determine the action $\theta_{t,n}^i$ so as to minimize the within-task regret after m rounds:

$$\mathbf{R}_{t,n} = \sum\nolimits_{i=1}^{m} l_{t,n}^i(\theta_{t,n}^i) - \sum\nolimits_{i=1}^{m} l_{t,n}^i(\theta_{t,n}^*).$$

For a convex and G-Lipschitz loss function, it is well-known that the best upper bound for $\mathbf{R}_{t,n}$ of online mirror descent (OMD), regularized by Bregman divergence, is given as follows [299]:

$$\mathbf{R}_{t,n} \leq \frac{1}{\alpha_{t,n}} \mathcal{B}_R(\theta_{t,n}^* || \phi_{t,n}) + \alpha_{t,n} G^2 m = \hat{\mathbf{R}}_{t,n}, \tag{8.1}$$

where for a continuously-differentiable strictly convex function $g : \Theta \to \mathbb{R}$, the Bregman divergence is defined as

$$\mathcal{B}_R(\theta || \phi) = g(\theta) - g(\phi) - \langle \nabla g(\phi), \theta - \phi \rangle.$$

This step corresponds to the within-task adaptation from the initial model $\phi_{t,n}$ using gradient descent regularized by the Bregman divergence, i.e., the inner loop of meta-learning. In order to use OCO for the meta-update of initial model $\phi_{t,n}$, we only consider the regularization as the set of Bregman divergence that is *convex and smooth in the second argument*, i.e., $\mathcal{B}_R(\theta || \cdot)$ is convex and smooth for any fixed $\theta \in \Theta$. For example, when $g(\cdot)$ is the negative generalized entropy function defined for the expected loss of convex proper loss functions, the corresponding Bregman divergence satisfies the above condition [254]. The widely used L_2 regularization also satisfies this condition.

8.3.2.2 Network-Level OCO

Based on the definition of ATAR, it is clear that ATAR can be bounded above by the average of $\{\hat{\mathbf{R}}_{t,n}\}$:

$$\mathbf{R}_a = \frac{1}{NT} \sum_{n=1}^{N} \sum_{t=1}^{T} \mathbf{R}_{t,n} \leq \frac{1}{NT} \sum_{n=1}^{N} \sum_{t=1}^{T} \hat{\mathbf{R}}_{t,n} \triangleq \bar{\mathbf{R}}_a, \tag{8.2}$$

which indicates that the ATAR is small if the average regret-upper-bound $\bar{\mathbf{R}}_a$ is small. Observe that each agent chooses one action pair $(\phi_{t,n}, \alpha_{t,n})$ and incurs the loss $\hat{\mathbf{R}}_{t,n}$ for each task $\mathcal{T}_{t,n}$. It follows that the outer loop of multi-agent online meta-learning, i.e., meta-update of the model initialization $\phi_{t,n}$ and the learning rate $\alpha_{t,n}$, can be cast as a distributed network-level OCO among all N agents. The objective here is to learn good meta-models $(\phi_{t,n}, \alpha_{t,n})$ for each agent via the multi-agent collaboration so as to minimize the following regret:

$$\mathbf{R}_{out} = \frac{1}{NT} \sum_{n=1}^{N} \sum_{t=1}^{T} \left[\hat{\mathbf{R}}_{t,n}(\phi_{t,n}, \alpha_{t,n}) - \hat{\mathbf{R}}_{t,n}(\phi^*, \alpha^*) \right] \tag{8.3}$$

where $(\phi^*, \alpha^*) = \arg\min \mathbb{E}_{\mathcal{P}_T}[\hat{\mathbf{R}}_{t,n}(\phi, \alpha)]$. This distributed network-level OCO enables the task-similarity to be learned on-the-fly, which is encapsulated in an adaptive learning rate by utilizing the information across the multi-agent network.

Note that the average regret-upper-bound $\bar{\mathbf{R}}_a$ corresponds to the average loss in the distributed network-level OCO for updating the meta-models. It is clear that $\bar{\mathbf{R}}_a$ is small if the regret $\mathbf{R}_{out}$ is small for the distributed network-level OCO, which consequently results in a small ATAR based on (8.2). This is intuitive as the performance of online meta-learning directly depends on how good the meta-models are. In other words, if we could quickly learn good meta-models, i.e., the model initialization and learning rate, by utilizing the knowledge across the multi-agent network, good performance can be guaranteed for each task in online meta-learning, without the need of learning over many tasks at a single agent. Therefore, the problem of accelerating distributed online meta-learning boils down to the problem of improving the performance per agent of distributed network-level OCO, i.e., quickly learn good meta-models, via limited communication.

8.4 Distributed Network-Level Online Convex Optimization

As alluded to earlier, it remains unclear a priori if any distributed OCO algorithms can achieve significant improvement in terms of the average regret per agent, with only one communication step per iteration. To tackle this challenge and also accelerate online meta-learning, we take a closer look to the distributed network-level OCO in this section, and devise a distributed OGD algorithm with gradient tracking.

For ease of exposition, we consider a more general formulation [55, 77, 138, 363] for the distributed network-level OCO (8.3): In iteration t the agent i makes a local model prediction $x_{t,i}$ from a convex compact set $\mathcal{K} \subset \mathbb{R}^d$ and incurs convex loss $f_{t,i}(x_{t,i})$ that follows some unknown distribution $\mathcal{P}$, i.e., $f_{t,i} \sim \mathcal{P}$, for any t and $i \in \mathcal{N}$. The stochastic assumption about the loss function corresponds to the underlying task distribution $\mathcal{P}_T$ of meta-learning in an implicit manner. The objective here is to make a sequence of predictions $\{x_{t,i}\}$ given the knowledge of previous ones and possibly additional information so as to minimize the average regret (achieved at each agent) compared with the best predictor, given as:

$$\mathbf{R} = \frac{1}{N} \left[\sum_{i=1}^{N} \sum_{t=1}^{T} f_{t,i}(x_{t,i}) - \sum_{i=1}^{N} \sum_{t=1}^{T} f_{t,i}(x^*) \right], \tag{8.4}$$

where $x^* = \arg\min \mathbb{E}_{f_{t,i} \sim \mathcal{P}}[f_{t,i}(x)]$. Note that the above problem formulation is closely related to but different from the classical stochastic optimization in the following sense [77]: Stochastic optimization is primarily concerned with finding the optimal solution efficiently, for a given underlying model distribution. In stark contrast, for the (stochastic) online convex optimization, each agent makes a sequence of decisions in a real-time manner when new data arrives, and the objective is to make a sequence of model predictions that results in a small cumulative loss along the way. In this study, distributed OCO algorithms are devised to reduce the average regret per agent with limited communication, compared with the single agent case.

Since the regret depends on the distribution of $f_{t,i}$, we focus on the expected regret $\mathbb{E}[\mathbf{R}]$, which is the same across agents because $\{f_{t,i}\}$ follow the same unknown distribution $\mathcal{P}$. It is well known that in the centralized case OGD can achieve the optimal regret $\mathbb{E}[\mathbf{R}] = O(\sqrt{T/N})$ after totally NT iterations are executed sequentially. In the distributed case where each agent runs OGD alone with no communication, it is clear that the regret $\mathbb{E}[\mathbf{R}]$ at each agent has the order of $O(\sqrt{T})$, which is a factor of $\sqrt{N}$ worse than the centralized case. This performance gap points to the need of the collaboration among agents in order to obtain the optimal regret per agent.

8.4.1 Distributed OGD with Gradient Tracking

Gradient tracking has shown great potentials in distributed optimization to improve the convergence rate through the collaboration among agents [203, 262, 267, 325]. Particularly, by taking advantage of the smoothness of the local functions, an accurate estimation of the global gradient can be obtained as a better descent direction based on the history information, in contrast to gradient descent with local gradients. Nevertheless, the benefit of gradient tracking, especially the acceleration capability, is not well understood in distributed online learning where one cares about the learning process. To fully unleash the potential of gradient tracking, we explore a distributed OGD algorithm with gradient tracking (DOGD-GT) in order to achieve the performance speedup at each agent for distributed OCO, as outlined in Algorithm 8.1.

More specifically, an auxiliary variable $s_{t,i}$ is introduced for each agent to track the average gradients over the network by leveraging history information:

$$s_{t,i} = \sum_{j \in \mathcal{N}_i} w_{ij} s_{t-1,j} + \nabla f_{t,i}(x_{t,i}) - \nabla f_{t-1,i}(x_{t-1,i}),$$

which serves as a more accurate estimation of the global gradient $\frac{1}{N}\sum_i \nabla f_{t,i}(x_{t,i})$, in contrast to the local gradient $\nabla f_{t,i}(x_{t,i})$. As a result, the local model at each agent is updated based on $s_{t,i}$ using the gradient descent:

$$x_{t+1,i} = \sum_{j \in \mathcal{N}_i} w_{ij} x_{t,j} - \eta s_{t,i}.$$

Compared with the standard distributed OGD (DOGD) algorithms, DOGD-GT has the same order of the communication cost, which is much smaller than that in the distributed mini-batch algorithm proposed in Dekel et al. [77], where additional consensus steps are needed in the network after every iteration.

Algorithm 8.1 Distributed OGD with gradient tracking

Initialize $x_{1,i} = 0$ for all $i \in \mathcal{N}$;
for $t = 1, 2, ..., T$ **do**
    for each agent i **do**
        Apply local model $x_{t,i}$ and incur loss $f_{t,i}(x_{t,i})$;
        Compute gradient $\nabla f_{t,i}(x_{t,i})$;
        if $t = 1$ **then**
            Query the local model $x_{t,j}$ from neighbors $j \in \mathcal{N}_i$;
            Compute $s_{t,i} = \nabla f_{t,i}(x_{t,i})$;
        else
            Query the local model $x_{t,j}$ and $s_{t-1,j}$ from neighbors $j \in \mathcal{N}_i$;
            Update $s_{t,i} = \sum_{j \in \mathcal{N}_i} w_{ij} s_{t-1,j} + \nabla f_{t,i}(x_{t,i}) - \nabla f_{t-1,i}(x_{t-1,i})$;
        end if
        Update $x_{t+1,i} = \sum_{j \in \mathcal{N}_i} w_{ij} x_{t,j} - \eta s_{t,i}$;
    end for
end for

8.4.2 Performance Analysis

We next quantify the performance speedup brought by the limited collaboration among agents in DOGD-GT. We first impose the following standard assumptions.

Assumption 8.1 Each $f_{t,i}(x)$ is convex and L-smooth. And there exists some constant D such that $\mathbb{E}[\|\nabla f_{t,i}(x)\|^2] \leq D$.

Assumption 8.2 Let $F(x) = \mathbb{E}[f_{t,i}(x)]$. The stochastic gradient $\nabla f_{t,i}(x)$ has a σ^2-bounded variance, i.e., there exists a constant $\sigma \geq 0$ such that

$$\mathbb{E}[\|\nabla f_{t,i}(x) - \nabla F(x)\|^2] \leq \sigma^2.$$

Let ρ denote the spectral norm of $W - \frac{1}{N}\mathbf{1}\mathbf{1}^T$ where $\mathbf{1}$ denotes an N-dimensional all one column vector, then $\rho \in (0, 1)$. Moreover, it can be shown that [267]

$$\|Wx - \mathbf{1}\bar{x}\| \leq \rho\|x - \mathbf{1}\bar{x}\| \tag{8.5}$$

where $\bar{x} = \frac{1}{N}\mathbf{1}^T x$.

Let $\bar{x}_t = \frac{1}{N}\sum_{i=1}^{N} x_{t,i}$, and $x_t = [x'_{t,1}, x'_{t,2}, \ldots, x'_{t,N}]'$. To analyze the regret of DOGD-GT, we note that the techniques in stochastic optimization [262] cannot be directly applied here, because it is necessary to track the regret accumulated within the learning process instead of the optimality gap $\lim_{t\to\infty}(f_{t,i}(x_{t,i}) - f_{t,i}(x^*))$. In light of this, we decompose the regret into two parts: (a) the regret $\sum_{i=1}^{N}\sum_{t=1}^{T}[f_{t,i}(x_{t,i}) - f_{t,i}(\bar{x}_t)]$ resulted from the consensus error among agents, and (b) the regret $\sum_{i=1}^{N}\sum_{t=1}^{T}[f_{t,i}(\bar{x}_t) - f_{t,i}(x^*)]$ accumulated over the iterations of $\bar{x}_t$.

For (a), we first have the following lemma to characterize the relationship between the regret and the consensus gap between model parameters.

Lemma 8.1 *Under Assumption 1, the following inequality holds:*

$$\mathbb{E}\left[\sum_{i=1}^{N}\sum_{t=1}^{T} f_{t,i}(x_{t,i}) - \sum_{i=1}^{N}\sum_{t=1}^{T} f_{t,i}(\bar{x}_t)\right] \leq 2L\sum_{t=1}^{T}\mathbb{E}\left[\|x_t - \mathbf{1}\bar{x}_t\|^2\right].$$

Next, we follow a similar way as in Pu and Nedić [262] to build a linear system to bound the consensus error $E[\|x_{t,i} - \bar{x}_t\|^2]$.

Lemma 8.2 *Let* $\alpha = \frac{3+\rho^2}{4}$. *Under Assumptions 1 and 2, the following inequality holds for some constant* A_1 *and* A_2*:*

$$\sum_{t=1}^{T}\mathbb{E}[\|x_t - \mathbf{1}\bar{x}_t\|^2] \leq A_1\frac{\alpha - \alpha^T}{1-\alpha} + \|x_1 - \mathbf{1}\bar{x}_1\|^2$$
$$+ A_2\eta^2\frac{1+\rho^2}{1-\rho^2}[18\eta^2\sigma^2L^2 + 18N\eta^2L^2D + (1 + \eta LN + N)\sigma^2]T.$$

The challenge lies in the characterization of the convergence rate of the consensus error, which needs a careful manipulation and analysis of the coefficient matrices in the linear system.

For (b), the key question is how to analyze this regret term without strong convexity. The techniques from Qu and Li [267] and Pu and Nedić [262] cannot be applied, as the former considers that each agent has the same loss function in the entire learning process and the later assumes the strong convexity. To resolve this issue, we quantify both the optimality gap at iteration $t+1$, i.e., $f_{t,i}(\bar{x}_{t+1}) - f_{t,i}(x^*)$, and the one-iteration gap between iteration t and iteration $t+1$, i.e., $f_{t,i}(\bar{x}_t) - f_{t,i}(\bar{x}_{t+1})$. In this way, we can characterize the relationship between the optimality gap and the consensus error, and bound the one-iteration gap by the norm of global gradients, which leads to the following result.

Lemma 8.3 *Under Assumptions 1 and 2, the following inequality holds:*

$$
\begin{aligned}
&\mathbb{E}\left[\sum_{i=1}^{N}\sum_{t=1}^{T} f_{t,i}(\bar{x}_t) - \sum_{i=1}^{N}\sum_{t=1}^{T} f_{t,i}(x^*)\right] \\
\leq &\frac{4N\|\bar{x}_1 - x^*\|^2}{\eta} + 26L\sum_{t=1}^{T}\mathbb{E}[\|x_t - \mathbf{1}\bar{x}_t\|^2] + \frac{N\eta}{2}\mathbb{E}[\|\nabla F(\bar{x}_{T+1})\|^2] \\
&+ 2\sigma^2\eta T + 24L\mathbb{E}[\|x_{T+1} - \mathbf{1}\bar{x}_{T+1}\|^2].
\end{aligned}
$$

Based on Lemmas 8.1, 8.2 and 8.3, we have the following result about the average regret per agent.

Theorem 8.1 *Under Assumptions 1 and 2, when η satisfies that*

$$
\eta \leq \min\left\{\frac{(1-\rho^2)^{1.5}}{32L\sqrt{1+\rho^2}}, \frac{1}{2L}\sqrt{\frac{N}{T}}\right\}
$$

with $N = o(T^{1/3})$, the DOGD-GT algorithm attains the following regret bound:

$$
\mathbb{E}[\mathbf{R}] = O\left(\eta^2 T + \frac{\eta T}{N} + \frac{1}{\eta}\right) = O\left(\sqrt{\frac{T}{N}}\right).
$$

Remark 8.1 (1) Theorem 8.1 indicates that each agent can achieve a factor of $\sqrt{1/N}$ speedup in terms of the average regret $\mathbb{E}[\mathbf{R}]$, through only one communication step per iteration by leveraging gradient tracking, compared to the case where a single agent can achieve a regret of order $O(\sqrt{T})$ without collaboration with other agents. (2) The overall regret obtained by DOGD-GT, i.e., $O(\sqrt{NT})$, also matches the optimal regret in the centralized case where NT iterations are processed sequentially. (3) Note that the learning rate η requires the knowledge of the time horizon T, which however can be relaxed by applying a standard doubling trick [48].

Remark 8.2 The classical DOGD algorithm [402] cannot achieve such performance gain in the setting here, because essentially DOGD performs a consensus step followed by a gradient descent along the local gradient $\nabla f_{t,i}(x_{t,i})$. For a fixed learning rate, DOGD only converges to a neighborhood of the optimizer x^*, because the local gradient is data-driven and hence random. Such an oscillation around x^* slows down the convergence and results in a larger regret, calling for a more elegant consensus algorithm. This is also corroborated by the consensus schemes in the work on distributed multi-armed bandits [187, 296]. In contrast, gradient tracking provides an efficient way to communicate local estimations of the global gradient with the neighbors, and each agent is able to quickly construct a more accurate estimate of the global gradient $\frac{1}{N}\sum_i \nabla f_{t,i}(x_{t,i})$ with only one communication step per iteration as the information diffuses in the network until consensus. And the global gradient estimation clearly serves as a better direction than the local gradient no matter the stochasticity is in place or not, leading to a better regret bound.

8.5 MAOML

Thanks to gradient tracking, the proposed DOGD-GT algorithm clearly showcases the potential for accelerating the learning process in distributed OCO through limited collaboration among agents. To reap the potential benefits, we next devise a multi-agent online meta-learning (MAOML) algorithm based on DOGD-GT, to mitigate the cold-start problem.

Algorithm 8.2 MAOML

Initialize $\phi_{1,n}$ and $\alpha_{1,n}$ for all $n \in \mathcal{N}$;
for $t = 1, 2, ..., T$ **do**
  for each agent n **do**
    Receive task $\mathcal{T}_{t,n}$ which would be learnt for m rounds;
    for round $i \in [m]$ **do**
      Run online mirror descent with $\phi_{t,n}$ and $\alpha_{t,n} = \frac{v_{t,n}}{G\sqrt{m}}$ to obtain $\theta^i_{t,n}$; // **(within-task adaptation)**
      Incur loss $l^i_{t,n}(\theta^i_{t,n})$;
    end for
    Run DOGD-GT with all agents to update $\phi_{t+1,n}$ and $\alpha_{t+1,n} = \frac{v_{t+1,n}}{G\sqrt{m}}$; // **(multi-agent meta-update of OMD initialization and learning rate)**
  end for
end for

As shown in (8.1), $\hat{\mathbf{R}}_{t,n}$ is a joint function for $\phi_{t,n}$ and $\alpha_{t,n}$, and it would be easier to learn $\phi_{t,n}$ and $\alpha_{t,n}$ separately [167]. Specifically, the distributed network-level OCO can be decoupled as two separate distributed OCOs over the following two function sequences $\{f^{init}_{t,n}\}_{t,n}$ and $\{f^{rate}_{t,n}\}_{t,n}$ for every task at each agent:

$$f^{init}_{t,n}(\phi) = \mathcal{B}_R(\theta^*_{t,n}||\phi)G\sqrt{m},$$
$$f^{rate}_{t,n}(v) = \left(\frac{\mathcal{B}_R(\theta^*_{t,n}||\phi_{t,n})}{v} + v\right)G\sqrt{m}.$$

In what follows, we make a few further remarks on the algorithm design:

- Here for each agent n at every task $\mathcal{T}_{t,n}$, the model initialization $\phi_{t,n}$ is updated based on DOGD-GT over the function $f^{init}_{t,n}$ for $\phi_{t,n} \in \Theta$, and the learning rate $\alpha_{t,n} = \frac{v_{t,n}}{G\sqrt{m}}$ where $v_{t,n}$ is updated based on DODG-GT over the function $f^{rate}_{t,n}$ for $v_{t,n} \geq \epsilon > 0$. By assuming that $\mathcal{B}_R(\theta||\phi) \leq H^2$ for any $\theta, \phi \in \Theta$, it is easy to check that $f_{t,n}^{rate}(v)$ is convex and $\frac{2H^2}{\epsilon^3}$-smooth for $v \in [\epsilon, \infty)$.

- Note that although the global optimal ϕ^* and α^* exist for all tasks, at each iteration t different agents would have distinct model initialization $\phi_{t,n}$ and learning rate $\alpha_{t,n}$ for their current tasks.
- And for implementation, one can use the last iterate $\theta_{t,n}^m$ to replace the optimal $\theta_{t,n}^*$, which incurs an additional $o(\sqrt{m})$ regret term only for many practical settings [167].

The details are summarized in Algorithm 8.2.

8.5.1 Performance Analysis

Based on Theorem 8.1, we have the following result about the performance of MAOML.

Theorem 8.2 *Suppose that the model initialization $\phi_{t,n}$ and $v_{t,n}$ are updated based on DOGD-GT with $\alpha_{t,n} = \frac{v_{t,n}}{G\sqrt{m}}$. Then, the ATAR achieved by each agent in the multi-agent online meta-learning satisfies that*

$$\mathbb{E}[\mathbf{R}_a] \leq \mathbb{E}[\bar{\mathbf{R}}_a] = O\left(\frac{1+\frac{1}{V_\phi}}{\sqrt{NT}} + V_\phi\right)\sqrt{m}$$

where $V_\phi^2 = \min_{\phi\in\Theta} \mathbb{E}[\mathcal{B}_R(\theta_{t,n}^*||\phi)]$.

To obtain a more concrete sense about the performance improvement of MAOML, we compare it with the single-agent online meta-learning, i.e., $N = 1$ (thus ignore the subscript n). In particular, we apply the general algorithm (Algorithm 1 therein) [167] to our setting here, which yields the following proposition.

Proposition 8.1 *Suppose that the model initialization is updated based on $\phi_t = \frac{1}{t-1}\sum_{j=1}^{t-1}\theta_j^*$, and the learning rate $\alpha_t = \frac{v_t}{G\sqrt{m}}$ where v_t is updated using simplified exponentially-weighted online-optimization (EWOO) [139] with parameter $\epsilon = \frac{1}{T^{1/4}}$. Then, the ATAR achieved by the single-agent online meta-learning satisfies that*

$$\mathbb{E}[\mathbf{R}_a] \leq \mathbb{E}[\bar{\mathbf{R}}_a] = \tilde{O}\left(\min\left\{\frac{1+\frac{1}{V_\phi}}{\sqrt{T}}, \frac{1}{T^{1/4}}\right\} + V_\phi\right)\sqrt{m}$$

where $V_\phi^2 = \min_{\phi\in\Theta} \mathbb{E}[\mathcal{B}_R(\theta_t^*||\phi)]$.

Remark 8.3 (1) It can be seen from Proposition 8.1 that for the single-agent online meta-learning, if V_ϕ, the average deviation of $\theta_{t,n}^*$, is $\Omega_T(1)$, then the ATAR approaches

$O(V_\phi\sqrt{m})$ at rate $\tilde{O}(1/\sqrt{T})$. In contrast, with the same number T of online learning tasks, each agent in multi-agent online meta-learning can achieve a clear performance gain by utilizing the task similarity across multiple agents through the limited collaboration, i.e., the ATAR approaches $O(V_\phi\sqrt{m})$ at a faster rate of $O(1/\sqrt{NT})$. Although we consider $m_{t,n} = m$ for simplicity, it is worth to note that the results still hold as long as all $\{m_{t,n}\}$ follow some distribution across all tasks.

(2) Moreover, the result shown in Theorem 8.2 also matches the optimal performance in the centralized case with NT tasks in total. It is worth to note that, for the set Θ with diameter H, the single-task regret achieved by OGD is $O(H\sqrt{m})$, whereas in online meta-learning the optimal regret for each task is smaller, i.e., $O(V_\phi\sqrt{m})$ when the optimal $\theta^*_{t,n}$ are close, especially for the few-shot setting of a small m [167].

(3) Built on joint learning of the model initialization and the learning rate from all past tasks, online meta-learning is intimately related to the regularization-based methods, particularly the prior-focused methods, in continual learning [76]. This strong connection indicates that the multi-agent online meta-learning methods can be used to speed up learning in continual learning, in particular, few-shot continual learning where each task only has a few data samples.

8.6 Experiments

In what follows, we present extensive experiments on both DOGD-GT and MAOML which corroborate the theoretic results in previous sections, respectively.

We first introduce the setup of the communication graph for the multi-agent network. More specifically, we consider that N agents communicate in a random network [262, 363], where each two agents are linked with probability 0.5 (discard the graphs that are not connected). And the weight matrix W is defined based on the Metropolis rule [290]:

$$w_{ij} = \begin{cases} 1/\max\{\deg(i), \deg(j)\} & \text{if } j \in \mathcal{N}_i, \\ 1 - \sum_{j\in\mathcal{N}_i} w_{ij} & \text{if } i = j, \\ 0 & \text{otherwise,} \end{cases}$$

where $\deg(i)$ is the degree of agent i. We also consider a complete communication graph where all agents are connected with each other for evaluating the performance of MAOML.

8.6.1 Performance of DOGD-GT

As in [262, 267], we study the online Ridge regression problem, where each agent i at each iteration t incurs the following loss:

$$f_{t,i}(x_{t,i}) = (u_{t,i}^T x_{t,i} - v_{t,i})^2 + \rho \|x_{t,i}\|^2$$

for a given model $x_{t,i}$ and the data sample $(u_{t,i}, v_{t,i})$. Here $\rho > 0$ is a penalty parameter.

In the experiments, each $u_{t,i}$ is uniformly sampled from $[0.3, 0.4]^p$ with dimension p, and $v_{t,i}$ is generated according to $v_{t,i} = u_{t,i}^T \tilde{x}_{t,i} + \epsilon_{t,i}$, where $\tilde{x}_{t,i}$ is a predefined parameter, and $\epsilon_{t,i}$ are independent Gaussian random noises with mean 0 and variance 0.5. For completeness, we evaluate the performance of DOGD-GT in both stochastic and adversarial setups:

(1) *Stochastic setup*: all $\tilde{x}_{t,i}$ are the same in this case, set as a constant from $[0, 5]^p$;
(2) *Adversarial setup*: $\tilde{x}_{t,i}$ are randomly and independently located in $[0, 10]^p$ in this case. Moreover, $\rho = 0.001$, $p = 10$, and the learning rate $\eta = 0.001$. We evaluate the average learning performance by measuring the average loss $\frac{1}{NT} \sum_{i=1}^{N} \sum_{t=1}^{T} f_{t,i}(x_{t,i})$ as in [402] over multiple simulations.

To demonstrate the performance gain achieved by gradient tracking, we compare the performance of DOGD-GT with both DOGD and the single agent approach. Clearly, as shown in Fig. 8.2a and c, DOGD-GT outperforms DOGD and the single agent approach in both stochastic and adversarial setups, indicating the benefits brought by collaborating with neighbors to track the global gradient via limited communication. We also evaluate the impact of the network size N on the performance of DOGD-GT. As expected, it can be seen from Fig. 8.2b and d that the learning performance improves with N for both stochastic and adversarial setups, validating the results in Theorem 8.1.

To further validate the performance of DOGD-GT, we study the online multiclass logistic regression on the MNIST dataset. For a batch $\mathcal{B}_{t,i}$ of data samples $j = (u_{t,i}^j, v_{t,i}^j) \in \mathbb{R}^d \times \{0, \ldots, 9\}$ where $u_{t,i}^j$ is the feature and $v_{t,i}^j$ is the label, the logistic loss function for $x_{t,i} \in \mathbb{R}^{d \times 10}$ is defined as:

$$f_{t,i}(x_{t,i}) = \frac{-1}{|\mathcal{B}_{t,i}|} \sum_{j \in \mathcal{B}_{t,i}} \sum_{v=0}^{9} \mathbf{1}\{v_{t,i}^j = v\} \log \frac{\exp(x_{t,i}^T u_{t,i}^j)}{\sum_{k=0}^{9} \exp(x_{t,i}^T (k) u_{t,i}^j)}.$$

Specifically, each batch $\mathcal{B}_{t,i}$ is randomly and independently sampled from the entire MNIST dataset. We set $d = 784$ and $\eta = 0.0001$. As shown in Fig. 8.4a, DOGD-GT outperforms both DOGD and the single agent case. When N increases, the performance of DOGD-GT will be better, as illustrated in Fig. 8.4b.

8.6.2 Performance of MAOML

Next, we evaluate the performance of MAOML in both convex and nonconvex setups: (1) *Convex setup*: we consider the online multiclass logistic regression [363] on the MNIST

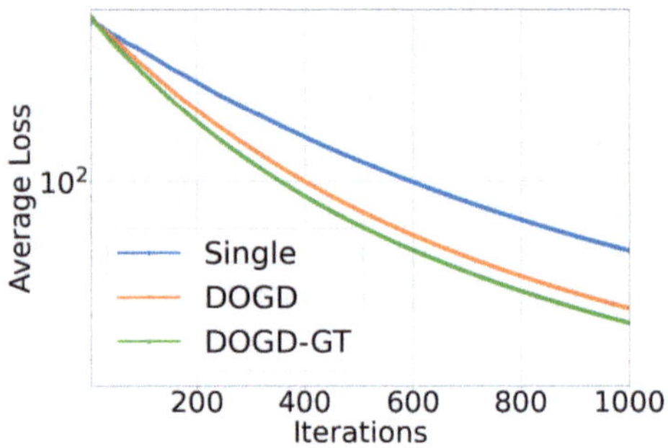

(a) Performance comparison in stochastic setup for $N = 8$.

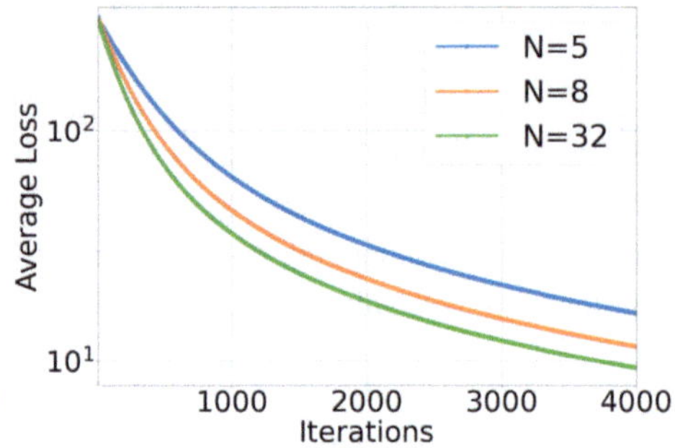

(b) Impact of N in stochastic setup on DOGD-GT.

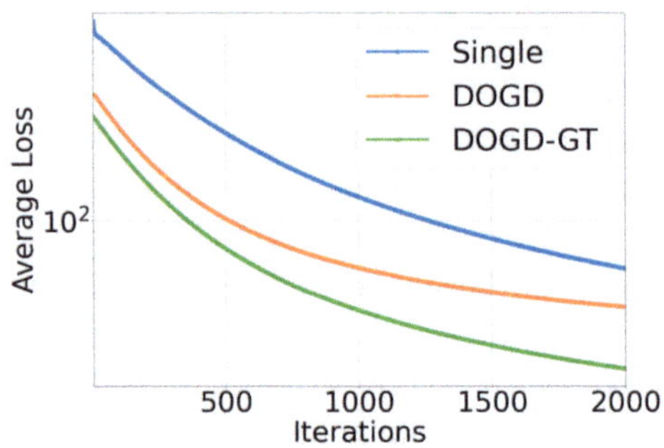

(c) Performance comparison in adversarial setup for $N = 12$.

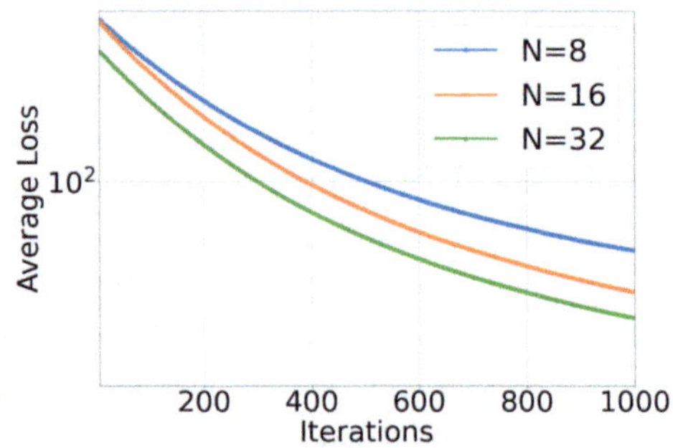

(d) Impact of N in adversarial setup on DOGD-GT.

Fig. 8.2 Performance evaluations of DOGD-GT on synthetic data

dataset as an online learning task $\mathcal{T}_{t,n}$ at each agent. For a batch $\mathcal{B}_{t,n}^i$ of data points $j = (u_{t,n}^j, v_{t,n}^j) \in \mathbb{R}^d \times \{0, \ldots, 9\}$ where $u_{t,n}^j$ is the feature and $v_{t,n}^j$ is the label, the logistic loss function for $\theta \in \mathbb{R}^{d\times 10}$ is defined as:

$$l_{t,n}^i(\theta) = \frac{-1}{|\mathcal{B}_{t,n}^i|} \sum_{j\in\mathcal{B}_{t,n}^i} \sum_{v=0}^{9} \mathbf{1}_{\{v_{t,n}^j = v\}} \log \frac{\exp(\theta^T u_{t,n}^j)}{\sum_{k=0}^{9} \exp(\theta^T(k) u_{t,n}^j)}.$$

And for each agent we consider 5-way 10-shot classification with the dataset randomly sampled from the entire dataset. (2) *Nonconvex setup*: we study 5-way 5-shot classification on Omniglot [185] as the online learning task $\mathcal{T}_{t,n}$ using a deep neural network (DNN). The DNN architecture for each task consists of two $2D$ convolutional layers (first with 6 output channels and second with 16 output channels) with kernel sizes 5×5. Each convolution operation is followed first by ReLu non-linearity, and then by $2D$ max-pooling operation with stride of 2. The final layer is a fully connected layer with input of size $16 \times 4 \times 4$ and

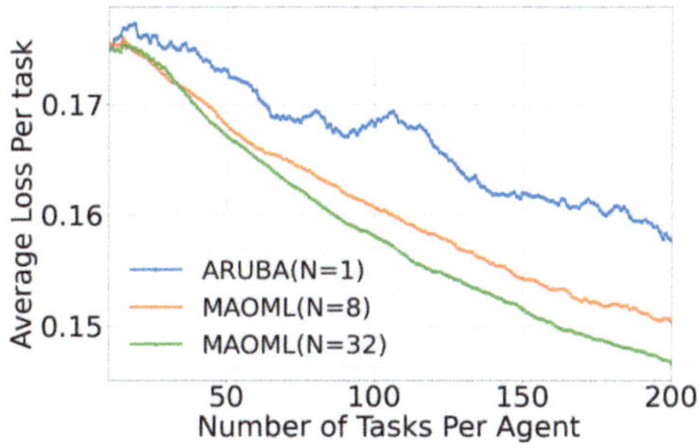

(a) Performance comparison for 5-way 10-shot MNIST.

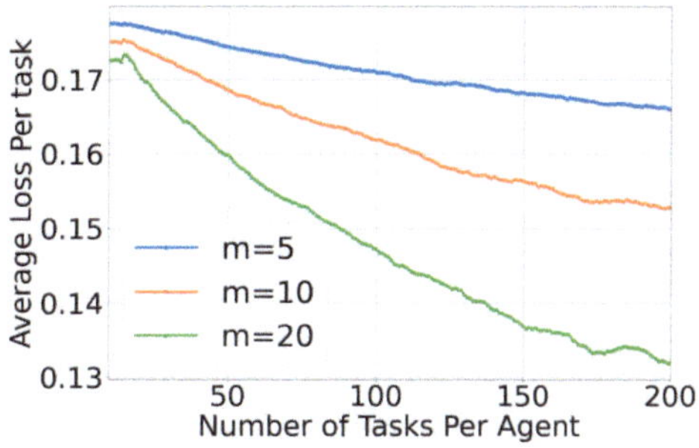

(b) Impact of m on MAOML for 5-way 10-shot MNIST.

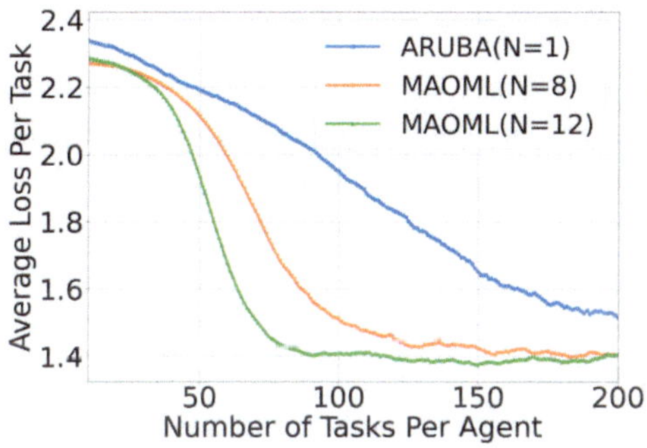

(c) Performance comparison for 5-way 5-shot Omniglot.

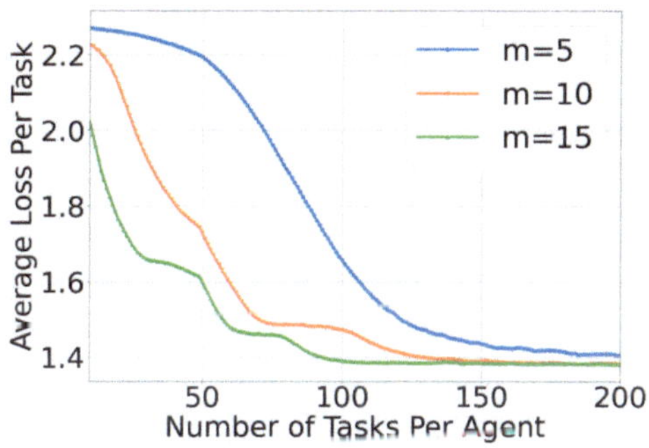

(d) Impact of m on MAOML for 5-way 5-shot Omniglot.

Fig. 8.3 Performance evaluations of MAOML on MNIST and Omniglot

output of size 10. We deploy the cross entropy to quantify the loss with respect to a single sample. In the experiments, we evaluate the average learning performance by measuring the average loss $\frac{1}{NTm}\sum_{n=1}^{N}\sum_{t=1}^{T}\sum_{i=1}^{m} l_{t,n}^{i}(\theta_{t,n}^{i})$. Along the same line in Khodak et al. [166], we use OGD as the learning algorithm within each task.

When applying DOGD-GT to update the model initialization $\phi_{t,n}$ and $v_{t,n}$, in the experiments, we set the learning rate $\eta = 0.001$ for the outer loop meta-update with DOGD-GT. For the selection of G, we test different values and choose the one with the best performance for every experimental setup. For example, when we use the logistic regression for the few-shot classification on MNIST, we set $G = 80$. We further clarify the parameters used in different experiments: (1) For Fig. 8.3a, we set $m = 10$; (2) For Fig. 8.3b and d, we set $N = 8$; (3) For Fig. 8.3c, we set $m = 5$.

We first compare the performance of MAOML under different number of agents with the single-agent general algorithm ARUBA in Khodak et al. [167]. As shown in Fig. 8.3a and c, MAOML clearly outperforms ARUBA, by utilizing the task similarity across multiple agents

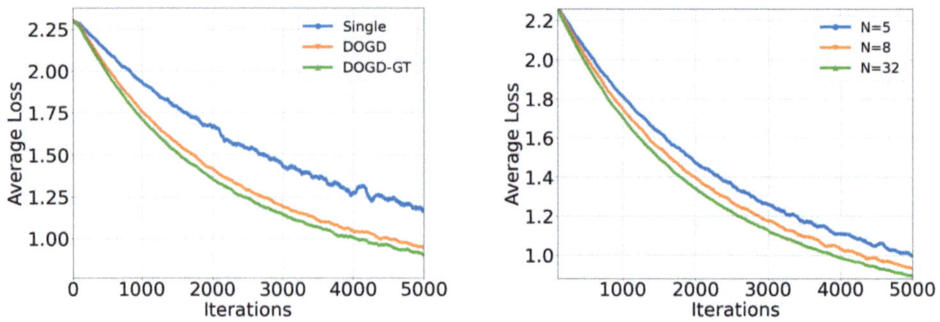

(a) Performance comparison for $N = 8$. (b) Impact of N on DOGD-GT.

Fig. 8.4 Performance evaluation of DOGD-GT on MNIST

through limited communication in both convex and nonconvex setups. More specifically, compared with ARUBA, MAOML learns good model priors at a faster rate, and performs significantly better after each agent learns over the same number of tasks. Moreover, with more agents collaborating in the network, the performance of MAOML increases further, corroborating the results in Theorem 8.2. We next examine the impact of m, i.e., the number of iterations within each task, on the learning performance of MAOML. As expected, the average loss per task decreases with m because each task has a sublinear regret on average, as illustrated in Fig. 8.3b and d.

Following the same line as in Khodak et al. [166, 167], we also evaluate the performance of MAOML in a meta-testing setup. More specifically, for each pair of $(\phi_{t,n}, v_{t,n})$ obtained at each iteration t, we test its performance on a set of testing tasks. For each testing task $\mathcal{T}_{t,n}^{te}$ at each agent with a training dataset $\mathcal{D}_{t,n}^{tr}$ and a testing dataset $\mathcal{D}_{t,n}^{te}$, we first run online gradient descent from the model initial $\phi_{t,n}$ with the learning rate $\alpha_{t,n} = \frac{v_{t,n}}{G\sqrt{m}}$ for m iterations using the training dataset $\mathcal{D}_{t,n}^{tr}$, and obtain the task specific model parameter $\theta_{t,n}^{te}$. Next, we evaluate the accuracy of $\theta_{t,n}^{te}$ on the testing dataset $\mathcal{D}_{t,n}^{te}$ for each testing task $\mathcal{T}_{t,n}^{te}$.

We run the experiments for 5-way 2-shot classification and 5-way 5-shot classification on Omniglot, and evaluate the average testing accuracy over 10 testing tasks after each iteration t for every agent. Particularly, we consider a complete graph where all agents are connected with each other, and set $m = 50$. As shown in Figs. 8.5 and 8.6, MAOML clearly achieves a better meta-testing accuracy compared with ARUBA, and its performance further increases as the number of agents N increases. Therefore, by utilizing the task similarity across different agents through limited collaboration among them, each agent can achieve good testing performance in MAOML after learning over a smaller number of tasks, in contrast to learning alone by itself.

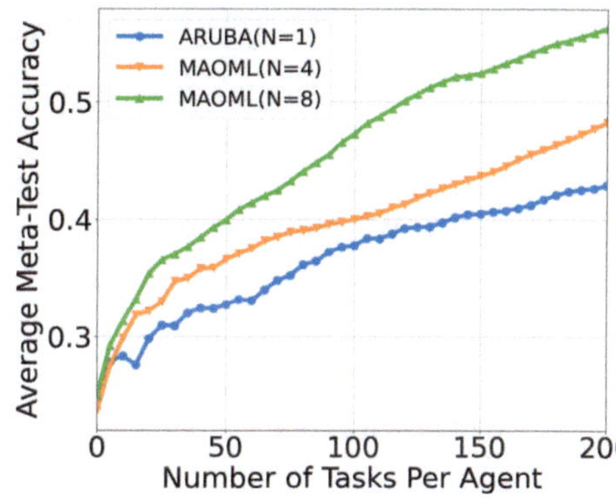

Fig. 8.5 Meta-testing performance evaluation of MAOML on 5-way 2-shot Omniglot

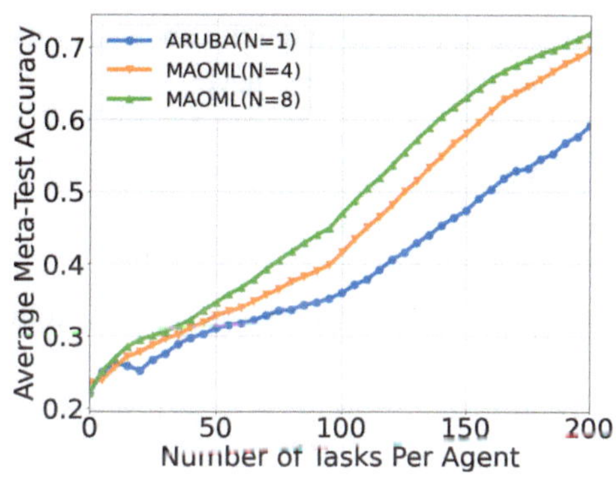

Fig. 8.6 Meta-testing performance evaluation of MAOML on 5-way 5-shot Omniglot

8.7 Proofs

For ease of exposition, we define the following average sequences:

$$\bar{x}_t = \frac{1}{N}\sum_{i=1}^{N} x_{t,i}, \quad \bar{s}_t = \frac{1}{N}\sum_{i=1}^{N} s_{t,i}, \quad g_t = \frac{1}{N}\sum_{i=1}^{N} \nabla f_{t,i}(x_{t,i}),$$

and further, rewrite $\{x_{t,i}\}$, $\{s_{t,i}\}$ and $\{\nabla f_{t,i}(x_{t,i})\}$ in vector form, i.e.,

$$x_t = \begin{bmatrix} x_{t,1} \\ x_{t,2} \\ \vdots \\ x_{t,N} \end{bmatrix}, \quad s_t = \begin{bmatrix} s_{t,1} \\ s_{t,2} \\ \vdots \\ s_{t,n} \end{bmatrix}, \quad \nabla_t = \begin{bmatrix} \nabla f_{t,1}(x_{t,1}) \\ \nabla f_{t,2}(x_{t,2}) \\ \vdots \\ \nabla f_{t,n}(x_{t,n}) \end{bmatrix}.$$

Based on Algorithm 8.1, the update rule can be reformulated as

$$s_t = W s_{t-1} + \nabla_t - \nabla_{t-1}, \tag{8.6}$$

$$x_{t+1} = W x_t - \eta s_t, \tag{8.7}$$

where $s_1 = \nabla_1$. We define $G_t = \begin{bmatrix} \nabla F(x_{t,1}) \\ \vdots \\ \nabla F(x_{t,N}) \end{bmatrix}$ as the expected gradient at x_t.

8.7.1 Preliminaries

To facilitate the regret analysis, we first restate some useful results in the literature. More specifically, to understand the updates of the average sequences $\bar{s}_t$ and $\bar{x}_t$, based on Lemma 7 in Qu and Li [267], we have

Lemma 8.4 *The following equalities hold. (a)* $\bar{s}_{t+1} = \bar{s}_t + g_{t+1} - g_t = g_{t+1}$; *(b)* $\bar{x}_{t+1} = \bar{x}_t - \eta \bar{s}_t = \bar{x}_t - \eta g_t$.

Proof Since W is doubly stochastic, it follows that $\mathbf{1}^T W = \mathbf{1}^T$. To prove (a), we have

$$\begin{aligned} \bar{s}_{t+1} =& \frac{1}{N}\mathbf{1}^T s_{t+1} \\ =& \frac{1}{N}\mathbf{1}^T (W s_t + \nabla_{t+1} - \nabla_t) \\ =& \bar{s}_t + g_{t+1} - g_t. \end{aligned}$$

Telescoping the above equation, we have $\bar{s}_{t+1} = \bar{s}_1 + g_{t+1} - g_1$. Since $\bar{s}_1 = g_1$, we can obtain $\bar{s}_{t+1} = g_{t+1}$.

To prove (b), we have

$$\begin{aligned} \bar{x}_{t+1} =& \frac{1}{N}\mathbf{1}^T x_{t+1} \\ =& \frac{1}{N}\mathbf{1}^T W x_t - \eta s_t \\ =& \bar{x}_t - \eta \bar{s}_t \\ =& \bar{x}_t - \eta g_t. \end{aligned}$$

□

Denote $\mathcal{F}_t$ as the σ-algebra generated by the sequence $\{f_1, f_2, \ldots, f_{t-1}\}$ where $f_t = [f_{t,1}^T, \ldots, f_{t,N}^T]^T$, and define $\mathbb{E}[\cdot|\mathcal{F}_t]$ as the conditional expectation given $\mathcal{F}_t$. Based on[262], we have the following two lemmas.

Lemma 8.5 *The following inequality holds:*

$$\mathbb{E}[\langle Ws_t - \mathbf{1}g_t, G_t - \nabla_t\rangle | \mathcal{F}_t] \leq \sigma^2.$$

Lemma 8.6 *The following inequality holds:*

$$\mathbb{E}[\langle G_{t+1}, G_t - \nabla_t\rangle | \mathcal{F}_t] \leq \eta L N \sigma^2.$$

We also need the following standard [52, 77, 331].

Lemma 8.7 *Let V be a closed convex set, ϕ be a convex function on V, and h be a differentiable, strongly convex function on V. Let d be the Bregman divergence generated by h. Given $u \in V$, if*

$$w^+ = \arg\min_{w \in V}\{\phi(w) + d(w, u)\},$$

then

$$\phi(w) + d(w, u) \geq \phi(w^+) + d(w^+, u) + d(w, w^+).$$

8.7.2 Regret Analysis

In order to prove Theorem 8.1, based on (8.4), we can first rewrite the regret as

$$\begin{aligned} N\mathbf{R} = \mathbf{R}_s &= \sum_{i=1}^{N}\sum_{t=1}^{T} f_{t,i}(x_{t,i}) - \sum_{i=1}^{N}\sum_{t=1}^{T} f_{t,i}(x^*) \\ &= \underbrace{\sum_{i=1}^{N}\sum_{t=1}^{T} f_{t,i}(x_{t,i}) - \sum_{i=1}^{N}\sum_{t=1}^{T} f_{t,i}(\bar{x}_t)}_{R_1} + \underbrace{\sum_{i=1}^{N}\sum_{t=1}^{T} f_{t,i}(\bar{x}_t) - \sum_{i=1}^{N}\sum_{t=1}^{T} f_{t,i}(x^*)}_{R_2}, \end{aligned} \tag{8.8}$$

where $x^* = \arg\min F(x)$. It can be seen from (8.8) that the regret can be decomposed into two terms: 1) R_1, the regret resulted by the difference between local model $x_{t,i}$ and the global average $\bar{x}_t$, and 2) R_2, the regret accumulated over the iteration of the global average $\bar{x}_t$.

8.7.2.1 Analysis of R_1

To analyze R_1, we first have the following lemma to characterize the relationship between the regret and the consensus gap between model parameters.

Lemma 8.8 *Under Assumption 1, the following inequality holds:*

$$\mathbb{E}\left[\sum_{i=1}^{N}\sum_{t=1}^{T} f_{t,i}(x_{t,i}) - \sum_{i=1}^{N}\sum_{t=1}^{T} f_{t,i}(\bar{x}_t)\right] \le 2L\sum_{t=1}^{T}\mathbb{E}\left[\|x_t - \mathbf{1}\bar{x}_t\|^2\right].$$

Proof For any x, we can have

$$\begin{aligned}
\frac{1}{N}\sum_{i=1}^{N} f_{t,i}(x) \le& \frac{1}{N}\sum_{i=1}^{N}\left\{f_{t,i}(x_{t,i}) + \langle\nabla f_{t,i}(x_{t,i}), x - x_{t,i}\rangle + \frac{L}{2}\|x - x_{t,i}\|^2\right\}\\
=& \frac{1}{N}\sum_{i=1}^{N}\left\{f_{t,i}(x_{t,i}) + \langle\nabla f_{t,i}(x_{t,i}), \bar{x}_t - x_{t,i}\rangle + \langle\nabla f_{t,i}(x_{t,i}), x - \bar{x}_t\rangle + \frac{L}{2}\|x - x_{t,i}\|^2\right\}\\
=& \frac{1}{N}\sum_{i=1}^{N}\{f_{t,i}(x_{t,i}) + \langle\nabla f_{t,i}(x_{t,i}), \bar{x}_t - x_{t,i}\rangle\} + \langle g_t, x - \bar{x}_t\rangle + \frac{L}{2N}\sum_{i=1}^{N}\|x - x_{t,i}\|^2\\
=& \frac{1}{N}\sum_{i=1}^{N}\{f_{t,i}(x_{t,i}) + \langle\nabla f_{t,i}(x_{t,i}), \bar{x}_t - x_{t,i}\rangle\} + \langle g_t, x - \bar{x}_t\rangle + \frac{L}{2N}\sum_{i=1}^{N}\|x - \bar{x}_t + \bar{x}_t - x_{t,i}\|^2\\
\le& \frac{1}{N}\sum_{i=1}^{N}\{f_{t,i}(x_{t,i}) + \langle\nabla f_{t,i}(x_{t,i}), \bar{x}_t - x_{t,i}\rangle\} + \langle g_t, x - \bar{x}_t\rangle + L\|x - \bar{x}_t\|^2 + \frac{L}{N}\|x_t - \mathbf{1}\bar{x}_t\|^2.
\end{aligned} \tag{8.9}$$

Moreover, based on the convexity and smoothness of $f_{t,i}$, it can be shown that

$$\begin{aligned}
f_{t,i}(\bar{x}_t) \ge& f_{t,i}(x_{t,i}) + \langle\nabla f_{t,i}(x_{t,i}), \bar{x}_t - x_{t,i}\rangle + \frac{1}{2L}\|\nabla f_{t,i}(\bar{x}_t) - \nabla f_{t,i}(x_{t,i})\|^2\\
\ge& f_{t,i}(x_{t,i}) + \langle\nabla f_{t,i}(x_{t,i}), \bar{x}_t - x_{t,i}\rangle.
\end{aligned}$$

Continuing with (8.9) and taking expectation at both sides, it follows that

$$\mathbb{E}\left[\frac{1}{N}\sum_{i=1}^{N} f_{t,i}(x)\right] \le \mathbb{E}\left[\frac{1}{N}\sum_{i=1}^{N} f_{t,i}(\bar{x}_t) + \langle g_t, x - \bar{x}_t\rangle + L\|x - \bar{x}_t\|^2 + \frac{L}{N}\|x_t - \mathbf{1}\bar{x}_t\|^2\right],$$

such that

$$\mathbb{E}[F(x)] \le \mathbb{E}[F(\bar{x}_t)] + \mathbb{E}[\langle g_t, x - \bar{x}_t\rangle] + L\mathbb{E}[\|x - \bar{x}_t\|^2] + \frac{L}{N}\mathbb{E}[\|x_t - \mathbf{1}\bar{x}_t\|^2]. \tag{8.10}$$

Since (8.10) holds for any x, we can have

$$\mathbb{E}[F(x_{t,i})] \le \mathbb{E}[F(\bar{x}_t)] + \mathbb{E}[\langle g_t, x_{t,i} - \bar{x}_t\rangle] + L\mathbb{E}[\|x_{t,i} - \bar{x}_t\|^2] + \frac{L}{N}\mathbb{E}[\|x_t - \mathbf{1}\bar{x}_t\|^2],$$

which indicates that

$$
\begin{aligned}
&\frac{1}{N}\sum_{i=1}^{N}\mathbb{E}[F(x_{t,i})]\\
\leq&\frac{1}{N}\sum_{i=1}^{N}\left\{\mathbb{E}[F(\bar{x}_t)]+\mathbb{E}[\langle g_t, x_{t,i}-\bar{x}_t\rangle]+L\mathbb{E}[\|x_{t,i}-\bar{x}_t\|^2]+\frac{L}{N}\mathbb{E}[\|x_t-\mathbf{1}\bar{x}_t\|^2]\right\}\\
=&\frac{1}{N}\sum_{i=1}^{N}\mathbb{E}[F(\bar{x}_t)]+\frac{1}{N}\sum_{i=1}^{N}\mathbb{E}[\langle g_t, x_{t,i}-\bar{x}_t\rangle]+\frac{L}{N}\sum_{i=1}^{N}\mathbb{E}[\|x_{t,i}-\bar{x}_t\|^2]+\frac{L}{N}\mathbb{E}[\|x_t-\mathbf{1}\bar{x}_t\|^2]\\
=&\frac{1}{N}\sum_{i=1}^{N}\mathbb{E}[F(\bar{x}_t)]+\mathbb{E}\left[\langle g_t, \frac{1}{N}\sum_{i=1}^{N}(x_{t,i}-\bar{x}_t)\rangle\right]+\mathbb{E}\left[\frac{L}{N}\sum_{i=1}^{N}\|x_{t,i}-\bar{x}_t\|^2\right]+\frac{L}{N}\mathbb{E}[\|x_t-\mathbf{1}\bar{x}_t\|^2]\\
=&\frac{1}{N}\sum_{i=1}^{N}\mathbb{E}[F(\bar{x}_t)]+\frac{2L}{N}\mathbb{E}[\|x_t-\mathbf{1}\bar{x}_t\|^2].
\end{aligned}
$$

Therefore, R_1 can be bounded above as follows:

$$
\begin{aligned}
\mathbb{E}\left[\sum_{i=1}^{N}\sum_{t=1}^{T}f_{t,i}(x_{t,i})-\sum_{i=1}^{N}\sum_{t=1}^{T}f_{t,i}(\bar{x}_t)\right]=&\sum_{i=1}^{N}\sum_{t=1}^{T}\mathbb{E}[F(x_{t,i})]-\sum_{i=1}^{N}\sum_{t=1}^{T}\mathbb{E}[F(\bar{x}_t)]\\
\leq&2L\sum_{t=1}^{T}\mathbb{E}[\|x_t-\mathbf{1}\bar{x}_t\|^2],
\end{aligned}
$$

thereby completing the proof of Lemma 8.1. □

Based on Lemma 8.1, to analyze R_1, it suffices to analyze the consensus error $\mathbb{E}[\|x_t - \mathbf{1}\bar{x}_t\|^2]$. To this end, it can be first seen that the average of $s_{t,i}$, i.e., $\bar{s}_t$, is equal to the global stochastic gradient average $g_t = \frac{1}{N}\sum_{i=1}^{N} f_{t,i}(x_{t,i})$ from Lemma 8.4. Since $s_{t,i}$ is designed to estimate the global gradient average g_t, it is necessary to quantify the estimation gap $\|s_t - \mathbf{1}g_t\|$. Through careful manipulations, we have the following result regarding the consensus error.

Lemma 8.9 *For any $\beta > 0$, we have the following result:*

$$
\begin{aligned}
\begin{bmatrix}\mathbb{E}[\|x_{t+1}-\mathbf{1}\bar{x}_{t+1}\|^2]\\ \mathbb{E}[\|s_{t+1}-\mathbf{1}g_{t+1}\|^2]\end{bmatrix}\leq&\begin{bmatrix}\frac{1+\rho^2}{2} & \eta^2\frac{1+\rho^2}{1-\rho^2}\\ \left(2+\frac{1}{\beta}\right)L^2\|W-I\|^2 & \rho^2+2\beta\rho^2+2L\rho\eta+2\eta^2L^2\end{bmatrix}\cdot\begin{bmatrix}\mathbb{E}[\|x_t-\mathbf{1}\bar{x}_t\|^2]\\ \mathbb{E}[\|s_t-\mathbf{1}g_t\|^2]\end{bmatrix}\\
&+\begin{bmatrix}0\\ \left(2+\frac{1}{\beta}\right)N\eta^2L^2\mathbb{E}[\|g_t\|^2]+2\sigma^2+2\eta LN\sigma^2+2N\sigma^2\end{bmatrix}.
\end{aligned}
$$

Proof Based on the update rule, we can have

$$
\begin{aligned}
\|x_{t+1} - \mathbf{1}\bar{x}_{t+1}\|^2 \leq & \| W x_t - \eta s_t - \mathbf{1}(\bar{x}_t - \eta g_t)\|^2 \\
= & \| W x_t - \mathbf{1}\bar{x}_t\|^2 - 2\eta \langle W x_t - \mathbf{1}\bar{x}_t, s_t - \mathbf{1} g_t \rangle + \eta^2 \|s_t - \mathbf{1} g_t\|^2 \\
\leq & \rho^2 \|x_t - \mathbf{1}\bar{x}_t\|^2 + \eta \left[\frac{1-\rho^2}{2\eta\rho^2} \| W x_t - \mathbf{1}\bar{x}_t\|^2 + \frac{2\eta\rho^2}{1-\rho^2} \|s_t - \mathbf{1} g_t\|^2 \right] + \eta^2 \|s_t - \mathbf{1} g_t\|^2 \\
\leq & \rho^2 \|x_t - \mathbf{1}\bar{x}_t\|^2 + \frac{(1-\rho^2)\rho^2}{2\rho^2} \|x_t - \mathbf{1}\bar{x}_t\|^2 + \frac{2\eta^2\rho^2}{1-\rho^2} \|s_t - \mathbf{1} g_t\|^2 + \eta^2 \|s_t - \mathbf{1} g_t\|^2 \\
\leq & \frac{1+\rho^2}{2} \|x_t - \mathbf{1}\bar{x}_t\|^2 + \eta^2 \frac{1+\rho^2}{1-\rho^2} \|s_t - \mathbf{1} g_t\|^2.
\end{aligned} \tag{8.11}
$$

Besides, for the global gradient estimation gap, it follows that

$$
\begin{aligned}
& \|s_{t+1} - \mathbf{1} g_{t+1}\|^2 \\
= & \| W s_t + \nabla_{t+1} - \nabla_t - \mathbf{1} g_{t+1}\|^2 \\
= & \| W s_t - \mathbf{1} g_t + \nabla_{t+1} - \nabla_t + \mathbf{1} g_t - \mathbf{1} g_{t+1}\|^2 \\
= & \| W s_t - \mathbf{1} g_t\|^2 + 2\langle W s_t - \mathbf{1} g_t, \nabla_{t+1} - \nabla_t + \mathbf{1} g_t - \mathbf{1} g_{t+1} \rangle + \|\nabla_{t+1} - \nabla_t + \mathbf{1} g_t - \mathbf{1} g_{t+1}\|^2 \\
= & \| W s_t - \mathbf{1} g_t\|^2 + \|\nabla_{t+1} - \nabla_t + \mathbf{1} g_t - \mathbf{1} g_{t+1}\|^2 + 2\langle W s_t - \mathbf{1} g_t, \nabla_{t+1} - \nabla_t \rangle \\
& + 2\langle W s_t - \mathbf{1} g_t, \mathbf{1} g_t - \mathbf{1} g_{t+1} \rangle \\
\leq & \rho^2 \|s_t - \mathbf{1} g_t\|^2 + \|\nabla_{t+1} - \nabla_t\|^2 + 2\langle W s_t - \mathbf{1} g_t, \nabla_{t+1} - \nabla_t \rangle
\end{aligned} \tag{8.12}
$$

where the last inequality holds because

$$
\begin{aligned}
& \|\nabla_{t+1} - \nabla_t + \mathbf{1} g_t - \mathbf{1} g_{t+1}\|^2 \\
= & \|\nabla_{t+1} - \nabla_t\|^2 + n\|g_{t+1} - g_t\|^2 - 2\langle \nabla_{t+1} - \nabla_t, \mathbf{1} g_{t+1} - \mathbf{1} g_t \rangle \\
\leq & \|\nabla_{t+1} - \nabla_t\|^2 - n\|g_{t+1} - g_t\|^2 \\
\leq & \|\nabla_{t+1} - \nabla_t\|^2,
\end{aligned}
$$

and

$$
\begin{aligned}
\langle W s_t - \mathbf{1} g_t, \mathbf{1} g_t - \mathbf{1} g_{t+1} \rangle = & \sum_{i=1}^{N} \left\langle \sum_{j=1}^{N} w_{ij} s_{t,j} - g_t, g_t - g_{t+1} \right\rangle \\
= & \left\langle \sum_{i=1}^{N} \sum_{j=1}^{N} w_{ij} s_{t,j} - N g_t, g_t - g_{t+1} \right\rangle \\
= & \left\langle \sum_{i=1}^{N} \sum_{j=1}^{N} w_{ij} s_{t,j} - \sum_{i=1}^{N} s_{t,i}, g_t - g_{t+1} \right\rangle \\
= & 0.
\end{aligned}
$$

We next bound the three terms in (8.12) separately. To bound $\|\nabla_{t+1} - \nabla_t\|^2$, we consider the conditional expectation $\mathbb{E}[\|\nabla_{t+1} - \nabla_t\|^2 | \mathcal{F}_t]$. It is clear that

$$\begin{aligned}\mathbb{E}[\|\nabla_{t+1}-\nabla_t\|^2|\mathcal{F}_t] =&\mathbb{E}[\|G_{t+1}-G_t+\nabla_{t+1}-G_{t+1}+G_t-\nabla_t\|^2|\mathcal{F}_t]\\ =&\mathbb{E}[\|G_{t+1}-G_t\|^2|\mathcal{F}_t]+2\mathbb{E}[\langle G_{t+1}-G_t,\nabla_{t+1}-G_{t+1}-\nabla_t+G_t\rangle|\mathcal{F}_t]\\ &+\mathbb{E}[\|\nabla_{t+1}-G_{t+1}-\nabla_t+G_t\|^2|\mathcal{F}_t]\\ \le&\mathbb{E}[\|G_{t+1}-G_t\|^2|\mathcal{F}_t]+2\mathbb{E}[\langle G_{t+1},G_t-\nabla_t\rangle|\mathcal{F}_t]+2N\sigma^2.\end{aligned}$$

Since

$$\begin{aligned}\|G_{t+1}-G_t\|^2\le&L^2\|x_{t+1}-x_t\|^2\\ \le&L^2\|Wx_t-\eta s_t-x_t\|^2\\ =&L^2(\|(W-I)(x_t-\mathbf{1}\bar{x}_t)\|^2-2\langle(W-I)(x_t-\mathbf{1}\bar{x}_t),\eta s_t\rangle+\eta^2\|s_t\|^2)\\ \le&2\|W-I\|^2L^2\|x_t-\mathbf{1}\bar{x}_t\|^2+2\eta^2L^2\|s_t-\mathbf{1}g_t+\mathbf{1}g_t\|^2\\ \le&2\|W-I\|^2L^2\|x_t-\mathbf{1}\bar{x}_t\|^2+2\eta^2L^2\|s_t-\mathbf{1}g_t\|^2+2N\eta^2L^2\|g_t\|^2,\end{aligned}$$

we use Lemma 8.6 to conclude that

$$\begin{aligned}&\mathbb{E}[\|\nabla_{t+1}-\nabla_t\|^2|\mathcal{F}_t]\\ \le&2\|W-I\|^2L^2\mathbb{E}[\|x_t-\mathbf{1}\bar{x}_t\|^2|\mathcal{F}_t]+2\eta^2L^2\mathbb{E}[\|s_t-\mathbf{1}g_t\|^2|\mathcal{F}_t]\\ &+2N\eta^2L^2\mathbb{E}[\|g_t\|^2|\mathcal{F}_t]+2\eta LN\sigma^2+2N\sigma^2.\end{aligned}\tag{8.13}$$

To bound $\langle Ws_t-\mathbf{1}g_t,\nabla_{t+1}-\nabla_t\rangle$, we consider the conditional expectation $\mathbb{E}[\langle Ws_t-\mathbf{1}g_t,\nabla_{t+1}-\nabla_t\rangle|\mathcal{F}_t]$ given $\mathcal{F}_t$, such that

$$\begin{aligned}&\mathbb{E}[\langle Ws_t-\mathbf{1}g_t,\nabla_{t+1}-\nabla_t\rangle|\mathcal{F}_t]\\ =&\mathbb{E}[\mathbb{E}[\langle Ws_t-\mathbf{1}g_t,\nabla_{t+1}-\nabla_t\rangle|\mathcal{F}_{t+1}]|\mathcal{F}_t]\\ =&\mathbb{E}[\langle Ws_t-\mathbf{1}g_t,G_{t+1}-\nabla_t\rangle|\mathcal{F}_t]\\ =&\mathbb{E}[\langle Ws_t-\mathbf{1}g_t,G_{t+1}-G_t\rangle|\mathcal{F}_t]+\mathbb{E}[\langle Ws_t-\mathbf{1}g_t,G_t-\nabla_t\rangle|\mathcal{F}_t].\end{aligned}\tag{8.14}$$

For the term $\mathbb{E}[\langle Ws_t-\mathbf{1}g_t,G_{t+1}-G_t\rangle|\mathcal{F}_t]$, we can obtain

$$\begin{aligned}&\mathbb{E}[\langle Ws_t-\mathbf{1}g_t,G_{t+1}-G_t\rangle|\mathcal{F}_t]\\ \le&L\rho\mathbb{E}[\|s_t-\mathbf{1}g_t\|\|x_{t+1}-x_t\||\mathcal{F}_t]\\ \le&L\rho\mathbb{E}[\|s_t-\mathbf{1}g_t\|\|Wx_t-\eta s_t-x_t\||\mathcal{F}_t]\\ =&L\rho\mathbb{E}[\|s_t-\mathbf{1}g_t\|\|(W-I)(x_t-\mathbf{1}\bar{x}_t)-\eta(s_t-\mathbf{1}g_t+\mathbf{1}g_t)\||\mathcal{F}_t]\\ \le&L\rho\mathbb{E}[\|s_t-\mathbf{1}g_t\|(\|W-I\|\|x_t-\mathbf{1}\bar{x}_t\|+\eta\|s_t-\mathbf{1}g_t\|+\eta\|\mathbf{1}g_t\|)|\mathcal{F}_t]\\ =&L\rho\mathbb{E}[\|W-I\|\|s_t-\mathbf{1}g_t\|\|x_t-\mathbf{1}\bar{x}_t\|+\eta\|s_t-\mathbf{1}g_t\|^2+\eta\sqrt{N}\|s_t-\mathbf{1}g_t\|\|g_t\||\mathcal{F}_t].\end{aligned}\tag{8.15}$$

For the term $\mathbb{E}[\langle Ws_t-\mathbf{1}g_t,G_t-\nabla_t\rangle|\mathcal{F}_t]$, it follows from Lemma 8.5 that

$$\mathbb{E}[\langle W s_t - \mathbf{1} g_t, G_t - \nabla_t \rangle | \mathcal{F}_t] \leq \sigma^2. \tag{8.16}$$

Therefore, by substituting (8.15) and (8.16) into (8.14), we have

$$\begin{aligned}
&\mathbb{E}[\langle W s_t - \mathbf{1} g_t, \nabla_{t+1} - \nabla_t \rangle | \mathcal{F}_t] \\
\leq & L\rho \mathbb{E}[\|W - I\| \|s_t - \mathbf{1} g_t\| \|x_t - \mathbf{1}\bar{x}_t\| | \mathcal{F}_t] + L\rho\eta \mathbb{E}[\|s_t - \mathbf{1} g_t\|^2 | \mathcal{F}_t] \\
& + L\rho\eta\sqrt{N}\mathbb{E}[\|s_t - \mathbf{1} g_t\| \|g_t\| | \mathcal{F}_t] + \sigma^2.
\end{aligned} \tag{8.17}$$

Combining (8.17) with (8.12) and (8.13), we conclude that for any $\beta > 0$,

$$\begin{aligned}
&\mathbb{E}[\|s_{t+1} - \mathbf{1} g_{t+1}\|^2 | \mathcal{F}_t] \\
\leq & \rho^2 \mathbb{E}[\|s_t - \mathbf{1} g_t\|^2 | \mathcal{F}_t] + \mathbb{E}[2\|W - I\|^2 L^2 \|x_t - \mathbf{1}\bar{x}_t\|^2 + 2\eta^2 L^2 \|s_t - \mathbf{1} g_t\|^2 + 2N\eta^2 L^2 \|g_t\|^2 | \mathcal{F}_t] \\
& + 2L\rho \mathbb{E}[\|W - I\| \|s_t - \mathbf{1} g_t\| \|x_t - \mathbf{1}\bar{x}_t\| | \mathcal{F}_t] + 2L\rho\eta \mathbb{E}[\|s_t - \mathbf{1} g_t\|^2 | \mathcal{F}_t] \\
& + 2L\rho\eta\sqrt{N}\mathbb{E}[\|s_t - \mathbf{1} g_t\| \|g_t\| | \mathcal{F}_t] + 2\sigma^2 + 2\eta L N\sigma^2 + 2N\sigma^2 \\
\leq & (\rho^2 + \beta\rho^2 + 2L\rho\eta + 2\eta^2 L^2)\mathbb{E}[\|s_t - \mathbf{1} g_t\|^2 | \mathcal{F}_t] + \left(2 + \frac{1}{\beta}\right) L^2 \|W - I\|^2 \mathbb{E}[\|x_t - \mathbf{1}\bar{x}_t\|^2 | \mathcal{F}_t] \\
& + 2N\eta^2 L^2 \mathbb{E}[\|g_t\|^2 | \mathcal{F}_t] + 2L\rho\eta\sqrt{N}\mathbb{E}[\|s_t - \mathbf{1} g_t\| \|g_t\| | \mathcal{F}_t] + 2\sigma^2 + 2\eta L N\sigma^2 + 2N\sigma^2 \\
\leq & (\rho^2 + 2\beta\rho^2 + 2L\rho\eta + 2\eta^2 L^2)\mathbb{E}[\|s_t - \mathbf{1} g_t\|^2 | \mathcal{F}_t] + \left(2 + \frac{1}{\beta}\right) L^2 \|W - I\|^2 \mathbb{E}[\|x_t - \mathbf{1}\bar{x}_t\|^2 | \mathcal{F}_t] \\
& + \left(2 + \frac{1}{\beta}\right) N\eta^2 L^2 \mathbb{E}[\|g_t\|^2 | \mathcal{F}_t] + 2\sigma^2 + 2\eta L N\sigma^2 + 2N\sigma^2
\end{aligned}$$

Taking expectation and combining with (8.11), we can obtain the following result:

$$\begin{aligned}
\begin{bmatrix} \mathbb{E}[\|x_{t+1} - \mathbf{1}\bar{x}_{t+1}\|^2] \\ \mathbb{E}[\|s_{t+1} - \mathbf{1} g_{t+1}\|^2] \end{bmatrix} \leq & \begin{bmatrix} \frac{1+\rho^2}{2} & \eta^2 \frac{1+\rho^2}{1-\rho^2} \\ \left(2 + \frac{1}{\beta}\right) L^2 \|W - I\|^2 & \rho^2 + 2\beta\rho^2 + 2L\rho\eta + 2\eta^2 L^2 \end{bmatrix} \cdot \begin{bmatrix} \mathbb{E}[\|x_t - \mathbf{1}\bar{x}_t\|^2] \\ \mathbb{E}[\|s_t - \mathbf{1} g_t\|^2] \end{bmatrix} \\
& + \begin{bmatrix} 0 \\ \left(2 + \frac{1}{\beta}\right) N\eta^2 L^2 \mathbb{E}[\|g_t\|^2] + 2\sigma^2 + 2\eta L N\sigma^2 + 2N\sigma^2 \end{bmatrix} \\
\triangleq & A_t \begin{bmatrix} \mathbb{E}[\|x_t - \mathbf{1}\bar{x}_t\|^2] \\ \mathbb{E}[\|s_t - \mathbf{1} g_t\|^2] \end{bmatrix} + B_t
\end{aligned} \tag{8.18}$$

where

$$A_t = \begin{bmatrix} \frac{1+\rho^2}{2} & \eta^2 \frac{1+\rho^2}{1-\rho^2} \\ \left(2 + \frac{1}{\beta}\right) L^2 \|W - I\|^2 & \rho^2 + 2\beta\rho^2 + 2L\rho\eta + 2\eta^2 L^2 \end{bmatrix}$$

and

$$B_t = \begin{bmatrix} 0 \\ \left(2 + \frac{1}{\beta}\right) N\eta^2 L^2 \mathbb{E}[\|g_t\|^2] + 2\sigma^2 + 2\eta L N\sigma^2 + 2N\sigma^2 \end{bmatrix},$$

completing the proof of Lemma 8.9. □

Telescoping (8.18), we have for $t \geq 2$

$$\begin{bmatrix} \mathbb{E}[\|x_t - \mathbf{1}\bar{x}_t\|^2] \\ \mathbb{E}[\|s_t - \mathbf{1}g_t\|^2] \end{bmatrix} \leq \left(\prod_{l=1}^{t-1} A_l\right) \begin{bmatrix} \mathbb{E}[\|x_1 - \mathbf{1}\bar{x}_1\|^2] \\ \mathbb{E}[\|s_1 - \mathbf{1}g_1\|^2] \end{bmatrix} + \sum_{l=1}^{t-1} \left(B_l \prod_{j=l+1}^{t-1} A_j \right) \tag{8.19}$$

where $\mathbb{E}[\|x_t - \mathbf{1}\bar{x}_t\|^2]$ and $\mathbb{E}[\|s_t - \mathbf{1}g_t\|^2]$ would converge to a neighborhood of 0, if the spectral radius of A_t, i.e., $\rho_s(A_t)$, is smaller than 1 for any $t \geq 1$. The next lemma characterizes the conditions for $\rho_s(A_t) < 1$.

Lemma 8.10 *When* $\eta \leq \frac{(1-\rho^2)^{1.5}}{32L\sqrt{1+\rho^2}}$ *and* $\beta = \frac{1-\rho^2}{4\rho^2} - \frac{L\eta}{\rho} - \frac{L^2\eta^2}{\rho^2}$*, we can have*

$$\rho_s(A_t) \leq \frac{3+\rho^2}{4} < 1.$$

Proof To ensure $\rho_s(A_t) < 1$, the eigenvalues λ of A_t, i.e., the solutions of $\det(A_t - \lambda I) = 0$, must be smaller than 1. By computing $\det(A_t - \lambda I)$, we first need the following holds for some $\beta > 0$:

$$\frac{1+\rho^2}{2} = \rho^2 + 2\beta\rho^2 + 2L\rho\eta + 2\eta^2 L^2 < 1.$$

Clearly, when $\eta \leq \frac{(1-\rho^2)^{1.5}}{32L\sqrt{1+\rho^2}} < \frac{1-\rho^2}{8L}$, it can be seen that the selection of β satisfying the above equality is positive:

$$\begin{aligned} \beta =& \frac{1-\rho^2}{4\rho^2} - \frac{L\eta}{\rho} - \frac{L^2\eta^2}{\rho^2} \\ \geq& \frac{1-\rho^2}{4\rho^2} - \frac{1-\rho^2}{8\rho} - \frac{(1-\rho^2)^2}{64\rho^2} \\ \geq& \frac{3}{32\rho^2} > 0. \end{aligned}$$

Therefore, for any λ satisfying $\det(A_t - \lambda I) = 0$, the following is true:

$$\begin{aligned} \lambda \leq& \frac{1}{2}\left(\frac{1+\rho^2}{2} + \rho^2 + 2\beta\rho^2 + 2L\rho\eta + 2L^2\eta^2 + 2L\eta\sqrt{\frac{1+\rho^2}{1-\rho^2}\left(8 + \frac{4}{\beta}\right)} \right) \\ =& \frac{1+\rho^2}{2} + L\eta\sqrt{\frac{1+\rho^2}{1-\rho^2}\left(8 + \frac{4}{\beta}\right)} \\ \leq& \frac{1+\rho^2}{2} + L\eta\sqrt{\frac{1+\rho^2}{1-\rho^2}(8 + 44\rho^2)} \end{aligned}$$

$$
\begin{aligned}
&\leq \frac{1+\rho^2}{2} + L\frac{(1-\rho^2)^{1.5}}{32L\sqrt{1+\rho^2}}\sqrt{\frac{1+\rho^2}{1-\rho^2}(8+44\rho^2)} \\
&\leq \frac{3+\rho^2}{4} < 1.
\end{aligned}
$$

□

Next, we are ready to prove Lemma 8.2.

Lemma 8.11 *Let* $\alpha = \frac{3+\rho^2}{4}$. *Under Assumptions 1 and 2, the following inequality holds for some constant* A_1 *and* A_2*:*

$$
\begin{aligned}
&\sum_{t=1}^{T} \mathbb{E}[\|x_t - \mathbf{1}\bar{x}_t\|^2] \leq A_1 \frac{\alpha - \alpha^T}{1-\alpha} + \|x_1 - \mathbf{1}\bar{x}_1\|^2 \\
&+ A_2\eta^2 \frac{1+\rho^2}{1-\rho^2}[18\eta^2\sigma^2L^2 + 18N\eta^2L^2D + (1+\eta LN + N)\sigma^2]T.
\end{aligned}
$$

Proof Let $\alpha = \frac{3+\rho^2}{4}$. With the same spirit in Qu and Li [267], we can show that

$$
\mathbb{E}[\|x_t - \mathbf{1}\bar{x}_t\|^2] \leq A_1\alpha^{t-1} + A_2\eta^2\frac{1+\rho^2}{1-\rho^2}(6N\eta^2L^2\mathbb{E}[\|g_t\|^2] + \sigma^2 + \eta LN\sigma^2 + N\sigma^2), \tag{8.20}
$$

for some A_1 and A_2. This can be achieved through diagonalization of A, which needs tedious calculations. To start with, for the selection of β, we first have

$$
\begin{aligned}
&\mathbb{E}[\|s_{t+1} - \mathbf{1}g_{t+1}\|^2|\mathcal{F}_t] \\
\leq &\frac{1+\rho^2}{2}\mathbb{E}[\|s_t - \mathbf{1}g_t\|^2|\mathcal{F}_t] + \left(2+\frac{1}{\beta}\right)L^2\|W-I\|^2\mathbb{E}[\|x_t - \mathbf{1}\bar{x}_t\|^2|\mathcal{F}_t] \\
&+ \left(2+\frac{1}{\beta}\right)N\eta^2L^2\mathbb{E}[\|g_t\|^2|\mathcal{F}_t] + 2\sigma^2 \\
\leq &\left[\frac{1+\rho^2}{2} + 2L^2\eta^2\frac{1+\rho^2}{1-\rho^2}\left(2+\frac{1}{\beta}\right)\right]\mathbb{E}[\|s_t - \mathbf{1}g_t\|^2|\mathcal{F}_t] + \left(2+\frac{1}{\beta}\right)L^2\|W-I\|^2\mathbb{E}[\|x_t - \mathbf{1}\bar{x}_t\|^2|\mathcal{F}_t] \\
&+ \left(2+\frac{1}{\beta}\right)N\eta^2L^2\mathbb{E}[\|g_t\|^2|\mathcal{F}_t] + 2\sigma^2 + 2\eta LN\sigma^2 + 2N\sigma^2.
\end{aligned} \tag{8.21}
$$

Combining (8.21) and (8.11), we can obtain

$$\begin{bmatrix}\mathbb{E}[\|s_{t+1}-\mathbf{1}g_{t+1}\|^2]\\ \mathbb{E}[\|x_{t+1}-\mathbf{1}\bar{x}_{t+1}\|^2]\end{bmatrix} \le \begin{bmatrix}\frac{1+\rho^2}{2}+2L^2\eta^2\frac{1+\rho^2}{1-\rho^2}(2+\frac{1}{\beta}) & 4L^2(2+\frac{1}{\beta})\\ \eta^2\frac{1+\rho^2}{1-\rho^2} & \frac{1+\rho^2}{2}\end{bmatrix}\cdot\begin{bmatrix}\mathbb{E}[\|s_t-\mathbf{1}g_t\|^2]\\ \mathbb{E}[\|x_t-\mathbf{1}\bar{x}_t\|^2]\end{bmatrix}$$
$$+\begin{bmatrix}(2+\frac{1}{\beta})N\eta^2L^2\mathbb{E}[\|g_t\|^2]+2\sigma^2+2\eta LN\sigma^2+2N\sigma^2\\ 0\end{bmatrix}.$$

For the matrix $\tilde{A}=\begin{bmatrix}\frac{1+\rho^2}{2}+2L^2\eta^2\frac{1+\rho^2}{1-\rho^2}\left(2+\frac{1}{\beta}\right) & 4L^2\left(2+\frac{1}{\beta}\right)\\ \eta^2\frac{1+\rho^2}{1-\rho^2} & \frac{1+\rho^2}{2}\end{bmatrix}$, we can diagonalize it as $\tilde{A}=V\Lambda V^{-1}$, where $\Sigma=\begin{bmatrix}\lambda_1 & 0\\ 0 & \lambda_2\end{bmatrix}$ with

$$\lambda_1=\frac{1+\rho^2+2L^2\eta^2\frac{1+\rho^2}{1-\rho^2}\left(2+\frac{1}{\beta}\right)-\sqrt{4L^4\eta^4\left(\frac{1+\rho^2}{1-\rho^2}\right)^2(2+\frac{1}{\beta})^2+16L^2\eta^2\frac{1+\rho^2}{1-\rho^2}\left(2+\frac{1}{\beta}\right)}}{2},$$
$$\lambda_2=\frac{1+\rho^2+2L^2\eta^2\frac{1+\rho^2}{1-\rho^2}\left(2+\frac{1}{\beta}\right)+\sqrt{4L^4\eta^4\left(\frac{1+\rho^2}{1-\rho^2}\right)^2\left(2+\frac{1}{\beta}\right)^2+16L^2\eta^2\frac{1+\rho^2}{1-\rho^2}\left(2+\frac{1}{\beta}\right)}}{2}.$$

And matrix V and V^{-1} are

$$V=\begin{bmatrix}\frac{2L^2\eta\sqrt{\frac{1+\rho^2}{1-\rho^2}}\left(2+\frac{1}{\beta}\right)-L\sqrt{2\left(2+\frac{1}{\beta}\right)\left[8+2L^2\eta^2\frac{1+\rho^2}{1-\rho^2}\left(2+\frac{1}{\beta}\right)\right]}}{2\eta\sqrt{\frac{1+\rho^2}{1-\rho^2}}} & \frac{2L^2\eta\sqrt{\frac{1+\rho^2}{1-\rho^2}}\left(2+\frac{1}{\beta}\right)+L\sqrt{2\left(2+\frac{1}{\beta}\right)\left[8+2L^2\eta^2\frac{1+\rho^2}{1-\rho^2}\left(2+\frac{1}{\beta}\right)\right]}}{2\eta\sqrt{\frac{1+\rho^2}{1-\rho^2}}}\\ 1 & 1\end{bmatrix}$$

and

$$V^{-1}=\begin{bmatrix}-\frac{\eta\sqrt{\frac{1+\rho^2}{1-\rho^2}}}{L\sqrt{2\left(2+\frac{1}{\beta}\right)\left[8+2L^2\eta^2\frac{1+\rho^2}{1-\rho^2}\left(2+\frac{1}{\beta}\right)\right]}} & \frac{1}{2}+\frac{1}{2}\sqrt{\frac{2L^2\eta^2\frac{1+\rho^2}{1-\rho^2}\left(2+\frac{1}{\beta}\right)}{8+2L^2\eta^2\frac{1+\rho^2}{1-\rho^2}\left(2+\frac{1}{\beta}\right)}}\\ \frac{\eta\sqrt{\frac{1+\rho^2}{1-\rho^2}}}{L\sqrt{2\left(2+\frac{1}{\beta}\right)\left[8+2L^2\eta^2\frac{1+\rho^2}{1-\rho^2}\left(2+\frac{1}{\beta}\right)\right]}} & \frac{1}{2}-\frac{1}{2}\sqrt{\frac{2L^2\eta^2\frac{1+\rho^2}{1-\rho^2}\left(2+\frac{1}{\beta}\right)}{8+2L^2\eta^2\frac{1+\rho^2}{1-\rho^2}\left(2+\frac{1}{\beta}\right)}}\end{bmatrix}.$$

Let $\tilde{B}_1=\begin{bmatrix}\mathbb{E}[\|s_t-\mathbf{1}g_t\|^2]\\ \mathbb{E}[\|x_t-\mathbf{1}\bar{x}_t\|^2]\end{bmatrix}$. We can show that the second row of $\tilde{A}^p\tilde{B}_1$ is smaller than $C\alpha^p$ for some constant C.

Moreover, it can be shown that

$$\sum_{p=0}^{t-1}\tilde{A}^p=(I-\tilde{A})^{-1}(I-\tilde{A}^{t-1})=\frac{1}{\det(I-\tilde{A})}\mathrm{adj}(I-\tilde{A})(I-\tilde{A}^{t-1})$$

where $\det(I-\tilde{A})>0$ since $\rho(\tilde{A})<1$, and $\mathrm{adj}(I-\tilde{A})$ is the adjugate matrix of $I-\tilde{A}$.

Let $\tilde{B}_t = \begin{bmatrix} \left(2+\frac{1}{\beta}\right) N\eta^2 L^2 \mathbb{E}[\|g_t\|^2] + 2\sigma^2 + 2\eta L N \sigma^2 + 2N\sigma^2 \\ 0 \end{bmatrix}$.

The following holds for the second row of $\sum_{p=0}^{t-1} \tilde{A}^p \tilde{B}_l$:

$$\sum_{p=0}^{t-1} \tilde{A}^p \tilde{B}_l \leq \frac{1}{\det(I-\tilde{A})}\eta^2 \frac{1+\rho^2}{1-\rho^2}\left[\left(2+\frac{1}{\beta}\right) N\eta^2 L^2 \mathbb{E}[\|g_t\|^2] + 2\sigma^2 + 2\eta L N\sigma^2 + 2N\sigma^2\right].$$

Let $A_1 = C$ and $A_2 = \frac{1}{\det(I-\tilde{A})}$, we can obtain (8.20).

Next, we need to bound the term $\mathbb{E}[\|g_t\|^2]$. It follows that

$$\begin{aligned} \mathbb{E}[\|g_t\|^2] =& \mathbb{E}\left[\left\|\frac{1}{N}\sum_{i=1}^{N} \nabla f_{t,i}(x_{t,i})\right\|^2\right] \\ \leq& \frac{1}{N}\sum_{i=1}^{N} \mathbb{E}[\|\nabla f_{t,i}(x_{t,i})\|^2] \\ \leq& D. \end{aligned}$$

For a constant learning rate η, from (8.20), it follows that

$$\begin{aligned} &\sum_{t=1}^{T} \mathbb{E}[\|x_t - \mathbf{1}\bar{x}_t\|^2] \\ \leq& \sum_{t=2}^{T}\left[A_1\alpha^{t-1} + A_2[18\eta^2L^2\sigma^2 + 18N\eta^2L^2D + (1+\eta LN+N)\sigma^2]\eta^2\frac{1+\rho^2}{1-\rho^2}\right] + \mathbb{E}[\|x_1 - \mathbf{1}\bar{x}_1\|^2] \\ =& A_1\frac{\alpha-\alpha^T}{1-\alpha} + A_2(18\eta^2L^2\sigma^2 + 18N\eta^2L^2D + (1+\eta LN+N)\sigma^2)\eta^2\frac{1+\rho^2}{1-\rho^2}(T-1) + \mathbb{E}[\|x_1 - \mathbf{1}\bar{x}_1\|^2] \\ \leq& A_2\eta^2\frac{1+\rho^2}{1-\rho^2}[18\eta^2\sigma^2L^2 + 18N\eta^2L^2D + (1+\eta LN+N)\sigma^2]T \\ &+ A_1\frac{\alpha-\alpha^T}{1-\alpha} + \mathbb{E}[\|x_1 - \mathbf{1}\bar{x}_1\|^2]. \end{aligned}$$

□

Based on Lemma 8.1, we can obtain the upper bound of R_1.

8.7.2.2 Analysis of R_2

Next, we analyze R_2. First, denote $f_t(\cdot) = \frac{1}{N}\sum_{i=1}^{N} f_{t,i}(\cdot)$, then we can have

$$\sum_{i=1}^{N}\sum_{t=1}^{T} f_{t,i}(\bar{x}_t) - \sum_{i=1}^{N}\sum_{t=1}^{T} f_{t,i}(x^*) = N\sum_{t=1}^{T}\left[f_t(\bar{x}_t) - f_t(x^*)\right].$$

And the following lemma gives an upper bound on R_2.

Lemma 8.12 *Under Assumptions 1 and 2, the following inequality holds:*

$$\begin{aligned}
&\mathbb{E}\left[\sum_{i=1}^{N}\sum_{t=1}^{T} f_{t,i}(\bar{x}_t) - \sum_{i=1}^{N}\sum_{t=1}^{T} f_{t,i}(x^*)\right] \\
\leq &\frac{4N\|\bar{x}_1 - x^*\|^2}{\eta} + 26L\sum_{t=1}^{T}\mathbb{E}[\|x_t - \mathbf{1}\bar{x}_t\|^2] + \frac{N\eta}{2}\mathbb{E}[\|\nabla F(\bar{x}_{T+1})\|^2] \\
&+ 2\sigma^2\eta T + 24L\mathbb{E}[\|x_{T+1} - \mathbf{1}\bar{x}_{T+1}\|^2].
\end{aligned}$$

Proof Following the same line as in Qu and Li [267], we can first have

$$f_t(x) \geq \hat{f}_t + \langle g_t, x - \bar{x}_t\rangle \tag{8.22}$$

and

$$f_t(x) \leq \hat{f}_t + \langle g_t, x - \bar{x}_t\rangle + L\|x - \bar{x}_t\|^2 + \frac{L}{N}\|x_t - \mathbf{1}\bar{x}_t\|^2 \tag{8.23}$$

where

$$\hat{f}_t = \frac{1}{N}\sum_{i=1}^{N}[f_{t,i}(x_{t,i}) + \langle\nabla f_{t,i}(x_{t,i}), x_t - x_{t,i}\rangle].$$

To show this, for (8.22), we have

$$\begin{aligned}
f_t(x) =&\frac{1}{N}\sum_{i=1}^{N} f_{t,i}(x) \\
\geq&\frac{1}{N}\sum_{i=1}^{N}[f_{t,i}(x_{t,i}) + \langle\nabla f_{t,i}(x_{t,i}), x - x_{t,i}\rangle] \\
=&\frac{1}{N}\sum_{i=1}^{N}[f_{t,i}(x_{t,i}) + \langle\nabla f_{t,i}(x_{t,i}), \bar{x}_t - x_{t,i}\rangle] + \frac{1}{N}\sum_{i=1}^{N}\langle\nabla f_{t,i}(x_{t,i}), x - \bar{x}_t\rangle \\
=&\hat{f}_t + \langle g_t, x - \bar{x}_t\rangle,
\end{aligned}$$

and for (8.23), it follows that

$$
\begin{aligned}
f_t(x) =& \frac{1}{N}\sum_{i=1}^{N} f_{t,i}(x) \\
\leq& \frac{1}{N}\sum_{i=1}^{N}[f_{t,i}(x_{t,i}) + \langle \nabla f_{t,i}(x_{t,i}), x - x_{t,i}\rangle + \frac{L}{2}\|x - x_{t,i}\|^2] \\
=& \frac{1}{N}\sum_{i=1}^{N}[f_{t,i}(x_{t,i}) + \langle \nabla f_{t,i}(x_{t,i}), \bar{x}_t - x_{t,i}\rangle] + \frac{1}{N}\sum_{i=1}^{N}\langle \nabla f_{t,i}(x_{t,i}), x - \bar{x}_t\rangle + \frac{L}{2N}\sum_{i=1}^{N}\|x - x_{t,i}\|^2 \\
\leq& \hat{f}_t + \langle g_t, x - \bar{x}_t\rangle + L\|x - \bar{x}_t\|^2 + \frac{L}{N}\|x_t - \mathbf{1}\bar{x}_t\|^2.
\end{aligned}
$$

Next, we can show that

$$
\begin{aligned}
&\|\bar{x}_t - x^*\|^2 \\
=& \|\bar{x}_{t+1} - x^* - \bar{x}_{t+1} + \bar{x}_t\|^2 \\
=& \|\bar{x}_{t+1} - x^*\|^2 - 2\langle \bar{x}_{t+1} - \bar{x}_t, \bar{x}_{t+1} - x^*\rangle + \|\bar{x}_{t+1} - \bar{x}_t\|^2 \\
\overset{(a)}{=}& \|\bar{x}_{t+1} - x^*\|^2 + 2\eta\langle g_t, \bar{x}_{t+1} - x^*\rangle + \eta^2\|g_t\|^2 \\
=& \|\bar{x}_{t+1} - x^*\|^2 + 2\eta\langle g_t, \bar{x}_t - x^*\rangle + 2\eta\langle g_t, \bar{x}_{t+1} - \bar{x}_t\rangle + \eta^2\|g_t\|^2 \\
\overset{(b)}{\geq}& \|\bar{x}_{t+1} - x^*\|^2 + 2\eta[\hat{f}_t - f_t(x^*)] + 2\eta[\langle g_t, \bar{x}_{t+1} - \bar{x}_t\rangle + \frac{\eta}{2}\|g_t\|^2] \\
\overset{(c)}{\geq}& \|\bar{x}_{t+1} - x^*\|^2 + 2\eta[\hat{f}_t - f_t(x^*)] + 2\eta\left[f_t(\bar{x}_{t+1}) - \hat{f}_t + \left(\frac{\eta}{2} - \eta^2 L\right)\|g_t\|^2 - \frac{L}{N}\|x_t - \mathbf{1}\bar{x}_t\|^2\right] \\
=& \|\bar{x}_{t+1} - x^*\|^2 + 2\eta[f_t(\bar{x}_{t+1}) - f_t(x^*)] + 2\eta\left[\left(\frac{\eta}{2} - \eta^2 L\right)\|g_t\|^2 - \frac{L}{N}\|x_t - \mathbf{1}\bar{x}_t\|^2\right]
\end{aligned}
$$

Here (a) is based on the update rule, (b) is based on (8.22), and (c) is based on (8.23) by setting $x = x^*$. Therefore,

$$
\sum_{t=1}^{T}[f_t(\bar{x}_{t+1}) - f_t(x^*)] \leq \frac{\|\bar{x}_1 - x^*\|^2}{2\eta} + \sum_{t=1}^{T}\left[\frac{L}{N}\|x_t - \mathbf{1}\bar{x}_t\|^2 - \left(\frac{\eta}{2} - \eta^2 L\right)\|g_t\|^2\right] \tag{8.24}
$$

which indicates that

$$
\sum_{t=1}^{T}\mathbb{E}[F(\bar{x}_{t+1}) - F(x^*)] \leq \frac{\|\bar{x}_1 - x^*\|^2}{2\eta} + \frac{L}{N}\sum_{t=1}^{T}\mathbb{E}[\|x_t - \mathbf{1}\bar{x}_t\|^2]. \tag{8.25}
$$

Moreover, it can be seen that

$$\sum_{t=1}^{T}[F(\bar{x}_t) - F(\bar{x}_{t+1})] \leq \sum_{t=1}^{T}\langle \nabla F(\bar{x}_{t+1}), \bar{x}_t - \bar{x}_{t+1}\rangle + \frac{L}{2}\sum_{t=1}^{T}\|\bar{x}_t - \bar{x}_{t+1}\|^2$$
$$= \eta\sum_{t=1}^{T}\langle \nabla F(\bar{x}_{t+1}), g_t\rangle + \frac{\eta^2 L}{2}\sum_{t=1}^{T}\|g_t\|^2$$
$$\leq \frac{\eta}{2}\sum_{t=1}^{T}\|\nabla F(\bar{x}_{t+1})\|^2 + \frac{\eta}{2}\sum_{t=1}^{T}\|g_t\|^2 + \frac{\eta^2 L}{2}\sum_{t=1}^{T}\|g_t\|^2. \quad (8.26)$$

Combing (8.25) and (8.26), we can obtain that

$$\sum_{t=1}^{T}\mathbb{E}[F(\bar{x}_t) - F(x^*)]$$
$$\leq \frac{\|\bar{x}_1 - x^*\|^2}{2\eta} + \frac{L}{N}\sum_{t=1}^{T}\mathbb{E}[\|x_t - \mathbf{1}\bar{x}_t\|^2] + \frac{\eta}{2}\sum_{t=1}^{T}\mathbb{E}[\|\nabla F(\bar{x}_{t+1})\|^2] + \frac{\eta+\eta^2 L}{2}\sum_{t=1}^{T}\mathbb{E}[\|g_t\|^2]$$
$$\leq \frac{\|\bar{x}_1 - x^*\|^2}{2\eta} + \frac{L}{N}\sum_{t=1}^{T}\mathbb{E}[\|x_t - \mathbf{1}\bar{x}_t\|^2] + \frac{\eta}{2}\mathbb{E}[\|\nabla F(\bar{x}_{T+1})\|^2] + \frac{\eta}{2}\sum_{t=1}^{T}\mathbb{E}[\|\nabla F(\bar{x}_t)\|^2]$$
$$+ \frac{\eta+\eta^2 L}{2}\sum_{t=1}^{T}\mathbb{E}[\|g_t\|^2]$$
$$\leq \frac{\|\bar{x}_1 - x^*\|^2}{2\eta} + \frac{L}{N}\sum_{t=1}^{T}\mathbb{E}[\|x_t - \mathbf{1}\bar{x}_t\|^2] + \frac{\eta}{2}\mathbb{E}[\|\nabla F(\bar{x}_{T+1})\|^2] + \eta\sum_{t=1}^{T}\mathbb{E}[\|\nabla F(\bar{x}_t) - g_t\|^2]$$
$$+ \frac{3\eta+\eta^2 L}{2}\sum_{t=1}^{T}\mathbb{E}[\|g_t\|^2]$$
$$\leq \frac{\|\bar{x}_1 - x^*\|^2}{2\eta} + \frac{L}{N}\sum_{t=1}^{T}\mathbb{E}[\|x_t - \mathbf{1}\bar{x}_t\|^2] + \frac{\eta}{2}\mathbb{E}[\|\nabla F(\bar{x}_{T+1})\|^2] + 2\eta\sum_{t=1}^{T}\mathbb{E}[\|\nabla F(\bar{x}_t) - \nabla f_t(\bar{x}_t)\|^2]$$
$$+ 2\eta\sum_{t=1}^{T}\mathbb{E}[\|\nabla f_t(\bar{x}_t) - g_t\|^2] + \frac{3\eta+\eta^2 L}{2}\sum_{t=1}^{T}\mathbb{E}[\|g_t\|^2]$$
$$\leq \frac{\|\bar{x}_1 - x^*\|^2}{2\eta} + \frac{L}{N}\sum_{t=1}^{T}\mathbb{E}[\|x_t - \mathbf{1}\bar{x}_t\|^2] + \frac{\eta}{2}\mathbb{E}[\|\nabla F(\bar{x}_{T+1})\|^2] + \frac{2\sigma^2\eta T}{N} + 2\eta\sum_{t=1}^{T}\mathbb{E}[\|\nabla f_t(\bar{x}_t) - g_t\|^2]$$
$$+ \frac{3\eta+\eta^2 L}{2}\sum_{t=1}^{T}\mathbb{E}[\|g_t\|^2],$$

where the last inequality holds because

$$\mathbb{E}[\|\nabla f_t(\bar{x}_t) - \nabla F(\bar{x}_t)\|^2] = \frac{1}{N^2}\sum_{i=1}^{N}\mathbb{E}[\|\nabla f_{t,i}(\bar{x}_t) - \nabla F(\bar{x}_t)\|^2] \leq \frac{\sigma^2}{N}.$$

For the term $\mathbb{E}[\|\nabla f_t(\bar{x}_t) - g_t\|^2]$, it is clear that

$$\begin{aligned}\|g_t - \nabla f_t(\bar{x}_t)\| &= \| \sum_{i=1}^{N} \frac{\nabla f_{t,i}(x_{t,i}) - \nabla f_{t,i}(\bar{x}_t)}{N} \| \\ &\leq L \sum_{i=1}^{N} \frac{\|x_{t,i} - \bar{x}_t\|}{N} \\ &\leq \frac{L}{\sqrt{N}} \|x_t - \mathbf{1}\bar{x}_t\|. \end{aligned} \tag{8.27}$$

Therefore,

$$\begin{aligned}\mathbb{E}\left[\sum_{t=1}^{T} [\|g_t - \nabla f_t(\bar{x}_t)\|^2]\right] &\leq \mathbb{E}\left[\sum_{t=1}^{T} \frac{L^2}{N} \|x_t - \mathbf{1}\bar{x}_t\|^2\right] \\ &= \frac{L^2}{N} \mathbb{E}\left[\sum_{t=1}^{T} \|x_t - \mathbf{1}\bar{x}_t\|^2\right]. \end{aligned} \tag{8.28}$$

To obtain an upper bound on $\sum_{t=1}^{T} \mathbb{E}[F(\bar{x}_t) - F(x^*)]$, it suffices to bound $\sum_{t=1}^{T} \mathbb{E}[\|g_t\|^2]$ from above. To this end, based on (8.23), we have

$$\begin{aligned} f_t(x_{t,i}) &\leq \hat{f}_t + \langle g_t, x_{t,i} - \bar{x}_t \rangle + L\|x_{t,i} - \bar{x}_t\|^2 + \frac{L}{N}\|x_t - \mathbf{1}\bar{x}_t\|^2 \\ &\leq f_t(\bar{x}_t) + \langle g_t, x_{t,i} - \bar{x}_t \rangle + L\|x_{t,i} - \bar{x}_t\|^2 + \frac{L}{N}\|x_t - \mathbf{1}\bar{x}_t\|^2. \end{aligned}$$

Therefore,

$$\mathbb{E}[F(x_{t,i})] \leq \mathbb{E}[F(\bar{x}_t)] + \mathbb{E}[\langle g_t, x_{t,i} - \bar{x}_t \rangle] + L\mathbb{E}[|x_{t,i} - \bar{x}_t\|^2] + \frac{L}{N}\mathbb{E}[\|x_t - \mathbf{1}\bar{x}_t\|^2],$$

and

$$\begin{aligned} &\sum_{t=1}^{T} [\frac{1}{N} \sum_{i=1}^{N} \mathbb{E}[F(x_{t+1,i}) - F(x^*)]] \\ &\leq \sum_{t=1}^{T} \mathbb{E}[F(\bar{x}_{t+1}) - F(x^*)] + \sum_{t=1}^{T} \mathbb{E}[\frac{1}{N} \sum_{i=1}^{N} \langle g_{t+1}, x_{t+1,i} - \bar{x}_{t+1} \rangle] + \frac{L}{N} \sum_{i=1}^{N} \sum_{t=1}^{T+1} \mathbb{E}[\|x_{t,i} - \bar{x}_t\|^2] \\ &+ \frac{L}{N} \sum_{t=1}^{T+1} \mathbb{E}[\|x_t - \mathbf{1}\bar{x}_t\|^2] \\ &\leq \sum_{t=1}^{T} \mathbb{E}[F(\bar{x}_{t+1}) - F(x^*)] + \frac{2L}{N} \sum_{t=1}^{T+1} \mathbb{E}[\|x_t - \mathbf{1}\bar{x}_t\|^2]. \end{aligned}$$

Based on (8.24), we can obtain that

$$\sum_{t=1}^{T}\mathbb{E}[F(\bar{x}_{t+1})-F(x^*)]\leq\frac{\|\bar{x}_1-x^*\|^2}{2\eta}+\frac{L}{N}\sum_{t=1}^{T}\mathbb{E}[\|x_t-\mathbf{1}\bar{x}_t\|^2]-\left(\frac{\eta}{2}-\eta^2L\right)\sum_{t=1}^{T}\mathbb{E}[\|g_t\|^2].$$

Continuing with $\sum_{t=1}^{T}\left[\frac{1}{N}\sum_{i=1}^{N}\mathbb{E}[F(x_{t+1,i})-F(x^*)]\right]$, we can have that

$$\begin{aligned}&\sum_{t=1}^{T}\left[\frac{1}{N}\sum_{i=1}^{N}\mathbb{E}[F(x_{t+1,i})-F(x^*)]\right]\\&\leq\frac{\|\bar{x}_1-x^*\|^2}{2\eta}+\frac{3L}{N}\sum_{t=1}^{T+1}\mathbb{E}[\|x_t-\mathbf{1}\bar{x}_t\|^2]-\left(\frac{\eta}{2}-\eta^2L\right)\sum_{t=1}^{T}\mathbb{E}[\|g_t\|^2].\end{aligned}$$

Since $\sum_{t=1}^{T}\left[\frac{1}{N}\sum_{i=1}^{N}\mathbb{E}[F(x_{t+1,i})-F(x^*)]\right]\geq 0$, it follows that

$$\left(\frac{\eta}{2}-\eta^2L\right)\sum_{t=1}^{T}\mathbb{E}[\|g_t\|^2]\leq\frac{\|\bar{x}_1-x^*\|^2}{2\eta}+\frac{3L}{N}\sum_{t=1}^{T+1}\mathbb{E}[\|x_t-\mathbf{1}\bar{x}_t\|^2].$$

For $\eta<\frac{1}{4L}$, it is clear that

$$\sum_{t=1}^{T}\mathbb{E}[\|g_t\|^2]\leq\frac{2\|x_1-x^*\|^2}{\eta^2}+\frac{12L}{\eta N}\sum_{t=1}^{T+1}\mathbb{E}[\|x_t-\mathbf{1}\bar{x}_t\|^2]$$

In a nutshell, we can obtain the upper bound for R_2:

$$\begin{aligned}&\mathbb{E}\left[\sum_{i=1}^{N}\sum_{t=1}^{T}f_{t,i}(\bar{x}_t)-\sum_{i=1}^{N}\sum_{t=1}^{T}f_{t,i}(x^*)\right]\\&=\mathbb{E}\left[N\sum_{t=1}^{T}[f_t(\bar{x}_t)-f_t(x^*)]\right]\\&=N\mathbb{E}\left[\sum_{t=1}^{T}[F(\bar{x}_t)-F(x^*)]\right]\\&\leq\frac{N\|\bar{x}_1-x^*\|^2}{2\eta}+L\sum_{t=1}^{T}\mathbb{E}[\|x_t-\mathbf{1}\bar{x}_t\|^2]+\frac{N\eta}{2}\mathbb{E}[\|\nabla F(\bar{x}_{T+1})\|^2]+2\sigma^2\eta T+2\eta L^2\mathbb{E}\left[\sum_{t=1}^{T}\|x_t-\mathbf{1}\bar{x}_t\|^2\right]\\&\quad+\frac{N(3\eta+\eta^2L)}{2}\left(\frac{2\|\bar{x}_1-x^*\|^2}{\eta^2}+\frac{12L}{\eta N}\sum_{t=1}^{T+1}\mathbb{E}[\|x_t-\mathbf{1}\bar{x}_t\|^2]\right)\\&\leq\frac{4N\|\bar{x}_1-x^*\|^2}{\eta}+26L\sum_{t=1}^{T}\mathbb{E}[\|x_t-\mathbf{1}\bar{x}_t\|^2]+\frac{N\eta}{2}\mathbb{E}[\|\nabla F(\bar{x}_{T+1})\|^2]+2\sigma^2\eta T+24L\mathbb{E}[\|x_{T+1}-\mathbf{1}\bar{x}_{T+1}\|^2].\end{aligned}$$

□

8.7.2.3 Proof of Theorem 1

Based on the analysis of R_1 and R_2, we can obtain the regret as follows:

$$
\begin{aligned}
\mathbb{E}[\mathbf{R}_s] =& \mathbb{E}\left[\sum_{i=1}^{N}\sum_{t=1}^{T} f_{t,i}(x_{t,i}) - \sum_{i=1}^{N}\sum_{t=1}^{T} f_{t,i}(x^*)\right] \\
=& \mathbb{E}\left[\sum_{i=1}^{N}\sum_{t=1}^{T} f_{t,i}(x_{t,i}) - \sum_{i=1}^{N}\sum_{t=1}^{T} f_{t,i}(\bar{x}_t)\right] + \mathbb{E}\left[\sum_{i=1}^{N}\sum_{t=1}^{T} f_{t,i}(\bar{x}_t) - \sum_{i=1}^{N}\sum_{t=1}^{T} f_{t,i}(x^*)\right] \\
\le& 28L\sum_{t=1}^{T}\mathbb{E}[\|x_t - \mathbf{1}\bar{x}_t\|^2] + \frac{4N\|\bar{x}_1 - x^*\|^2}{\eta} + \frac{N\eta}{2}\mathbb{E}[\|\nabla F(\bar{x}_{T+1})\|^2] + 2\sigma^2\eta T \\
&+ 24L\mathbb{E}[\|x_{T+1} - \mathbf{1}\bar{x}_{T+1}\|^2] \\
\le& 28LA_2\eta^2\frac{1+\rho^2}{1-\rho^2}[18\eta^2\sigma^2L^2 + 18N\eta^2L^2D + (1+\eta LN + N)\sigma^2]T + 28LA_1\frac{\alpha-\alpha^T}{1-\alpha} \\
&+ 28L\|x_1 - \mathbf{1}\bar{x}_1\|^2 + \frac{4N\|\bar{x}_1 - x^*\|^2}{\eta} + \frac{N\eta}{2}\mathbb{E}[\|\nabla F(\bar{x}_{T+1})\|^2] \\
&+ 2\sigma^2\eta T + 24L\mathbb{E}[\|x_{T+1} - \mathbf{1}\bar{x}_{T+1}\|^2] \\
=& O(\eta^2 T + \eta^2 NT + \frac{N}{\eta} + \eta T) \\
=& O(\sqrt{NT})
\end{aligned}
$$

where $\eta \le \frac{1}{2L}\sqrt{\frac{N}{T}}$ and $N = o(T^{1/3})$. Therefore, we conclude that the optimal regret can be achieved and the average regret per agent $\mathbb{E}[\mathbf{R}] = \frac{1}{N}\mathbb{E}[\mathbf{R}_s] \le O(\sqrt{\frac{T}{N}})$.

8.7.3 Distributed Convex Stochastic Optimization

As a byproduct, we can achieve the following convergence guarantee of DOGD-GT for distributed convex stochastic optimization.

Corollary 8.1 *Suppose Assumptions 1, 2, and 3 hold, and let $\hat{x}_i = \frac{1}{T}\sum_{t=1}^{T} x_{t,i}$ be the final output of DOGD-GT for each agent i. It follows that*

$$\frac{1}{N}\sum_{i=1}^{N}\mathbb{E}[F(\hat{x}_i) - F(x^*)] = O\left(\frac{1}{\sqrt{NT}}\right).$$

Proof Based on the convexity of $F(\cdot)$ and Jensen's inequality, we can have

$$\begin{aligned}\frac{1}{N}\sum_{i=1}^{N}\mathbb{E}[F(\hat{x}_i)-F(x^*)] \leq & \frac{1}{N}\sum_{i=1}^{N}\mathbb{E}\left[\frac{1}{T}\sum_{t=1}^{T}F(x_{t,i})-F(x^*)\right]\\ = & \frac{1}{N}\sum_{i=1}^{N}\mathbb{E}\left[\frac{1}{T}\sum_{t=1}^{T}F(x_{t,i})-\frac{1}{T}\sum_{t=1}^{T}F(x^*)\right]\\ = & \frac{1}{T}\mathbb{E}[\mathbf{R}]\\ = & O\left(\frac{1}{\sqrt{NT}}\right).\end{aligned}$$

□

Corollary 8.1 indicates that the optimal convergence rate of $O(1/\sqrt{NT})$ can be obtained by DOGD-GT for convex stochastic optimization problems. In contrast to standard stochastic gradient descent algorithms, it is clear that DOGD-GT can achieve a factor of $\sqrt{1/N}$ speedup compared with the single-agent case.

8.7.4 Proof of Theorem 8.2

Define the regret for the network-level OCO about $\phi_{t,n}$ with respect to any reference point ϕ as

$$\mathbf{R}^{init}(\phi)=\frac{1}{N}\sum_{n=1}^{N}\sum_{t=1}^{T}f_{t,n}^{init}(\phi_{t,n})-\frac{1}{N}\sum_{n=1}^{N}\sum_{t=1}^{T}f_{t,n}^{init}(\phi),$$

and the regret about $v_{t,n}$ with respect to any reference point v as

$$\mathbf{R}^{rate}(v)=\frac{1}{N}\sum_{n=1}^{N}\sum_{t=1}^{T}f_{t,n}^{rate}(v_{t,n})-\frac{1}{N}\sum_{n=1}^{N}\sum_{t=1}^{T}f_{t,n}^{rate}(v).$$

Therefore, according to Theorem 8.1 in Sect. 8.4, it follows that

$$\mathbb{E}[\mathbf{R}^{init}(\phi^*)]=O\left(\sqrt{\frac{mT}{N}}\right)$$

for $\phi^*=\arg\min_{\phi\in\Theta}\mathbb{E}[f_{t,n}^{init}(\phi)]$, and that

$$\mathbb{E}[\mathbf{R}^{rate}(\tilde{v}^*)]=O\left(\sqrt{\frac{mT}{N}}\right)$$

for $\tilde{v}^*=\arg\min_{v\geq\epsilon}\mathbb{E}[f_{t,n}^{rate}(v)]$.

Based on Theorem 3.1 in Khodak et al. [167], we can have

$$
\begin{aligned}
\mathbb{E}[\mathbf{R}_a] \leq & \mathbb{E}\left[\frac{1}{NT}\sum_{n=1}^{N}\sum_{t=1}^{T}\left(\frac{\mathcal{B}_R(\theta_{t,n}^*||\phi_{t,n})}{v_{t,n}} + v_{t,n}\right)G\sqrt{m}\right] \\
\leq & \frac{1}{T}\left\{\mathbb{E}[\mathbf{R}^{rate}(\tilde{v}^*)] + \min_v \mathbb{E}\left[\frac{1}{N}\sum_{n=1}^{N}\sum_{t=1}^{T}\left(\frac{\mathcal{B}_R(\theta_{t,n}^*||\phi_{t,n})}{v} + v\right)G\sqrt{m}\right]\right\} \\
\leq & \frac{\mathbb{E}[\mathbf{R}^{rate}(\tilde{v}^*)]}{T} + \min_v \frac{1}{T}\left\{\frac{\mathbb{E}[\mathbf{R}^{init}(\phi^*)]}{v} + \mathbb{E}\left[\frac{1}{N}\sum_{n=1}^{N}\sum_{t=1}^{T}\left(\frac{\mathcal{B}_R(\theta_{t,n}^*||\phi^*)}{v} + v\right)G\sqrt{m}\right]\right\} \\
\overset{(a)}{\leq} & \frac{\mathbb{E}[\mathbf{R}^{rate}(\tilde{v}^*)]}{T} + \frac{1}{T}\min\left\{\frac{\mathbb{E}[\mathbf{R}^{init}(\phi^*)]}{V_\phi}, 2\sqrt{\mathbb{E}[\mathbf{R}^{init}(\phi^*)]GT\sqrt{m}}\right\} + 2V_\phi GT\sqrt{m} \\
= & O\left(\frac{1+\frac{1}{V_\phi}}{\sqrt{NT}} + V_\phi\right)\sqrt{m}
\end{aligned}
$$

where (a) is true for $V_\phi = \sqrt{\mathbb{E}[\mathcal{B}_R(\theta_{t,n}^*||\phi^*)]}$ and $v = \max\left\{V_\phi, \sqrt{\frac{\mathbb{E}[\mathbf{R}^{init}(\phi^*)]}{GT\sqrt{m}}}\right\}$.

8.8 Conclusion

In single-agent online meta-learning, the agent has to learn over many tasks so as to obtain good meta-models, based on which within-task fast adaptation can be achieved. Nevertheless, this would inevitably lead to the cold-start problem. To address this problem, we propose a multi-agent online meta-learning framework to leverage the task similarity across multiple agents, and cast it into an equivalent two-level nested OCO problem. By pinpointing that the performance bottleneck lies in the distributed network-level OCO, where it still remains unclear that how much an agent can benefit from it through limited communication with neighboring agents, we further explore a DOGD algorithm with gradient tracking. We show that the average regret $O(\sqrt{T/N})$ can be achieved at each agent, thus revealing a linear speedup of the learning performance compared with the single-agent case. Building on the foundation of the agent-level performance speedup achieved in the distributed network-level OCO, we next propose a multi-agent online meta-learning algorithm MAOML, and show that the optimal within-task regret can be achieved at a faster rate of $O(1/\sqrt{NT})$ compared with the rate of $\tilde{O}(1/\sqrt{T})$ in the single agent case. The theoretic results have been clearly verified in the experimental studies on different datasets.

Part III
Applications and Future Directions

In Part III, we explore the real-world applications of edge AI technologies and delve into their potential future directions. This section highlights how advancements in AI are driving innovation across various fields, from autonomous vehicles to robotics, and foundational models that are shaping the next generation of intelligent systems.

In this part, we first introduce Embodied AI, which focuses on physical systems, such as autonomous vehicles and general-purpose robotics, that interact directly with the environment. We examine how these systems use reinforcement and continual learning to navigate complex tasks in dynamic settings. Next, we introduce Foundation Models and discuss the impact of large-scale models, including Large Language Models and World Models, on diverse AI applications. We also explore the growing importance of on-device foundation models, emphasizing the shift towards more localized, efficient, and personalized AI systems. Together, these chapters provide a comprehensive view of both current applications and the future possibilities for AI technologies across a variety of domains.

Applications and Future Directions

9

In this chapter, we explore the applications and future directions of continual and reinforcement learning in edge AI, offering our subjective perspective on recent trends and potential cross-disciplinary developments.

9.1 Embodied AI

Recent progress in deep learning has sparked explosive interest in creating versatile AI systems. This has led to a transition away from "Internet AI", which relies on analyzing datasets of images, videos, and text collected from the web, towards "embodied AI" [289, 311, 389]. Embodied AI emphasizes that genuine intelligence arises through an agent's active engagement with its environment, allowing it to *learn and adapt* from real-world interactions. Embodied AI's reliance on real-time interaction with its environment aligns closely with the principles of edge AI, where processing is performed locally on devices rather than relying on centralized cloud resources. This decentralized approach is ideal for embodied AI because it enables artificial agents to make decisions, learn, and adapt in real time, directly at the edge, where latency and bandwidth constraints are critical. Embodied AI is also well-suited for continual and reinforcement learning, as both learning paradigms depend on continuous feedback loops and interactions with dynamic environments. Continual learning allows agents to evolve and adapt without forgetting previous knowledge, and reinforcement learning focuses on optimizing decision-making based on rewards. Together, these learning paradigms enable embodied AI agents to achieve higher levels of adaptability and autonomy

H. Wang et al., *Continual and Reinforcement Learning for Edge AI*,
Synthesis Lectures on Learning, Networks, and Algorithms,
https://doi.org/10.1007/978-3-031-84363-1_9

in edge environments. In this section, we study two prominent applications of embodied AI that leverage continual and reinforcement learning at the edge: automated vehicles and artificial general robotics.

9.1.1 Autonomous Vehicles

A key application of embodied AI is in autonomous driving, which seeks to create vehicles capable of perceiving their environment, making real-time decisions, and operating autonomously without human intervention. The growing interest in this area is driven by the potential of automated vehicles (AVs) to offer safer and more environmentally sustainable transportation solutions. Studies have shown that automated vehicles (AVs) could significantly reduce road accidents [95, 252], shorten travel times [343], minimize the need for frequent road maintenance [38], reduce travel cost [68] and lower the overall number of vehicles on the road while improving urban road capacity [245].

In recent decades, AI advancements in areas such as computer vision have greatly accelerated the development and real-world implementations of autonomous driving systems [151, 257]. The conventional framework in autonomous driving is the modular architecture, which divides the design pipeline into separate sub-tasks handled by individual sensors and algorithms to produce control outputs. This system consists of interconnected modules, such as perception, planning, and control. However, it faces several limitations that hinder further progress in autonomous driving. One major drawback is the potential for error propagation; for example, a misclassification in the perception module can lead to errors in the subsequent planning and control stages, potentially causing unsafe behavior. Additionally, managing the complexity of these interconnected modules and the computational inefficiencies of processing data at each stage presents further challenges. To address these issues, the end-to-end learning approach has emerged as an alternative and drives the evolution of next-generation autonomous vehicles (AV2.0). By mapping sensory inputs directly to control outputs, this method simplifies the system, improving both efficiency and robustness. The end-to-end approach leverages integrated sensor suites, powerful computational resources like GPUs and TPUs, and deep learning techniques to support the navigation through adaptive learning, scaling, and generalization in complex driving environments.

AV2.0 sets higher standards for autonomous vehicles, requiring them to not only navigate complex traffic situations safely but also continuously improve in real-time. Edge computing plays a critical role by enabling on-vehicle computation, allowing autonomous vehicles to process data locally and make real-time decisions without solely relying on cloud servers. This reduces latency and enhances responsiveness, which is vital for tasks requiring immediate action, such as emergency braking, dynamic path planning, and obstacle detection. Meanwhile, introducing AI techniques at the edge is particularly essential for autonomous vehicles to accomplish tasks that require real-time decision making, such as emergency braking, dynamic path planning, and obstacle detection. Integrating AI at the edge is especially important for autonomous vehicles to handle these real-time demands. Specifically, continual learning enhances AV2.0 by allowing vehicles to adapt to new environments and driving conditions without requiring a complete retraining of models. This capability helps AVs evolve over time, learning from fresh data gathered in real-world scenarios such as changing weather, varying road conditions, and unusual traffic patterns, thereby improving safety and performance. Meanwhile, reinforcement learning complements this by enabling vehicles to learn optimal driving strategies through direct interaction with their surroundings. By receiving feedback in the form of rewards or penalties, autonomous vehicles refine their driving policies to navigate more efficiently and safely. This trial-and-error process helps AVs discover behaviors such as fuel-efficient driving, collision avoidance, and adaptive speed control, making them increasingly capable in complex, dynamic environments. Together, continual learning, and reinforcement learning empower AV2.0 to operate more autonomously and intelligently in real-world conditions.

Future Directions. Despite the impressive performance of end-to-end approaches in AV2.0, there are still many challenges towards realizing safe and self-adaptive autonomous driving in the real-world.

- Multi-Agent Collaboration and Traffic Coordination: Future autonomous vehicles (AVs) could communicate with each other and surrounding infrastructure to create a collaborative driving ecosystem. Research in vehicle-to-vehicle (V2V) and vehicle-to-infrastructure (V2I) communication can enable AVs to share real-time traffic data, road conditions, and driving strategies, optimizing overall traffic flow and reducing accidents through collective decision-making.
- Robustness in Adverse Conditions: Expanding AV capabilities to handle extreme weather conditions, low-visibility environments, and rare edge cases remains a major challenge. Future work can focus on improving the robustness of AVs through advanced sensor fusion, domain adaptation, and real-time learning to ensure safe driving in unpredictable or hazardous scenarios, such as icy roads, heavy rain, or construction zones.
- Enhanced Personalization and Human-AI Interaction: Research could explore how autonomous vehicles can adapt driving behavior based on passenger preferences, comfort levels, or specific needs. This includes integrating natural language processing (NLP) and personalized reinforcement learning to provide seamless interactions between humans and AVs, improving user trust and enhancing the passenger experience.
- Scalable Learning for Diverse Environments: As AVs are deployed globally, future research should focus on scaling learning across diverse and changing environments, including different urban, rural, and cultural settings. This could involve studying how AVs generalize their learned knowledge and adapt efficiently to different countries' traffic laws, driving styles, and infrastructure layouts.

- Energy Efficiency and Sustainability: As AVs become more widespread, optimizing energy consumption will be crucial. Future research could focus on developing energy-efficient driving strategies through reinforcement learning and edge AI, optimizing routes, driving behaviors, and energy consumption to extend battery life in electric AVs and reduce fuel consumption.
- Safety and Ethical Decision-Making Frameworks: Further research will be needed to establish clear frameworks for ethical decision-making in critical situations, such as how an AV should react in unavoidable accidents or moral dilemmas. Ensuring transparency and accountability in these decisions will be essential as AVs are integrated into society.
- Cross-Modal Learning and World Models: Integrating multimodal data–such as visual, auditory, and contextual inputs–could improve the situational awareness of AVs. Research could explore how world models enable autonomous vehicles to predict, simulate, and react to dynamic events, improving decision-making and planning in real time.
- Autonomous Vehicle Fleet Management: As AV fleets are deployed for ride-sharing or public transportation, research could focus on efficient fleet management and optimization using edge AI and reinforcement learning. Topics such as dynamic scheduling, maintenance prediction, and route optimization could be explored to maximize fleet performance and minimize costs.

9.1.2 Artificial General Robotics

In recent years, the use of mobile robots has grown significantly, with applications spanning industrial, household, educational, and healthcare settings. Traditionally, robots are designed to achieve "narrow AI", where they perform specific tasks in specific contexts. For example, a robotic assistant may assist in ultrasound scans [378], or a robotic cleaner may be developed for household chores [21]. These systems, however, are limited in their flexibility. Even minor changes to the task or environment often require human intervention for reprogramming or reconfiguration to maintain functionality.

This rigidity contrasts sharply with human intelligence, which can easily adapt to changing goals or environments without needing external adjustments. Artificial General Robotics (AGR) seeks to bridge this gap by developing robotic systems that not only perform a wide range of tasks but also self-adapt to new situations. The ultimate aim of AGR is to create systems with capabilities that rival or even surpass human adaptability and intelligence in performing everyday tasks [114]. By leveraging advanced learning algorithms, continual learning, and reinforcement learning, AGR aspires to build robots capable of independent, versatile, and intelligent behavior in a wide variety of real-world environments.

In real-world applications, edge AI plays a crucial role in advancing AGR by enabling robots to process data and make decisions locally, without relying on cloud-based systems. This on-device computation allows robots to perform real-time analysis and decision-making, which is essential for tasks that require split-second responses, such as navigating obstacles or interacting with humans. By reducing the latency associated with sending data to and from centralized servers, edge AI improves the robot's ability to operate autonomously in dynamic and unpredictable environments. This localized processing not only enhances the robot's responsiveness but also ensures continuous operation in areas with limited or unreliable internet connectivity, making it suitable for a wide range of real-world applications.

Furthermore, edge AI supports the scalability and efficiency of AGR by optimizing resource use on hardware-constrained devices. Advanced algorithms running on the edge allow robots to handle complex tasks, such as object recognition, speech processing, and decision-making, while consuming less power and computational resources. This efficiency is critical for deploying robots in environments where power and processing capacity are limited, such as in small household robots or in industrial settings with mobile units. By integrating edge AI with continual learning and reinforcement learning, robots can not only process and act on real-time data locally but also continuously improve their performance, achieving greater adaptability and autonomy in diverse, real-world tasks.

In particular, RL is instrumental in achieving artificial general robotics by enabling robots to learn through interaction with their environment. Instead of following pre-programmed instructions, RL allows robots to make decisions based on trial and error, receiving feedback in the form of rewards or penalties. Over time, this process helps the robot refine its decision-making abilities, discovering optimal behaviors for complex tasks such as navigation, object manipulation, and even social interactions. The adaptability of RL is key to creating more flexible robotic systems capable of handling varied tasks and adjusting to changing conditions without human intervention. This learning paradigm supports the development of robots that can operate in diverse, unstructured environments, such as homes, hospitals, and factories, where conditions can shift unpredictably.

Continual learning (CL) further enhances artificial general robotics by allowing robots to incrementally learn new tasks without forgetting previously acquired knowledge. In traditional robotic systems, introducing new tasks often requires retraining from scratch, which is time-consuming and inefficient. CL, on the other hand, enables robots to build on their existing knowledge, seamlessly integrating new information while retaining their previous skills. This approach mirrors human learning, where individuals continually learn and adapt to new experiences over time. By combining CL with RL, robots can not only improve their performance through real-world interactions but also maintain a broader, long-term understanding of various tasks. This dynamic adaptability is crucial for achieving the goal of artificial general robotics, where robots must demonstrate versatility and sustained competence across a wide range of tasks and environments.

9.2 Foundation Models

9.2.1 Large Language Models

Large Language Models (LLMs) are advanced machine learning systems trained on vast amounts of text data to understand and generate human-like language. These models, such as GPT [5] and BERT [82], have shown remarkable capabilities in tasks like language translation, summarization, question-answering, and content generation [401]. LLMs can process and produce human-like text across a variety of domains, offering versatility in applications such as chatbots, virtual assistants, and automated content creation. Their ability to generalize across languages and topics makes them a powerful tool in both consumer-facing applications and enterprise-level automation.

However, LLMs still face significant challenges, particularly with alignment and deployment [18]. Alignment refers to the challenge of ensuring that an LLM's outputs align with human values, objectives, and expectations. Since LLMs learn from vast and often unfiltered datasets, they can generate harmful, biased, or inaccurate content. Additionally, deploying these models on end devices, such as smartphones or edge devices, presents technical hurdles due to their model size and computational requirements. LLMs typically require significant cloud-based resources for inference, making real-time, on-device deployment impractical in most cases. The reliance on cloud processing can also lead to issues with latency, privacy, and reliability, particularly in areas with poor connectivity.

Edge AI, particularly when integrated with reinforcement learning (RL) and continual learning (CL), can address some of these challenges. By enabling LLMs to run locally on devices, edge AI reduces the dependence on cloud servers, enhancing privacy, responsiveness, and accessibility. RL can be employed to help LLMs adapt and optimize their responses based on real-time user feedback, improving alignment and personalization. Continual learning further allows LLMs to update and refine their knowledge over time, adjusting to changing user preferences and evolving language without requiring a full retraining process. This localized, adaptive learning on the edge makes LLMs more reliable, efficient, and tailored to specific environments and users.

In the future, advancements in model compression and quantization will likely make it possible to deploy even more powerful LLMs on resource-constrained edge devices. Further research in federated learning could enable LLMs to learn from multiple devices while maintaining privacy, ensuring that personal data remains secure. These advancements will help LLMs become more versatile and accessible, driving their application in areas such as personalized assistants, real-time language translation, and human-AI collaboration.

9.2.2 World Models

World models [126, 128, 129] are machine learning frameworks that attempt to create an internal representation or simulation of the environment based on sensory inputs, e.g., vision-based data. Vision-based world models are used to interpret and predict the dynamics of the surrounding world by learning the spatial and temporal relationships from visual information. These models enable systems, such as robots or autonomous vehicles, to understand their environment, simulate potential actions, and make decisions accordingly. One key advantage of world models is their ability to plan and predict future outcomes without requiring constant real-world interaction. This predictive capability allows systems to handle complex, real-world tasks such as navigation, object manipulation, and collision avoidance more effectively and efficiently.

One major challenge with world models is the fine-tuning process, which often requires extensive data and significant computational resources. Adapting these models to specific environments, such as different driving conditions in autonomous vehicles, is computationally intensive and time-consuming. Additionally, deploying world models on edge devices presents a technical hurdle. Automated vehicles, for instance, need to process data and make decisions in real-time. The large-scale computations required for world models are typically handled by cloud infrastructure, but relying on external resources increases latency and raises concerns about system reliability in areas with poor connectivity.

Edge AI addresses these challenges by enabling real-time, local computation directly on the device, reducing latency and improving the system's reliability, especially in critical applications like autonomous driving. Reinforcement learning (RL) allows world models to adapt dynamically by learning from interaction with the environment, refining decision-making processes without requiring constant cloud connectivity. Continual learning (CL) also plays a crucial role by allowing world models to evolve incrementally, learning from new data without the need to restart the training process. This ensures that the system remains adaptive and improves over time, even in constantly changing environments.

Future research in edge AI for world models will likely focus on making these systems more efficient and scalable. Advances in model compression and hardware acceleration could allow world models to run more effectively on devices with limited resources, such as mobile robots and autonomous vehicles. Additionally, the integration of federated learning could enable multiple edge devices to share learning experiences, making the models more robust and adaptable across different domains. By combining edge AI with RL and CL, world models will continue to push the boundaries of autonomy, safety, and real-world problem-solving.

9.2.3 On-Device Foundation Model

As noted before, the advancements in model compression and quantization will likely make it possible to deploy even more powerful LLMs on resource-constrained edge devices. "On-device foundation models", such as those (scaled down versions of foundation models) embedded in smartphones, represent a significant advancement in AI technology by shifting from cloud-based computation to localized processing. Apple's intelligence on iPhone [122, 232], for example, uses machine learning models that operate directly on the device, enabling personalized experiences like facial recognition, language suggestions, and image processing. The key advantage of on-device models over cloud-based solutions, such as ChatGPT, lies in their ability to operate without continuous internet connectivity, providing faster response times, greater privacy, and reduced dependence on cloud infrastructure. By handling data locally, these models eliminate the need to transfer sensitive personal information to external servers, addressing privacy concerns and enhancing security for the user.

However, deploying foundation models on end devices, such as mobile phones, presents significant challenges. These models typically require substantial computational resources and memory, which can strain the limited capabilities of mobile device hardware. Furthermore, the personalized use of foundation models poses an additional challenge: each user interacts with their device in a unique way, and the model must continually adapt to the preferences and behavior of that specific individual. Personalization at scale demands that these models not only be lightweight but also flexible enough to adjust based on real-time interactions while conserving device power and storage.

RL and continual learning offer promising solutions to these challenges. In particular, RL allows models to adapt based on user interactions, gradually optimizing performance for individual usage patterns. Continual learning enables models to acquire new knowledge over time without forgetting previously learned information. These techniques help create more personalized and efficient on-device models that evolve with user needs while operating within hardware constraints.

The field of edge AI for on-device foundation models has several exciting directions. Researchers are exploring more efficient model architectures and compression techniques to reduce the computational footprint. Additionally, there's growing interest in federated learning approaches that allow devices to collaboratively improve their models while maintaining user privacy. As hardware capabilities advance and AI algorithms become more sophisticated, we can expect increasingly powerful and personalized AI experiences running directly on our personal devices.

References

1. D. Abel, A. Barreto, B. Van Roy, D. Precup, H. P. van Hasselt, and S. Singh. 2024. A definition of continual reinforcement learning. *Advances in Neural Information Processing Systems*, 36.
2. J. D. Abernethy, E. Hazan, and A. Rakhlin. 2009. Competing in the dark: An efficient algorithm for bandit linear optimization.
3. H. G. Abreha, M. Hayajneh, and M. A. Serhani. 2022. Federated learning in edge computing: a systematic survey. *Sensors*, 22(2): 450.
4. D. A. E. Acar, R. Zhu, and V. Saligrama. 2021. Memory efficient online meta learning. In *International Conference on Machine Learning*, pp. 32–42. PMLR.
5. J. Achiam, S. Adler, S. Agarwal, L. Ahmad, I. Akkaya, F. L. Aleman, D. Almeida, J. Altenschmidt, S. Altman, S. Anadkat, et al. 2023. Gpt-4 technical report. *arXiv preprint* arXiv:2303.08774.
6. A. Agarwal, S. M. Kakade, J. D. Lee, and G. Mahajan. 2020a. Optimality and approximation with policy gradient methods in markov decision processes. In *Conference on Learning Theory*, pp. 64–66. PMLR.
7. R. Agarwal, D. Schuurmans, and M. Norouzi. 2020b. An optimistic perspective on offline reinforcement learning. In *International Conference on Machine Learning*, pp. 104–114. PMLR.
8. M. Agueh and G. Carlier. 2011. Barycenters in the wasserstein space. *SIAM J. on Math. Anal.*, 43(2): 904–924.
9. J. M. Aguiar-Pérez and M. Á. Pérez-Juárez. 2023. An insight of deep learning based demand forecasting in smart grids. *Sensors*, 23(3): 1467.
10. J.-H. Ahn, O. Simeone, and J. Kang. 2019. Wireless federated distillation for distributed edge learning with heterogeneous data. In *2019 IEEE 30th Annual International Symposium on Personal, Indoor and Mobile Radio Communications (PIMRC)*, pp. 1–6. IEEE.
11. A. Ajalloeian and S. U. Stich. 2020. On the convergence of SGD with biased gradients. *arXiv preprint* arXiv:2008.00051.
12. R. Aljundi, F. Babiloni, M. Elhoseiny, M. Rohrbach, and T. Tuytelaars. 2018. Memory aware synapses: Learning what (not) to forget. In *Proceedings of the European Conference on Computer Vision (ECCV)*, pp. 139–154.

H. Wang et al., *Continual and Reinforcement Learning for Edge AI*,
Synthesis Lectures on Learning, Networks, and Algorithms,
https://doi.org/10.1007/978-3-031-84363-1

13. L. Ambrosio, N. Gigli, and G. Savaré. 2008. *Gradient flows: in metric spaces and in the space of probability measures*. Springer Science & Business Media.
14. E. Anderes, S. Borgwardt, and J. Miller. 2016. Discrete wasserstein barycenters: Optimal transport for discrete data. *Math. Methods of Operations Res.*, 84(2): 389–409.
15. J. Andle and S. Yasaei Sekeh. 2022. Theoretical understanding of the information flow on continual learning performance. In *European Conference on Computer Vision*, pp. 86–101. Springer.
16. Andy Patrizio, 2018. IDC: Expect 175 zettabytes of data worldwide by 2025. https://www.networkworld.com/article/3325397/idc-expect-175-zettabytes-of-data-worldwide-by-2025.html. [Online; accessed 03-December-2018].
17. O. Anschel, N. Baram, and N. Shimkin. 2017. Averaged-dqn: Variance reduction and stabilization for deep reinforcement learning. In *International Conference on Machine Learning*, pp. 176–185. PMLR.
18. U. Anwar, A. Saparov, J. Rando, D. Paleka, M. Turpin, P. Hase, E. S. Lubana, E. Jenner, S. Casper, O. Sourbut, et al. 2024. Foundational challenges in assuring alignment and safety of large language models. *arXiv preprint* arXiv:2404.09932.
19. M. Arjovsky, S. Chintala, and L. Bottou. 2017. Wasserstein gan. *arXiv preprint* arXiv:1701.07875.
20. S. Arora, R. Ge, Y. Liang, T. Ma, and Y. Zhang. 2017. Generalization and equilibrium in generative adversarial nets (gans). In *Proc. of the 34th Int. Conf. on Mach. Learn.-Volume 70*, pp. 224–232. JMLR. org.
21. T. Asafa, T. Afonja, E. Olaniyan, and H. Alade. 2018. Development of a vacuum cleaner robot. *Alexandria engineering journal*, 57(4): 2911–2920.
22. H. Asanuma, S. Takagi, Y. Nagano, Y. Yoshida, Y. Igarashi, and M. Okada. 2021. Statistical mechanical analysis of catastrophic forgetting in continual learning with teacher and student networks. *Journal of the Physical Society of Japan*, 90(10): 104001.
23. K. J. Astrom. 1987. Adaptive feedback control. *Proceedings of the IEEE*, 75(2): 185–217.
24. J. Bagnell, S. M. Kakade, J. Schneider, and A. Ng. 2003. Policy search by dynamic programming. *Advances in neural information processing systems*, 16.
25. M.-F. Balcan, M. Khodak, and A. Talwalkar. 09–15 Jun 2019. Provable guarantees for gradient-based meta-learning. In K. Chaudhuri and R. Salakhutdinov, eds., *Proceedings of the 36th International Conference on Machine Learning*, volume 97 of *Proceedings of Machine Learning Research*, pp. 424–433. PMLR. https://proceedings.mlr.press/v97/balcan19a.html.
26. M. Belkin, S. Ma, and S. Mandal. 2018. To understand deep learning we need to understand kernel learning. In *International Conference on Machine Learning*, pp. 541–549. PMLR.
27. M. Belkin, D. Hsu, and J. Xu. 2020. Two models of double descent for weak features. *SIAM Journal on Mathematics of Data Science*, 2(4): 1167–1180.
28. S. J. Bell and N. D. Lawrence. 2022. The effect of task ordering in continual learning. *arXiv preprint* arXiv:2205.13323.
29. R. Bellman. 1958. Dynamic programming and stochastic control processes. *Information and control*, 1(3): 228–239.
30. Y. Bengio, J. Louradour, R. Collobert, and J. Weston. 2009. Curriculum learning. In *Proceedings of the 26th annual international conference on machine learning*, pp. 41–48.
31. M. A. Bennani, T. Doan, and M. Sugiyama. 2020. Generalisation guarantees for continual learning with orthogonal gradient descent. *arXiv preprint* arXiv:2006.11942.
32. D. Bertsekas. 2019. *Reinforcement and Optimal Control*. Athena Scientific.
33. D. Bertsekas. 2022a. *Abstract Dynamic Programming*. Athena Scientific.
34. D. Bertsekas. 2022b. *Lessons from Alphazero for Optimal, Model Predictive, and Adaptive Control*. Athena Scientific.

35. J. Bhandari, D. Russo, and R. Singal. 2018. A finite time analysis of temporal difference learning with linear function approximation. In *Conference on learning theory*, pp. 1691–1692. PMLR.
36. I. Bistritz, A. Mann, and N. Bambos. 2020. Distributed distillation for on-device learning. *Advances in Neural Information Processing Systems*, 33: 22593–22604.
37. E. Boissard, T. Le Gouic, J.-M. Loubes, et al. 2015. Distribution's template estimate with wasserstein metrics. *Bernoulli*, 21(2): 740–759.
38. P. M. Bösch, F. Becker, H. Becker, and K. W. Axhausen. 2018. Cost-based analysis of autonomous mobility services. *Transport Policy*, 64: 76–91.
39. S. Boyd, S. P. Boyd, and L. Vandenberghe. 2004. *Convex Optimization*. Cambridge university press.
40. L. Breiman. 1996. Bagging predictors. *Mach. Learn.*, 24(2): 123–140.
41. Y. Brenier. 1991. Polar factorization and monotone rearrangement of vector-valued functions. *Commun. on Pure and Appl. Math.*, 44(4): 375–417.
42. T. Brooks, B. Peebles, C. Holmes, W. DePue, Y. Guo, L. Jing, D. Schnurr, J. Taylor, T. Luhman, E. Luhman, C. Ng, R. Wang, and A. Ramesh. 2024. Video generation models as world simulators. https://openai.com/research/video-generation-models-as-world-simulators.
43. T. Brown, B. Mann, N. Ryder, M. Subbiah, J. D. Kaplan, P. Dhariwal, A. Neelakantan, P. Shyam, G. Sastry, A. Askell, et al. 2020. Language models are few-shot learners. *Advances in neural information processing systems*, 33: 1877–1901.
44. Y. Bulatov. 2011. Notmnist dataset. *Google (Books/OCR), Tech. Rep.[Online]. Available:* http://yaroslavvb blogspot. it/2011/09/notmnist-dataset. html, 2.
45. M. Caccia, P. R. López, O. Ostapenko, F. Normandin, M. Lin, L. Page-Caccia, I. H. Laradji, I. Rish, A. Lacoste, D. Vázquez, and L. Charlin. 2020. Online fast adaptation and knowledge accumulation (osaka): a new approach to continual learning. In *NeurIPS*.
46. H. Cai, C. Gan, L. Zhu, and S. Han. 2020. Tinytl: Reduce memory, not parameters for efficient on-device learning. *Advances in Neural Information Processing Systems*, 33: 11285–11297.
47. X. Cao, W. Liu, and S. Vempala. 2022. Provable lifelong learning of representations. In *International Conference on Artificial Intelligence and Statistics*, pp. 6334–6356. PMLR.
48. N. Cesa-Bianchi and G. Lugosi. 2006. *Prediction, learning, and games*. Cambridge university press.
49. A. Chaudhry, M. Ranzato, M. Rohrbach, and M. Elhoseiny. 2018. Efficient lifelong learning with a-gem. *arXiv preprint* arXiv:1812.00420.
50. A. Chaudhry, M. Rohrbach, M. Elhoseiny, T. Ajanthan, P. K. Dokania, P. H. Torr, and M. Ranzato. 2019. Continual learning with tiny episodic memories.
51. F. Chen, M. Luo, Z. Dong, Z. Li, and X. He. 2018a. Federated meta-learning with fast convergence and efficient communication. *arXiv preprint* arXiv:1802.07876.
52. G. Chen and M. Teboulle. 1993. Convergence analysis of a proximal-like minimization algorithm using bregman functions. *SIAM Journal on Optimization*, 3(3): 538–543.
53. J. Chen, Y. Zheng, Y. Liang, Z. Zhan, M. Jiang, X. Zhang, D. S. da Silva, W. Wu, and V. H. C. de Albuquerque. 2022a. Edge2analysis: a novel aiot platform for atrial fibrillation recognition and detection. *IEEE Journal of Biomedical and Health Informatics*, 26(12): 5772–5782.
54. J. Chen, O. Esrafilian, H. Bayerlein, D. Gesbert, and M. Caccamo. 2023. Model-aided federated reinforcement learning for multi-uav trajectory planning in iot networks. In *2023 IEEE Globecom Workshops (GC Wkshps)*, pp. 818–823. IEEE.
55. L. Chen, C. Harshaw, H. Hassani, and A. Karbasi. 2018b. Projection-free online optimization with stochastic gradient: From convexity to submodularity. *arXiv preprint* arXiv:1802.08183.
56. L. Chen, K. Bai, C. Tao, Y. Zhang, G. Wang, W. Wang, R. Henao, and L. Carin. Apr. 2020a. Sequence generation with optimal-transport-enhanced reinforcement learning. *Proc. of the AAAI Conf. on Artif. Intell.*, 34: 7512–7520.

57. L. Chen, Z. Gan, Y. Cheng, L. Li, L. Carin, and J. Liu. 13–18 Jul 2020b. Graph optimal transport for cross-domain alignment. In H. D. III and A. Singh, eds., *Proc. of the 37th Int. Conf. on Mach. Learn.*, volume 119 of *Proc. of Mach. Learn. Res.*, pp. 1542–1553. PMLR.
58. X. Chen, C. Wang, Z. Zhou, and K. Ross. 2021a. Randomized ensembled double q-learning: Learning fast without a model. *arXiv preprint* arXiv:2101.05982.
59. X. Chen, C. Papadimitriou, and B. Peng. 2022b. Memory bounds for continual learning. *arXiv preprint* arXiv:2204.10830.
60. Y. Chen, C. Hawkins, K. Zhang, Z. Zhang, and C. Hao. 2021b. 3u-edgeai: Ultra-low memory training, ultra-low bitwidth quantization, and ultra-low latency acceleration. In *Proceedings of the 2021 on Great Lakes Symposium on VLSI*, pp. 157–162.
61. Z. Cheng, X. Xia, M. Liwang, X. Fan, Y. Sun, X. Wang, and L. Huang. 2023. Cheese: Distributed clustering-based hybrid federated split learning over edge networks. *IEEE Transactions on Parallel and Distributed Systems*.
62. H.-Y. Chiang, N. Frumkin, F. Liang, and D. Marculescu. 2023. Mobiletl: on-device transfer learning with inverted residual blocks. In *Proceedings of the AAAI Conference on Artificial Intelligence*, volume 37, pp. 7166–7174.
63. J. J. Choi, D. Laibson, B. C. Madrian, and A. Metrick. 2009. Reinforcement learning and savings behavior. *The Journal of finance*, 64(6): 2515–2534.
64. M. J. Chong and D. Forsyth, 2019. Effectively unbiased fid and inception score and where to find them.
65. L. Collins, A. Mokhtari, and S. Shakkottai. 2020. Distribution-agnostic model-agnostic meta-learning. *arXiv preprint* arXiv:2002.04766.
66. G. Constantinou, G. S. Ramachandran, A. Alfarrarjeh, S. H. Kim, B. Krishnamachari, and C. Shahabi. 2019. A crowd-based image learning framework using edge computing for smart city applications. In *2019 IEEE Fifth International Conference on Multimedia Big Data (BigMM)*, pp. 11–20. IEEE.
67. A. Coronato, M. Naeem, G. De Pietro, and G. Paragliola. 2020. Reinforcement learning for intelligent healthcare applications: A survey. *Artificial intelligence in medicine*, 109: 101964.
68. T. J. Crayton and B. M. Meier. 2017. Autonomous vehicles: Developing a public health research agenda to frame the future of transportation policy. *Journal of Transport & Health*, 6: 245–252.
69. F.-A. Croitoru, V. Hondru, R. T. Ionescu, and M. Shah. 2023. Diffusion models in vision: A survey. *IEEE Transactions on Pattern Analysis and Machine Intelligence*, 45(9): 10850–10869.
70. M. Cuturi. 2013. Sinkhorn distances: Lightspeed computation of optimal transport. In *Advances in Neural Inf. Process. Syst. 26*, pp. 2292–2300. Curran Associates, Inc.
71. M. Cuturi and A. Doucet. 2014. Fast computation of wasserstein barycenters. In *Int. Conf. on Mach. Learn.*, pp. 685–693.
72. M. Cuturi and G. Peyré. 2016. A smoothed dual approach for variational wasserstein problems. *SIAM J. on Imag. Sci.*, 9(1): 320–343.
73. G. Dalal, G. Thoppe, B. Szörényi, and S. Mannor. 2018. Finite sample analysis of two-timescale stochastic approximation with applications to reinforcement learning. In *Conference On Learning Theory*, pp. 1199–1233. PMLR.
74. G. Dalal, B. Szorenyi, and G. Thoppe. 2020. A tale of two-timescale reinforcement learning with the tightest finite-time bound. In *Proceedings of the AAAI Conference on Artificial Intelligence*, volume 34, pp. 3701–3708.
75. Y. Dang, C. Benzaïd, B. Yang, T. Taleb, and Y. Shen. 2022. Deep-ensemble-learning-based gps spoofing detection for cellular-connected uavs. *IEEE Internet of Things Journal*, 9(24): 25068–25085.
76. M. De Lange, R. Aljundi, M. Masana, S. Parisot, X. Jia, A. Leonardis, G. Slabaugh, and T. Tuytelaars. 2019. A continual learning survey: Defying forgetting in classification tasks. *arXiv preprint* arXiv:1909.08383.

77. O. Dekel, R. Gilad-Bachrach, O. Shamir, and L. Xiao. 2012. Optimal distributed online prediction using mini-batches. *The Journal of Machine Learning Research*, 13: 165–202.
78. G. Denevi, D. Stamos, C. Ciliberto, and M. Pontil. 2019. Online-within-online meta-learning. *Advances in Neural Information Processing Systems*, 32.
79. L. Deng. 2012. The mnist database of handwritten digit images for machine learning research. *IEEE Signal Process. Mag.*, 29(6): 141–142.
80. S. Deng, H. Zhao, W. Fang, J. Yin, S. Dustdar, and A. Y. Zomaya. 2020. Edge intelligence: The confluence of edge computing and artificial intelligence. *IEEE Internet of Things Journal*, 7(8): 7457–7469.
81. C. D'Eramo, M. Restelli, and A. Nuara. 2016. Estimating maximum expected value through gaussian approximation. In *International Conference on Machine Learning*, pp. 1032–1040. PMLR.
82. J. Devlin, M.-W. Chang, K. Lee, and K. Toutanova. 2018. Bert: Pre-training of deep bidirectional transformers for language understanding. *arXiv preprint* arXiv:1810.04805.
83. P. Dhar, R. V. Singh, K.-C. Peng, Z. Wu, and R. Chellappa. 2019. Learning without memorizing. In *Proceedings of the IEEE/CVF Conference on Computer Vision and Pattern Recognition*, pp. 5138–5146.
84. M. Dhuheir, E. Baccour, A. Erbad, S. Sabeeh, and M. Hamdi. 2021. Efficient real-time image recognition using collaborative swarm of uavs and convolutional networks. In *2021 International Wireless Communications and Mobile Computing (IWCMC)*, pp. 1954–1959. IEEE.
85. T. G. Dietterich. 2000. Ensemble methods in machine learning. In *International workshop on multiple classifier systems*, pp. 1–15. Springer.
86. C. T. Dinh, N. H. Tran, and T. D. Nguyen. 2020. Personalized federated learning with moreau envelopes. *arXiv preprint* arXiv:2006.08848.
87. T. Doan, M. A. Bennani, B. Mazoure, G. Rabusseau, and P. Alquier. 2021. A theoretical analysis of catastrophic forgetting through the ntk overlap matrix. In *International Conference on Artificial Intelligence and Statistics*, pp. 1072–1080. PMLR.
88. J. Dong, L. Wang, Z. Fang, G. Sun, S. Xu, X. Wang, and Q. Zhu. 2022. Federated class-incremental learning. In *Proceedings of the IEEE/CVF conference on computer vision and pattern recognition*, pp. 10164–10173.
89. J. Dong, H. Li, Y. Cong, G. Sun, Y. Zhang, and L. Van Gool. 2023. No one left behind: Real-world federated class-incremental learning. *IEEE Transactions on Pattern Analysis and Machine Intelligence*.
90. M. Doshi and A. Varghese. 2022. Smart agriculture using renewable energy and ai-powered iot. In *AI, edge and IoT-based smart agriculture*, pp. 205–225. Elsevier.
91. I. Durugkar, I. Gemp, and S. Mahadevan. 2016. Generative multi-adversarial networks. *arXiv preprint* :1611.01673.
92. S. Elfwing, E. Uchibe, and K. Doya. 2018. Sigmoid-weighted linear units for neural network function approximation in reinforcement learning. *Neural Networks*, 107: 3–11.
93. B. Ermis, G. Zappella, M. Wistuba, A. Rawal, and C. Archambeau. 2022. Memory efficient continual learning with transformers. *Advances in Neural Information Processing Systems*, 35: 10629–10642.
94. I. Evron, E. Moroshko, R. Ward, N. Srebro, and D. Soudry. 2022. How catastrophic can catastrophic forgetting be in linear regression? In *Conference on Learning Theory*, pp. 4028–4079. PMLR.
95. D. J. Fagnant and K. Kockelman. 2015. Preparing a nation for autonomous vehicles: opportunities, barriers and policy recommendations. *Transportation Research Part A: Policy and Practice*, 77: 167–181.

96. A. Fallah, A. Mokhtari, and A. Ozdaglar. 2020a. On the convergence theory of gradient-based model-agnostic meta-learning algorithms. In *International Conference on Artificial Intelligence and Statistics (AISTATS)*, pp. 1082–1092. PMLR.
97. A. Fallah, A. Mokhtari, and A. Ozdaglar. 2020b. Personalized federated learning: A meta-learning approach. *arXiv preprint* arXiv:2002.07948.
98. J. Fan, B. Jiang, and Q. Sun. 2021. Hoeffding's inequality for general markov chains and its applications to statistical learning. *Journal of Machine Learning Research*, 22(139): 1–35.
99. A.-m. Farahmand, C. Szepesvári, and R. Munos. 2010. Error propagation for approximate policy and value iteration. *Advances in Neural Information Processing Systems*, 23.
100. M. Farajtabar, N. Azizan, A. Mott, and A. Li. 2020. Orthogonal gradient descent for continual learning. In *International Conference on Artificial Intelligence and Statistics*, pp. 3762–3773. PMLR.
101. Y. Feng, Y. Qi, H. Li, X. Wang, and J. Tian. 2024. Leveraging federated learning and edge computing for recommendation systems within cloud computing networks. In *Third International Symposium on Computer Applications and Information Systems (ISCAIS 2024)*, volume 13210, pp. 279–287. SPIE.
102. C. Fernando, D. Banarse, C. Blundell, Y. Zwols, D. Ha, A. A. Rusu, A. Pritzel, and D. Wierstra. 2017. Pathnet: Evolution channels gradient descent in super neural networks. *arXiv preprint* arXiv:1701.08734.
103. C. Finn, P. Abbeel, and S. Levine. 2017. Model-agnostic meta-learning for fast adaptation of deep networks. In *Proc. International Conference on Machine Learning (ICML)*, pp. 1126–1135.
104. C. Finn, K. Xu, and S. Levine. 2018. Probabilistic model-agnostic meta-learning. *Advances in neural information processing systems*, 31.
105. C. Finn, A. Rajeswaran, S. Kakade, and S. Levine. 2019. Online meta-learning. In *International Conference on Machine Learning*, pp. 1920–1930. PMLR.
106. S. Fu, F. Dong, D. Shen, and Q. He. 2023. Joint quality evaluation, model splitting and resource provisioning for split edge learning. In *2023 20th Annual IEEE International Conference on Sensing, Communication, and Networking (SECON)*, pp. 420–428. IEEE.
107. Z. Fu, Z. Yang, and Z. Wang. 2020. Single-timescale actor-critic provably finds globally optimal policy. In *International Conference on Learning Representations*.
108. S. Fujimoto, H. Hoof, and D. Meger. 2018. Addressing function approximation error in actor-critic methods. In *International conference on machine learning*, pp. 1587–1596. PMLR.
109. S. Fujimoto, D. Meger, and D. Precup. 2019. Off-policy deep reinforcement learning without exploration. In *International conference on machine learning*, pp. 2052–2062. PMLR.
110. Q. Gao, Z. Luo, D. Klabjan, and F. Zhang. 2022. Efficient architecture search for continual learning. *IEEE Trans. on Neural Netw. and Learn. Syst.*, pp. 1–11. https://doi.org/10.1109/TNNLS.2022.3151511.
111. J. Garcıa and F. Fernández. 2015. A comprehensive survey on safe reinforcement learning. *Journal of Machine Learning Research*, 16(1): 1437–1480.
112. M. Gargiani, A. Zanelli, D. Liao-McPherson, T. Summers, and J. Lygeros. 2022. Dynamic programming through the lens of semismooth newton-type methods. *IEEE Control Systems Letters*.
113. S. Gelly and D. Silver. 2007. Combining online and offline knowledge in UCT. In *Proceedings of the 24th international conference on Machine learning*, pp. 273–280.
114. B. Goertzel. 2014. Artificial general intelligence: concept, state of the art, and future prospects. *Journal of Artificial General Intelligence*, 5(1): 1.
115. D. Goldfarb and P. Hand. 2023. Analysis of catastrophic forgetting for random orthogonal transformation tasks in the overparameterized regime. In *International Conference on Artificial Intelligence and Statistics*, pp. 2975–2993. PMLR.

116. I. Goodfellow, J. Pouget-Abadie, M. Mirza, B. Xu, D. Warde-Farley, S. Ozair, A. Courville, and Y. Bengio. 2014a. Generative adversarial nets. In *Advances in neural Inf. Process. Syst.*, pp. 2672–2680.
117. I. Goodfellow, J. Pouget-Abadie, M. Mirza, B. Xu, D. Warde-Farley, S. Ozair, A. Courville, and Y. Bengio. 2014b. Generative adversarial nets. In *Advances in neural Inf. Process. Syst.*, pp. 2672–2680.
118. J. Grand-Clément. 2021. From convex optimization to MDPs: A review of first-order, second-order and quasi-newton methods for MDPs. *arXiv preprint* arXiv:2104.10677.
119. P. Grnarova, K. Y. Levy, A. Lucchi, N. Perraudin, I. Goodfellow, T. Hofmann, and A. Krause. 2019. A domain agnostic measure for monitoring and evaluating gans. In *Advances in Neural Inf. Process. Syst. 32*, pp. 12092–12102. Curran Associates, Inc.
120. I. Grondman, L. Busoniu, G. A. Lopes, and R. Babuska. 2012. A survey of actor-critic reinforcement learning: Standard and natural policy gradients. *IEEE Transactions on Systems, Man, and Cybernetics, Part C (Applications and Reviews)*, 42(6): 1291–1307.
121. S. Gunasekar, J. Lee, D. Soudry, and N. Srebro. 2018. Characterizing implicit bias in terms of optimization geometry. In *International Conference on Machine Learning*, pp. 1832–1841. PMLR.
122. T. Gunter, Z. Wang, C. Wang, R. Pang, A. Narayanan, A. Zhang, B. Zhang, C. Chen, C.-C. Chiu, D. Qiu, et al. 2024. Apple intelligence foundation language models. *arXiv preprint* arXiv:2407.21075.
123. Y. Guo, F. Liu, Z. Cai, L. Chen, and N. Xiao. 2020a. Feel: A federated edge learning system for efficient and privacy-preserving mobile healthcare. In *Proceedings of the 49th International Conference on Parallel Processing*, pp. 1–11.
124. Y. Guo, M. Liu, T. Yang, and T. Rosing. 2020b. Improved schemes for episodic memory based lifelong learning algorithm. In *Conference on Neural Information Processing Systems*.
125. A. Gupta, V. Kumar, C. Lynch, S. Levine, and K. Hausman. 2020. Relay policy learning: Solving long horizon tasks via imitation and reinforcement learning. In *Conference on Robot Learning*, pp. 1025–1037. PMLR.
126. D. Ha and J. Schmidhuber. 2018. World models. *arXiv preprint* arXiv:1803.10122.
127. T. Haarnoja, A. Zhou, P. Abbeel, and S. Levine. 2018. Soft actor-critic: Off-policy maximum entropy deep reinforcement learning with a stochastic actor. In *International Conference on Machine Learning*, pp. 1861–1870. PMLR.
128. D. Hafner, T. Lillicrap, J. Ba, and M. Norouzi. 2019. Dream to control: Learning behaviors by latent imagination. *arXiv preprint* arXiv:1912.01603.
129. D. Hafner, T. Lillicrap, M. Norouzi, and J. Ba. 2020. Mastering atari with discrete world models. *arXiv preprint* arXiv:2010.02193.
130. J. Han, Y. Ma, Q. Mei, and X. Liu. 2021. Deeprec: On-device deep learning for privacy-preserving sequential recommendation in mobile commerce. In *Proceedings of the Web Conference 2021*, pp. 900–911.
131. S. Han, H. Mao, and W. J. Dally. 2015. Deep compression: Compressing deep neural networks with pruning, trained quantization and huffman coding. *arXiv preprint* arXiv:1510.00149.
132. J. Hannan. 1957. Approximation to bayes risk in repeated play. *Contributions to the Theory of Games*, 3(2): 97–139.
133. C. Hardy, E. Le Merrer, and B. Sericola. 2019. Md-gan: Multi-discriminator generative adversarial networks for distributed datasets. In *2019 IEEE Int. Parallel and Distrib. Process. Symp. (IPDPS)*, pp. 866–877. IEEE.
134. J. Harrison, A. Sharma, C. Finn, and M. Pavone. 2020. Continuous meta-learning without tasks. *Advances in neural information processing systems*, 33: 17571–17581.
135. H. Hasselt. 2010. Double q-learning. *Advances in neural information processing systems*, 23: 2613–2621.

136. T. Hastie, A. Montanari, S. Rosset, and R. J. Tibshirani. 2022. Surprises in high-dimensional ridgeless least squares interpolation. *The Annals of Statistics*, 50(2): 949–986.
137. E. Hazan. 2019. Introduction to online convex optimization. *arXiv preprint* arXiv:1909.05207.
138. E. Hazan and S. Kale. 2012. Projection-free online learning. *arXiv preprint* arXiv:1206.4657.
139. E. Hazan, A. Agarwal, and S. Kale. 2007. Logarithmic regret algorithms for online convex optimization. *Machine Learning*, 69(2): 169–192.
140. E. Hazan et al. 2016. Introduction to online convex optimization. *Foundations and Trends® in Optimization*, 2(3-4): 157–325.
141. K. He, X. Zhang, S. Ren, and J. Sun. 2016. Deep residual learning for image recognition. In *Computer Vision and Pattern Recognition*, pp. 770–778.
142. Z. He and D. Fan. 2019. Simultaneously optimizing weight and quantizer of ternary neural network using truncated gaussian approximation. In *Proc. of the IEEE Conf. on Comput. Vision and Pattern Recognit.*, pp. 11438–11446.
143. M. Heusel, H. Ramsauer, T. Unterthiner, B. Nessler, and S. Hochreiter. 2017. Gans trained by a two time-scale update rule converge to a local nash equilibrium. In I. Guyon, U. V. Luxburg, S. Bengio, H. Wallach, R. Fergus, S. Vishwanathan, and R. Garnett, eds., *Advances in Neural Inf. Process. Syst. 30*, pp. 6626–6637. Curran Associates, Inc.
144. S. Hochreiter and J. Schmidhuber. 1997. Long short-term memory. *Neural computation*, 9(8): 1735–1780.
145. K. Hu, M. Lu, Y. Li, S. Gong, J. Wu, F. Zhou, S. Jiang, and Y. Yang. 2022. A federated incremental learning algorithm based on dual attention mechanism. *Applied Sciences*, 12(19): 10025.
146. Z. Hu, L. Shen, Z. Wang, T. Liu, C. Yuan, and D. Tao. 2023. Architecture, dataset and model-scale agnostic data-free meta-learning. In *Proceedings of the IEEE/CVF Conference on Computer Vision and Pattern Recognition*, pp. 7736–7745.
147. S. C. Hung, C.-H. Tu, C.-E. Wu, C.-H. Chen, Y.-M. Chan, and C.-S. Chen. 2019. Compacting, picking and growing for unforgetting continual learning. *arXiv preprint* arXiv:1910.06562.
148. N. N. T. Huu, L. Mai, and T. V. Minh. 2021. Detecting abnormal and dangerous activities using artificial intelligence on the edge for smart city application. In *2021 15th International Conference on Advanced Computing and Applications (ACOMP)*, pp. 85–92. IEEE.
149. J. Ibarz, J. Tan, C. Finn, M. Kalakrishnan, P. Pastor, and S. Levine. 2021. How to train your robot with deep reinforcement learning: lessons we have learned. *The International Journal of Robotics Research*, 40(4-5): 698–721.
150. A. Ijspeert, J. Nakanishi, and S. Schaal. 2002. Learning attractor landscapes for learning motor primitives. *Advances in neural information processing systems*, 15.
151. J. Janai, F. Güney, A. Behl, A. Geiger, et al. 2020. Computer vision for autonomous vehicles: Problems, datasets and state of the art. *Foundations and Trends® in Computer Graphics and Vision*, 12(1–3): 1–308.
152. M. Janner, J. Fu, M. Zhang, and S. Levine. 2019. When to trust your model: Model-based policy optimization. *Advances in neural information processing systems*, 32.
153. K. Ji, J. Yang, and Y. Liang. 2020. Multi-step model-agnostic meta-learning: Convergence and improved algorithms. *arXiv preprint* arXiv:2002.07836.
154. B. Jiang, Q. Sun, and J. Fan. 2018. Bernstein's inequality for general markov chains. *arXiv preprint* arXiv:1805.10721.
155. Y. Jiang, J. Konečný, K. Rush, and S. Kannan. 2019. Improving federated learning personalization via model agnostic meta learning. *arXiv preprint* arXiv:1909.12488.
156. Y. Jiang, S. Wang, V. Valls, B. J. Ko, W.-H. Lee, K. K. Leung, and L. Tassiulas. 2022. Model pruning enables efficient federated learning on edge devices. *IEEE Transactions on Neural Networks and Learning Systems*, 34(12): 10374–10386.

157. P. Ju, X. Lin, and J. Liu. 2020. Overfitting can be harmless for basis pursuit, but only to a degree. *Advances in Neural Information Processing Systems*, 33: 7956–7967.
158. L. P. Kaelbling, M. L. Littman, and A. W. Moore. 1996. Reinforcement learning: A survey. *Journal of artificial intelligence research*, 4: 237–285.
159. M. Kamp, M. Boley, D. Keren, A. Schuster, and I. Sharfman. 2014. Communication-efficient distributed online prediction by dynamic model synchronization. In *Joint European Conference on Machine Learning and Knowledge Discovery in Databases*, pp. 623–639. Springer.
160. M. Kamruzzaman. 2021. New opportunities, challenges, and applications of edge-ai for connected healthcare in smart cities. In *2021 IEEE Globecom Workshops (GC Wkshps)*, pp. 1–6. IEEE.
161. Y. Kang, J. Hauswald, C. Gao, A. Rovinski, T. Mudge, J. Mars, and L. Tang. 2017. Neurosurgeon: Collaborative intelligence between the cloud and mobile edge. *ACM SIGARCH Computer Architecture News*, 45(1): 615–629.
162. Z. Ke, B. Liu, and X. Huang. 2020. Continual learning of a mixed sequence of similar and dissimilar tasks. *Advances in Neural Information Processing Systems*, 33: 18493–18504.
163. R. Kemker, M. McClure, A. Abitino, T. Hayes, and C. Kanan. 2018. Measuring catastrophic forgetting in neural networks. In *Proc. of the AAAI Conf. on Artif. Intell.*, volume 32.
164. K. Khetarpal, M. Riemer, I. Rish, and D. Precup. 2022. Towards continual reinforcement learning: A review and perspectives. *Journal of Artificial Intelligence Research*, 75: 1401–1476.
165. S. Khodadadian, T. T. Doan, J. Romberg, and S. T. Maguluri. 2022. Finite sample analysis of two-time-scale natural actor-critic algorithm. *IEEE Transactions on Automatic Control.*
166. M. Khodak, M.-F. Balcan, and A. Talwalkar. 2019a. Provable guarantees for gradient-based meta-learning. *arXiv preprint* arXiv:1902.10644.
167. M. Khodak, M.-F. F. Balcan, and A. S. Talwalkar. 2019b. Adaptive gradient-based meta-learning methods. *Advances in Neural Information Processing Systems*, 32.
168. B. Kim, A.-m. Farahmand, J. Pineau, and D. Precup. 2013. Learning from limited demonstrations. *Advances in Neural Information Processing Systems*, 26.
169. Y. G. Kim and C.-J. Wu. 2022. Fedgpo: Heterogeneity-aware global parameter optimization for efficient federated learning. In *2022 IEEE International Symposium on Workload Characterization (IISWC)*, pp. 117–129. IEEE.
170. B. R. Kiran, I. Sobh, V. Talpaert, P. Mannion, A. A. Al Sallab, S. Yogamani, and P. Pérez. 2021. Deep reinforcement learning for autonomous driving: A survey. *IEEE Transactions on Intelligent Transportation Systems*, 23(6): 4909–4926.
171. J. Kirkpatrick, R. Pascanu, N. Rabinowitz, J. Veness, G. Desjardins, A. A. Rusu, K. Milan, J. Quan, T. Ramalho, A. Grabska-Barwinska, et al. 2017. Overcoming catastrophic forgetting in neural networks. *Proceedings of the national academy of sciences*, 114(13): 3521–3526.
172. B. Kizilkaya, E. Ever, H. Y. Yatbaz, and A. Yazici. 2022. An effective forest fire detection framework using heterogeneous wireless multimedia sensor networks. *ACM Transactions on Multimedia Computing, Communications, and Applications (TOMM)*, 18(2): 1–21.
173. G. Koch, R. Zemel, R. Salakhutdinov, et al. 2015. Siamese neural networks for one-shot image recognition. In *ICML deep learning workshop*, volume 2, p. 0. Lille.
174. A. Konar, I. G. Chakraborty, S. J. Singh, L. C. Jain, and A. K. Nagar. 2013. A deterministic improved q-learning for path planning of a mobile robot. *IEEE Transactions on Systems, Man, and Cybernetics: Systems*, 43(5): 1141–1153.
175. V. Konda and J. Tsitsiklis. 1999. Actor-critic algorithms. *Advances in neural information processing systems*, 12.
176. A. Korotin, V. Egiazarian, A. Asadulaev, A. Safin, and E. Burnaev. 2021. Wasserstein-2 generative networks. In *Int. Conf. on Learn. Representations*.
177. A. Krizhevsky and G. Hinton. 2009a. Learning multiple layers of features from tiny images. Technical Report 0, University of Toronto, Toronto, Ontario.

178. A. Krizhevsky and G. Hinton. 2009b. Learning multiple layers of features from tiny images.
179. A. Krizhevsky, I. Sutskever, and G. E. Hinton. 2012. Imagenet classification with deep convolutional neural networks. In *Proc. of NIPS*.
180. M. Krouka, A. Elgabli, C. B. Issaid, and M. Bennis. 2021. Energy-efficient model compression and splitting for collaborative inference over time-varying channels. In *2021 IEEE 32nd Annual International Symposium on Personal, Indoor and Mobile Radio Communications (PIMRC)*, pp. 1173–1178. IEEE.
181. A. Kumar, J. Fu, M. Soh, G. Tucker, and S. Levine. 2019. Stabilizing off-policy q-learning via bootstrapping error reduction. *Advances in Neural Information Processing Systems*, 32.
182. A. Kumar, A. Gupta, and S. Levine. 2020. Discor: Corrective feedback in reinforcement learning via distribution correction. *Advances in Neural Information Processing Systems*, 33: 18560–18572.
183. H. Kumar, A. Koppel, and A. Ribeiro. 2023. On the sample complexity of actor-critic method for reinforcement learning with function approximation. *Machine Learning*, pp. 1–35.
184. A. Kwasniewska, M. Szankin, M. Ozga, J. Wolfe, A. Das, A. Zajac, J. Ruminski, and P. Rad. 2019. Deep learning optimization for edge devices: Analysis of training quantization parameters. In *IECON 2019-45th Annual Conference of the IEEE Industrial Electronics Society*, volume 1, pp. 96–101. IEEE.
185. B. Lake, R. Salakhutdinov, J. Gross, and J. Tenenbaum. 2011. One shot learning of simple visual concepts. In *Proceedings of the annual meeting of the cognitive science society*, volume 33.
186. Q. Lan, Y. Pan, A. Fyshe, and M. White. 2020. Maxmin q-learning: Controlling the estimation bias of q-learning. *arXiv preprint* arXiv:2002.06487.
187. P. Landgren, V. Srivastava, and N. E. Leonard. 2016. Distributed cooperative decision-making in multiarmed bandits: Frequentist and bayesian algorithms. In *2016 IEEE 55th Conference on Decision and Control (CDC)*, pp. 167–172. IEEE.
188. A. Lazaric, M. Ghavamzadeh, and R. Munos. 2010. Analysis of a classification-based policy iteration algorithm. In *27th International Conference on Machine Learning*, pp. 607–614. Omnipress.
189. Y. LeCun. 1998. The mnist database of handwritten digits. *http://yann. lecun. com/exdb/mnist/*.
190. Y. LeCun, B. Boser, J. Denker, D. Henderson, R. Howard, W. Hubbard, and L. Jackel. 1989. Handwritten digit recognition with a back-propagation network. *Advances in neural information processing systems*, 2.
191. Y. Lecun, Y. Bengio, and G. Hinton. 2015. Deep learning. *Nature*, 521(7553): 436.
192. D. Lee, B. Defourny, and W. B. Powell. 2013. Bias-corrected q-learning to control max-operator bias in q-learning. In *2013 IEEE Symposium on Adaptive Dynamic Programming and Reinforcement Learning (ADPRL)*, pp. 93–99. IEEE.
193. J. Lee, H. G. Hong, D. Joo, and J. Kim. 2020a. Continual learning with extended kronecker-factored approximate curvature. In *Proceedings of the IEEE/CVF Conference on Computer Vision and Pattern Recognition*, pp. 9001–9010.
194. K. Lee, M. Laskin, A. Srinivas, and P. Abbeel. 2020b. Sunrise: A simple unified framework for ensemble learning in deep reinforcement learning. *arXiv preprint* arXiv:2007.04938.
195. S. Lee, S. Goldt, and A. Saxe. 2021. Continual learning in the teacher-student setup: Impact of task similarity. In *International Conference on Machine Learning*, pp. 6109–6119. PMLR.
196. S. Lee, Y. Seo, K. Lee, P. Abbeel, and J. Shin. 2022. Offline-to-online reinforcement learning via balanced replay and pessimistic q-ensemble. In *Conference on Robot Learning*, pp. 1702–1712. PMLR.
197. S.-W. Lee, J.-H. Kim, J. Jun, J.-W. Ha, and B.-T. Zhang. 2017. Overcoming catastrophic forgetting by incremental moment matching. *arXiv preprint* arXiv:1703.08475.
198. M. I. Leontev, V. Islenteva, and S. V. Sukhov. 2020. Non-iterative knowledge fusion in deep convolutional neural networks. *Neural Process. Letters*, 51(1): 1–22.

199. D. A. Levin and Y. Peres. 2017. *Markov Chains and Mixing Times*, volume 107. American Mathematical Soc.
200. S. Levine, A. Kumar, G. Tucker, and J. Fu. 2020. Offline reinforcement learning: Tutorial, review, and perspectives on open problems. *arXiv preprint* arXiv:2005.01643.
201. J. Leygonie, J. She, A. Almahairi, S. Rajeswar, and A. C. Courville. 2019. Adversarial computation of optimal transport maps. *CoRR*, abs/1906.09691.
202. P. Lezaud. 1998. Chernoff-type bound for finite markov chains. *Annals of Applied Probability*, pp. 849–867.
203. B. Li, S. Cen, Y. Chen, and Y. Chi. 2020. Communication-efficient distributed optimization in networks with gradient tracking and variance reduction. In *International Conference on Artificial Intelligence and Statistics*, pp. 1662–1672.
204. E. Li, L. Zeng, Z. Zhou, and X. Chen. 2019a. Edge ai: On-demand accelerating deep neural network inference via edge computing. *IEEE Transactions on Wireless Communications*, 19(1): 447–457.
205. H. Li, R. Wang, J. Wu, and W. Zhang. 2022a. Federated edge learning via reconfigurable intelligent surface with one-bit quantization. In *GLOBECOM 2022-2022 IEEE Global Communications Conference*, pp. 1055–1060. IEEE.
206. H. Li, S. Lin, L. Duan, Y. Liang, and N. B. Shroff. 2024. Theory on mixture-of-experts in continual learning. *arXiv preprint* arXiv:2406.16437.
207. Q. Li, Z. Wang, G. Li, J. Pang, and G. Xu. 2021. Hilbert sinkhorn divergence for optimal transport. In *2021 IEEE/CVF Conf. on Comput. Vision and Pattern Recognit. (CVPR)*, pp. 3834–3843. https://doi.org/10.1109/CVPR46437.2021.00383.
208. X. Li, Y. Zhou, T. Wu, R. Socher, and C. Xiong. 2019b. Learn to grow: A continual structure learning framework for overcoming catastrophic forgetting. In *International Conference on Machine Learning*, pp. 3925–3934. PMLR.
209. Y. Li. 2017. Deep reinforcement learning: An overview. *arXiv preprint* arXiv:1701.07274.
210. Y. Li, M. Li, M. S. Asif, and S. Oymak. 2022b. Provable and efficient continual representation learning. *arXiv preprint* arXiv:2203.02026.
211. Z. Li and D. Hoiem. 2017. Learning without forgetting. *IEEE transactions on pattern analysis and machine intelligence*, 40(12): 2935–2947.
212. S. Liang and R. Srikant. 2017. Why deep neural networks for function approximation? In *5th International Conference on Learning Representations, ICLR 2017.*
213. S. Lin, G. Yang, and J. Zhang. 2020. A collaborative learning framework via federated meta-learning. *arXiv preprint* arXiv:2001.03229.
214. S. Lin, M. Dedeoglu, and J. Zhang. 2021a. Accelerating distributed online meta-learning via multi-agent collaboration under limited communication. In *Proceedings of the Twenty-Second International Symposium on Theory, Algorithmic Foundations, and Protocol Design for Mobile Networks and Mobile Computing*, MobiHoc '21, p. 261-270. Association for Computing Machinery, New York, NY, USA. ISBN 9781450385589. https://doi.org/10.1145/3466772.3467055. 10.1145/3466772.3467055.
215. S. Lin, L. Yang, D. Fan, and J. Zhang. 2021b. Trgp: Trust region gradient projection for continual learning. In *International Conference on Learning Representations.*
216. S. Lin, L. Yang, Z. He, D. Fan, and J. Zhang. 2021c. Metagater: Fast learning of conditional channel gated networks via federated meta-learning. In *2021 IEEE 18th International Conference on Mobile Ad Hoc and Smart Systems (MASS)*, pp. 164–172. IEEE.
217. S. Lin, L. Yang, D. Fan, and J. Zhang. 2022a. Beyond not-forgetting: Continual learning with backward knowledge transfer. In *Thirty-Sixth Conference on Neural Information Processing Systems.*
218. S. Lin, L. Yang, D. Fan, and J. Zhang. 2022b. Trgp: Trust region gradient projection for continual learning. *Tenth International Conference on Learning Representations, ICLR 2022.*

219. T.-Y. Lin, P. Goyal, R. Girshick, K. He, and P. Dollár. 2017. Focal loss for dense object detection. In *Proceedings of the IEEE international conference on computer vision*, pp. 2980–2988.
220. H. Liu and H. Liu. 2022. Continual learning with recursive gradient optimization. *arXiv preprint* arXiv:2201.12522.
221. H. Liu, X. Gu, and D. Samaras. 2019. Wasserstein gan with quadratic transport cost. In *Proc. of the IEEE Int. Conf. on Comput. Vision*, pp. 4832–4841.
222. S. Liu, G. Yu, R. Yin, J. Yuan, L. Shen, and C. Liu. 2021a. Joint model pruning and device selection for communication-efficient federated edge learning. *IEEE Transactions on Communications*, 70(1): 231–244.
223. S. Liu, H. Qu, Q. Chen, W. Jian, R. Liu, and L. You. 2022. Afmeta: Asynchronous federated meta-learning with temporally weighted aggregation. In *2022 IEEE Smartworld, Ubiquitous Intelligence & Computing, Scalable Computing & Communications, Digital Twin, Privacy Computing, Metaverse, Autonomous & Trusted Vehicles (SmartWorld/UIC/ScalCom/DigitalTwin/PriComp/Meta)*, pp. 641–648. IEEE.
224. W. Liu, X. Wang, J. Owens, and Y. Li. 2020. Energy-based out-of-distribution detection. *Advances in Neural Information Processing Systems*, 33: 21464–21475.
225. X. Liu, M. Masana, L. Herranz, J. Van de Weijer, A. M. Lopez, and A. D. Bagdanov. 2018. Rotate your networks: Better weight consolidation and less catastrophic forgetting. In *2018 24th International Conference on Pattern Recognition (ICPR)*, pp. 2262–2268. IEEE.
226. Y. Liu, Y. Zhu, and J. James. 2021b. Resource-constrained federated edge learning with heterogeneous data: Formulation and analysis. *IEEE Transactions on Network Science and Engineering*, 9(5): 3166–3178.
227. D. Lopez-Paz and M. Ranzato. 2017. Gradient episodic memory for continual learning. *Advances in neural information processing systems*, 30: 6467–6476.
228. K. Luo, X. Li, Y. Lan, and M. Gao. 2023. Gradma: A gradient-memory-based accelerated federated learning with alleviated catastrophic forgetting. In *Proceedings of the IEEE/CVF Conference on Computer Vision and Pattern Recognition*, pp. 3708–3717.
229. H. Mao, M. Alizadeh, I. Menache, and S. Kandula. 2016. Resource management with deep reinforcement learning. In *Proceedings of the 15th ACM workshop on hot topics in networks*, pp. 50–56.
230. R. J. McCann. 1997. A convexity principle for interacting gases. *Advances in Math.*, 128(1): 153–179.
231. M. McCloskey and N. J. Cohen. 1989. Catastrophic interference in connectionist networks: The sequential learning problem. In *Psychology of learning and motivation*, volume 24, pp. 109–165. Elsevier.
232. B. McKinzie, Z. Gan, J.-P. Fauconnier, S. Dodge, B. Zhang, P. Dufter, D. Shah, X. Du, F. Peng, F. Weers, et al. 2024. Mm1: Methods, analysis & insights from multimodal llm pre-training. *arXiv preprint* arXiv:2403.09611.
233. B. McMahan, E. Moore, D. Ramage, S. Hampson, and B. A. y Arcas. 2017. Communication-efficient learning of deep networks from decentralized data. In *Artificial Intelligence and Statistics*, pp. 1273–1282.
234. V. Mnih, K. Kavukcuoglu, D. Silver, A. Graves, I. Antonoglou, D. Wierstra, and M. Riedmiller. 2013. Playing atari with deep reinforcement learning. *arXiv preprint* arXiv:1312.5602.
235. H. Modares, F. L. Lewis, and M.-B. Naghibi-Sistani. 2013. Adaptive optimal control of unknown constrained-input systems using policy iteration and neural networks. *IEEE transactions on neural networks and learning systems*, 24(10): 1513–1525.
236. J. Mori, I. Teranishi, and R. Furukawa. 2022. Continual horizontal federated learning for heterogeneous data. In *2022 International Joint Conference on Neural Networks (IJCNN)*, pp. 1–8. IEEE.

237. R. Munos. 2003. Error bounds for approximate policy iteration. In *ICML*, volume 3, pp. 560–567. Citeseer.
238. A. Nair, M. Dalal, A. Gupta, and S. Levine. 2020. Accelerating online reinforcement learning with offline datasets. *arXiv preprint* arXiv:2006.09359.
239. S. Nasiriany, V. H. Pong, S. Lin, and S. Levine. 2019. Planning with goal-conditioned policies. *arXiv preprint* arXiv:1911.08453.
240. Y. Nesterov. 2003. *Introductory Lectures on Convex Optimization: A Basic Course*, volume 87. Springer Science & Business Media.
241. Y. Netzer, T. Wang, A. Coates, A. Bissacco, B. Wu, and A. Y. Ng. 2011. Reading digits in natural images with unsupervised feature learning.
242. B. Neyshabur, S. Bhojanapalli, and A. Chakrabarti. 2017. Stabilizing GAN training with multiple random projections. *CoRR*, abs/1705.07831.
243. T. Nguyen, T. Le, N. Dam, Q. H. Tran, T. Nguyen, and D. Phung. 8 2021. Tidot: A teacher imitation learning approach for domain adaptation with optimal transport. In Z.-H. Zhou, ed., *Proc. of the 13th Int. Joint Conf. on Artif. Intell., IJCAI-21*, pp. 2862–2868. Int. Joint Conf. on Artif. Intell. Org. Main Track.
244. A. Nichol, J. Achiam, and J. Schulman. 2018. On first-order meta-learning algorithms. *arXiv preprint* arXiv:1803.02999.
245. I. Olayode, L. Tartibu, M. Okwu, and d. U. Uchechi. 2020. Intelligent transportation systems, un-signalized road intersections and traffic congestion in johannesburg: A systematic review. *Procedia CIRP*, 91: 844–850.
246. R. Ortega and Y. Tang. 1989. Robustness of adaptive controllers-a survey. *Automatica*, 25(5): 651–677.
247. R. Ortner. 2020. Regret bounds for reinforcement learning via markov chain concentration. *Journal of Artificial Intelligence Research*, 67: 115–128.
248. S. A. Osia, A. S. Shamsabadi, S. Sajadmanesh, A. Taheri, K. Katevas, H. R. Rabiee, N. D. Lane, and H. Haddadi. 2020. A hybrid deep learning architecture for privacy-preserving mobile analytics. *IEEE Internet of Things Journal*, 7(5): 4505–4518.
249. O. Ostapenko, M. Puscas, T. Klein, P. Jahnichen, and M. Nabi. 2019. Learning to remember: A synaptic plasticity driven framework for continual learning. In *Proc. of the IEEE Conf. on Comput. Vision and Pattern Recognit.*, pp. 11321–11329.
250. O. Ostapenko, T. Lesort, P. Rodriguez, M. R. Arefin, A. Douillard, I. Rish, and L. Charlin. 2022. Continual learning with foundation models: An empirical study of latent replay. In *Conference on lifelong learning agents*, pp. 60–91. PMLR.
251. M. H. Ostertag, S. Al-Doweesh, and T. Rosing. 2020. Efficient training on edge devices using online quantization. In *2020 Design, Automation & Test in Europe Conference & Exhibition (DATE)*, pp. 1011–1014. IEEE.
252. K. Othman. 2022. Exploring the implications of autonomous vehicles: A comprehensive review. *Innovative Infrastructure Solutions*, 7(2): 165.
253. S. Padakandla, P. KJ, and S. Bhatnagar. 2020. Reinforcement learning algorithm for non-stationary environments. *Applied Intelligence*, 50(11): 3590–3606.
254. A. Painsky and G. W. Wornell. 2019. Bregman divergence bounds and universality properties of the logarithmic loss. *IEEE Transactions on Information Theory*, 66(3): 1658–1673.
255. K. Palanisamy, V. Khimani, M. H. Moti, and D. Chatzopoulos. 2021. Spliteasy: A practical approach for training ml models on mobile devices. In *Proceedings of the 22nd International Workshop on Mobile Computing Systems and Applications*, pp. 37–43.
256. S. J. Pan and Q. Yang. 2009. A survey on transfer learning. *IEEE Transactions on knowledge and data engineering*, 22(10): 1345–1359.
257. D. Parekh, N. Poddar, A. Rajpurkar, M. Chahal, N. Kumar, G. P. Joshi, and W. Cho. 2022. A review on autonomous vehicles: Progress, methods and challenges. *Electronics*, 11(14): 2162.

258. G. I. Parisi, R. Kemker, J. L. Part, C. Kanan, and S. Wermter. 2019. Continual lifelong learning with neural networks: A review. *Neural Netw.*, 113: 54–71.
259. O. Peer, C. Tessler, N. Merlis, and R. Meir. 2021. Ensemble bootstrapping for q-learning. *arXiv preprint* arXiv:2103.00445.
260. F. Pelosin. 2022. Simpler is better: off-the-shelf continual learning through pretrained backbones. *arXiv preprint* arXiv:2205.01586.
261. J. Peters and S. Schaal. 2008. Natural actor-critic. *Neurocomputing*, 71(7-9): 1180–1190.
262. S. Pu and A. Nedić. 2020. Distributed stochastic gradient tracking methods. *Mathematical Programming*, pp. 1–49.
263. M. L. Puterman. 2014. *Markov decision processes: discrete stochastic dynamic programming*. John Wiley & Sons.
264. M. L. Puterman and S. L. Brumelle. 1979. On the convergence of policy iteration in stationary dynamic programming. *Mathematics of Operations Research*, 4(1): 60–69.
265. D. Qi, H. Zhao, and S. Li. 2023. Better generative replay for continual federated learning. *arXiv preprint* arXiv:2302.13001.
266. M. Qi, Y. Wang, J. Qin, and A. Li. 2019. Ke-gan: Knowledge embedded generative adversarial networks for semi-supervised scene parsing. In *Proc. of the IEEE Conf. on Comput. Vision and Pattern Recognit.*, pp. 5237–5246.
267. G. Qu and N. Li. 2017. Harnessing smoothness to accelerate distributed optimization. *IEEE Transactions on Control of Network Systems*, 5(3): 1245–1260.
268. A. Radford, L. Metz, and S. Chintala. 2015. Unsupervised representation learning with deep convolutional generative adversarial networks. *arXiv preprint* arXiv:1511.06434.
269. B. Ragavi, L. Pavithra, P. Sandhiyadevi, G. Mohanapriya, and S. Harikirubha. 2020. Smart agriculture with ai sensor by using agrobot. In *2020 Fourth International Conference on Computing Methodologies and Communication (ICCMC)*, pp. 1–4. IEEE.
270. A. Raghu, M. Raghu, S. Bengio, and O. Vinyals. 2019. Rapid learning or feature reuse? towards understanding the effectiveness of MAML. *International Conference on Learning Representations (ICLR)*.
271. J. Rajasegaran, C. Finn, and S. Levine. 2022. Fully online meta-learning without task boundaries. *arXiv preprint* arXiv:2202.00263.
272. V. V. Ramasesh, E. Dyer, and M. Raghu. 2020. Anatomy of catastrophic forgetting: Hidden representations and task semantics. *arXiv preprint* arXiv:2007.07400.
273. V. V. Ramasesh, A. Lewkowycz, and E. Dyer. 2021. Effect of scale on catastrophic forgetting in neural networks. In *International Conference on Learning Representations*.
274. S. Ravi and H. Larochelle. 2016. Optimization as a model for few-shot learning. In *International Conference on Learning Representations (ICLR)*.
275. T. Ravi Shanker Reddy and B. Beena. 2022. Ai integrated blockchain technology for secure health care-consent-based secured federated transfer learning for predicting covid-19 on wearable devices. In *International Conference on Innovative Computing and Communications: Proceedings of ICICC 2022, Volume 1*, pp. 345–356. Springer.
276. W. Rawat and Z. Wang. 2017. Deep convolutional neural networks for image classification: A comprehensive review. *Neural computation*, 29(9): 2352–2449.
277. S.-A. Rebuffi, A. Kolesnikov, G. Sperl, and C. H. Lampert. 2017. icarl: Incremental classifier and representation learning. In *Proceedings of the IEEE conference on Computer Vision and Pattern Recognition*, pp. 2001–2010.
278. M. Riemer, I. Cases, R. Ajemian, M. Liu, I. Rish, Y. Tu, and G. Tesauro. 2018. Learning to learn without forgetting by maximizing transfer and minimizing interference. *arXiv preprint* arXiv:1810.11910.

279. M. Riemer, T. Klinger, D. Bouneffouf, and M. Franceschini. 2019. Scalable recollections for continual lifelong learning. In *Proc. of the AAAI Conf. on Artif. Intell.*, volume 33, pp. 1352–1359.
280. D. Rolnick, A. Ahuja, J. Schwarz, T. Lillicrap, and G. Wayne. 2019. Experience replay for continual learning. In *Advances in Neural Inf. Process. Syst.*, pp. 350–360.
281. A. Rosenfeld and J. K. Tsotsos. 2018. Incremental learning through deep adaptation. *IEEE transactions on pattern analysis and machine intelligence*, 42(3): 651–663.
282. A. A. Rusu, N. C. Rabinowitz, G. Desjardins, H. Soyer, J. Kirkpatrick, K. Kavukcuoglu, R. Pascanu, and R. Hadsell. 2016. Progressive neural networks. *arXiv preprint* arXiv:1606.04671.
283. G. Saha, I. Garg, and K. Roy. 2021. Gradient projection memory for continual learning. In *International Conference on Learning Representations*. https://openreview.net/forum?id=3AOj0RCNC2.
284. A. Şahin. 2023. Over-the-air computation based on balanced number systems for federated edge learning. *IEEE Transactions on Wireless Communications*.
285. E. Samikwa, A. Di Maio, and T. Braun. 2022. Ares: Adaptive resource-aware split learning for internet of things. *Computer Networks*, 218: 109380.
286. A. Santoro, S. Bartunov, M. Botvinick, D. Wierstra, and T. Lillicrap. 2016. Meta-learning with memory-augmented neural networks. In *International conference on machine learning*, pp. 1842–1850. PMLR.
287. M. S. Santos and J. Rust. 2004. Convergence properties of policy iteration. *SIAM Journal on Control and Optimization*, 42(6): 2094–2115.
288. S. Savazzi, M. Nicoli, and V. Rampa. 2020. Federated learning with cooperating devices: A consensus approach for massive iot networks. *IEEE Internet of Things Journal*, 7(5): 4641–4654.
289. M. Savva, A. Kadian, O. Maksymets, Y. Zhao, E. Wijmans, B. Jain, J. Straub, J. Liu, V. Koltun, J. Malik, et al. 2019. Habitat: A platform for embodied ai research. In *Proceedings of the IEEE/CVF international conference on computer vision*, pp. 9339–9347.
290. A. H. Sayed. 2014. Adaptive networks. *Proceedings of the IEEE*, 102(4): 460–497.
291. R. E. Schapire. 1999. A brief introduction to boosting. In *Ijcai*, volume 99, pp. 1401–1406.
292. J. Schmidhuber. 1987. *Evolutionary principles in self-referential learning*. PhD thesis, Technische Universität München.
293. J. Schwarz, W. Czarnecki, J. Luketina, A. Grabska-Barwinska, Y. W. Teh, R. Pascanu, and R. Hadsell. 2018. Progress & compress: A scalable framework for continual learning. In *International Conference on Machine Learning*, pp. 4528–4537. PMLR.
294. J. Serra, D. Suris, M. Miron, and A. Karatzoglou. 2018. Overcoming catastrophic forgetting with hard attention to the task. In *International Conference on Machine Learning*, pp. 4548–4557. PMLR.
295. M. J. Shafiee, F. Li, B. Chwyl, and A. Wong. 2017. Squishednets: Squishing squeezenet further for edge device scenarios via deep evolutionary synthesis. *arXiv preprint* arXiv:1711.07459.
296. S. Shahrampour, A. Rakhlin, and A. Jadbabaie. 2017. Multi-armed bandits in multi-agent networks. In *2017 IEEE International Conference on Acoustics, Speech and Signal Processing (ICASSP)*, pp. 2786–2790. IEEE.
297. S. Shalev-Shwartz and Y. Singer. 2007. Online learning: Theory, algorithms, and applications.
298. S. Shalev-Shwartz, S. Shammah, and A. Shashua. 2016. Safe, multi-agent, reinforcement learning for autonomous driving. *arXiv preprint* arXiv:1610.03295.
299. S. Shalev-Shwartz et al. 2011. Online learning and online convex optimization. *Foundations and trends in Machine Learning*, 4(2): 107–194.
300. S. Shalev-Shwartz et al. 2012. Online learning and online convex optimization. *Foundations and Trends® in Machine Learning*, 4(2): 107–194.

301. H. Shin, J. K. Lee, J. Kim, and J. Kim. 2017. Continual learning with deep generative replay. *arXiv preprint* arXiv:1705.08690.
302. D. Silver, G. Lever, N. Heess, T. Degris, D. Wierstra, and M. Riedmiller. 2014. Deterministic policy gradient algorithms. In *International conference on machine learning*, pp. 387–395. PMLR.
303. D. Silver, T. Hubert, J. Schrittwieser, I. Antonoglou, M. Lai, A. Guez, M. Lanctot, L. Sifre, D. Kumaran, T. Graepel, et al. 2017. Mastering chess and shogi by self-play with a general reinforcement learning algorithm. *arXiv preprint* arXiv:1712.01815.
304. D. Silver, T. Hubert, J. Schrittwieser, I. Antonoglou, M. Lai, A. Guez, M. Lanctot, L. Sifre, D. Kumaran, T. Graepel, et al. 2018. A general reinforcement learning algorithm that masters chess, shogi, and go through self-play. *Science*, 362(6419): 1140–1144.
305. D. Simon and A. Aberdam. 2020. Barycenters of natural images constrained wasserstein barycenters for image morphing. In *Proc. of the IEEE/CVF Conf. on Comput. Vision and Pattern Recognit.*, pp. 7910–7919.
306. S. P. Singh and M. Jaggi. 2019. Model fusion via optimal transport. *arXiv preprint* arXiv:1910.05653.
307. A. Sinha, H. Namkoong, R. Volpi, and J. Duchi. 2017. Certifying some distributional robustness with principled adversarial training. *arXiv preprint* arXiv:1710.10571.
308. R. Sitharthan, M. Rajesh, S. Vimal, S. Kumar, S. Yuvaraj, A. Kumar, J. Raglend, K. Vengatesan, et al. 2023. A novel autonomous irrigation system for smart agriculture using ai and 6g enabled iot network. *Microprocessors and Microsystems*, 101: 104905.
309. J. Smith and M. Gashler. 2017. An investigation of how neural networks learn from the experiences of peers through periodic weight averaging. In *2017 16th IEEE Int. Conf. on Mach. Learn. and Appl. (ICMLA)*, pp. 731–736. IEEE.
310. J. S. Smith, L. Karlinsky, V. Gutta, P. Cascante-Bonilla, D. Kim, A. Arbelle, R. Panda, R. Feris, and Z. Kira. 2023. Coda-prompt: Continual decomposed attention-based prompting for rehearsal-free continual learning. In *Proceedings of the IEEE/CVF Conference on Computer Vision and Pattern Recognition*, pp. 11909–11919.
311. L. Smith and M. Gasser. 2005. The development of embodied cognition: Six lessons from babies. *Artificial life*, 11(1-2): 13–29.
312. V. Smith, C.-K. Chiang, M. Sanjabi, and A. S. Talwalkar. 2017. Federated multi-task learning. In *Advances in Neural Information Processing Systems*, pp. 4424–4434.
313. J. Snell, K. Swersky, and R. Zemel. 2017. Prototypical networks for few-shot learning. In *Advances in Neural Information Processing Systems (NIPS)*.
314. X. Song, W. Gao, Y. Yang, K. Choromanski, A. Pacchiano, and Y. Tang. 2019a. Es-maml: Simple hessian-free meta learning. In *International Conference on Learning Representations (ICLR)*.
315. Y. Song, Y. Zhou, A. Sekhari, J. A. Bagnell, A. Krishnamurthy, and W. Sun. 2022. Hybrid rl: Using both offline and online data can make rl efficient. *arXiv preprint* arXiv:2210.06718.
316. Z. Song, R. Parr, and L. Carin. 2019b. Revisiting the softmax bellman operator: New benefits and new perspective. In *International Conference on Machine Learning*, pp. 5916–5925. PMLR.
317. S. Srivastava, V. Cevher, Q. Dinh, and D. Dunson. 2015. Wasp: Scalable bayes via barycenters of subset posteriors. In *Artif. Intell. and Statist.*, pp. 912–920.
318. M. Staib, S. Claici, J. M. Solomon, and S. Jegelka. 2017. Parallel streaming wasserstein barycenters. In *Advances in Neural Inf. Process. Syst.*, pp. 2647–2658.
319. X. Su, S. Guo, T. Tan, and F. Chen. 2020. Generative memory for lifelong learning. *IEEE Trans. on Neural Netw. and Learn. Syst.*, 31(6): 1884–1898. https://doi.org/10.1109/TNNLS.2019.2927369.

320. F. Sung, Y. Yang, L. Zhang, T. Xiang, P. H. Torr, and T. M. Hospedales. 2018. Learning to compare: Relation network for few-shot learning. In *Proceedings of the IEEE conference on computer vision and pattern recognition*, pp. 1199–1208.
321. R. S. Sutton and A. G. Barto. 2018. *Reinforcement learning: An introduction*. MIT press.
322. R. S. Sutton, D. McAllester, S. Singh, and Y. Mansour. 1999. Policy gradient methods for reinforcement learning with function approximation. *Advances in neural information processing systems*, 12.
323. C. Szegedy, A. Toshev, and D. Erhan. 2013. Deep neural networks for object detection. In *Advances in neural information processing systems*, pp. 2553–2561.
324. A. Taik, B. Nour, and S. Cherkaoui. 2021. Empowering prosumer communities in smart grid with wireless communications and federated edge learning. *IEEE Wireless Communications*, 28(6): 26–33.
325. Y. Tang, J. Zhang, and N. Li. 2019. Distributed zero-order algorithms for nonconvex multi-agent optimization. *arXiv*, pp. arXiv–1908.
326. S. Thrun. 1995. A lifelong learning perspective for mobile robot control. In *Intelligent robots and Syst.*, pp. 201–214. Elsevier.
327. S. Thrun. 1998. Lifelong learning algorithms. In *Learning to learn*, pp. 181–209. Springer.
328. S. Thrun and A. Schwartz. 1993. Issues in using function approximation for reinforcement learning. In *Proceedings of the 1993 Connectionist Models Summer School Hillsdale, NJ. Lawrence Erlbaum*, volume 6.
329. E. Todorov, T. Erez, and Y. Tassa. 2012. Mujoco: A physics engine for model-based control. In *2012 IEEE/RSJ International Conference on Intelligent Robots and Systems*, pp. 5026–5033. IEEE.
330. C. Tsanikidis and J. Ghaderi. 2020. On the power of randomization for scheduling real-time traffic in wireless networks. In *IEEE INFOCOM 2020-IEEE Conference on Computer Communications*, pp. 59–68. IEEE.
331. P. Tseng. 2008. On accelerated proximal gradient methods for convex-concave optimization. *submitted to SIAM Journal on Optimization*, 2(3).
332. I. Uchendu, T. Xiao, Y. Lu, B. Zhu, M. Yan, J. Simon, M. Bennice, C. Fu, C. Ma, J. Jiao, et al. 2022. Jump-start reinforcement learning. *arXiv preprint* arXiv:2204.02372.
333. H. Van Hasselt, A. Guez, and D. Silver. 2016. Deep reinforcement learning with double q-learning. In *Proceedings of the AAAI conference on artificial intelligence*, volume 30.
334. H. Van Hasselt, Y. Doron, F. Strub, M. Hessel, N. Sonnerat, and J. Modayil. 2018. Deep reinforcement learning and the deadly triad. *arXiv preprint* arXiv:1812.02648.
335. A. Vaswani, N. Shazeer, N. Parmar, J. Uszkoreit, L. Jones, A. N. Gomez, Ł. Kaiser, and I. Polosukhin. 2017. Attention is all you need. *Advances in neural information processing systems*, 30.
336. T. Veniat, L. Denoyer, and M. Ranzato. 2020. Efficient continual learning with modular networks and task-driven priors. *arXiv preprint* arXiv:2012.12631.
337. S. Venkataramani, A. Ranjan, S. Banerjee, D. Das, S. Avancha, A. Jagannathan, A. Durg, D. Nagaraj, B. Kaul, P. Dubey, et al. 2017. Scaledeep: A scalable compute architecture for learning and evaluating deep networks. In *Proc. of the 44th Annu.Int. Symp. on Comput. Architecture*, pp. 13–26.
338. R. Vilalta and Y. Drissi. 2002. A perspective view and survey of meta-learning. *Artificial intelligence review*, 18: 77–95.
339. C. Villani. 2003. *Topics in optimal transportation*. Number 58. American Mathematical Society.
340. C. Villani. 2008. *Optimal transport: old and new*, volume 338. Springer Science & Business Media.
341. O. Vinyals, C. Blundell, T. Lillicrap, and D. Wierstra. 2016. Matching networks for one shot learning. In *Advances in Neural Information Processing Systems (NIPS)*.

342. R. Volpi, H. Namkoong, O. Sener, J. C. Duchi, V. Murino, and S. Savarese. 2018. Generalizing to unseen domains via adversarial data augmentation. In *Advances in Neural Inf. Process. Syst.*, pp. 5334–5344.
343. Z. Wadud, D. MacKenzie, and P. Leiby. 2016. Help or hindrance? the travel, energy and carbon impacts of highly automated vehicles. *Transportation Research Part A: Policy and Practice*, 86: 1–18.
344. A. Wagenmaker and A. Pacchiano. 2022. Leveraging offline data in online reinforcement learning. *arXiv preprint* arXiv:2211.04974.
345. J. Wang, J. Zhang, W. Bao, X. Zhu, B. Cao, and P. S. Yu. 2018a. Not just privacy: Improving performance of private deep learning in mobile cloud. In *Proc. of the 24th ACM SIGKDD Int. Conf. on Knowledge Discovery & Data Mining*, pp. 2407–2416.
346. J. Wang, W. Bao, L. Sun, X. Zhu, B. Cao, and S. Y. Philip. 2019a. Private model compression via knowledge distillation. In *Proc. of the AAAI Conf. on Artif. Intell.*, volume 33, pp. 1190–1197.
347. J. Wang, W. Zhou, G.-J. Qi, Z. Fu, Q. Tian, and H. Li. 2020a. Transformation gan for unsupervised image synthesis and representation learning. In *Proc. of the IEEE/CVF Conf. on Comput. Vision and Pattern Recognit.*, pp. 472–481.
348. L. Wang, Q. Cai, Z. Yang, and Z. Wang. 2020b. On the global optimality of model-agnostic meta-learning. *arXiv preprint* arXiv:2006.13182.
349. Q. Wang, Y. Li, B. Shao, S. Dey, and P. Li. 2017. Energy efficient parallel neuromorphic architectures with approximate arithmetic on fpga. *Neurocomputing*, 221: 146–158.
350. X. Wang, Y. Han, C. Wang, Q. Zhao, X. Chen, and M. Chen. 2019b. In-edge ai: Intelligentizing mobile edge computing, caching and communication by federated learning. *IEEE Network*.
351. Y. Wang, C. Wu, L. Herranz, J. van de Weijer, A. Gonzalez-Garcia, and B. Raducanu. 2018b. Transferring gans: generating images from limited data. In *Proc. of the Eur. Conf. on Comput. Vision (ECCV)*, pp. 218–234.
352. Y.-C. Wang and J. M. Usher. 2005. Application of reinforcement learning for agent-based production scheduling. *Engineering Applications of Artificial Intelligence*, 18(1): 73–82.
353. Z. Wang, X. Wang, L. Shen, Q. Suo, K. Song, D. Yu, Y. Shen, and M. Gao. 2022a. Meta-learning without data via wasserstein distributionally-robust model fusion. In *Uncertainty in Artificial Intelligence*, pp. 2045–2055. PMLR.
354. Z. Wang, Z. Zhang, C.-Y. Lee, H. Zhang, R. Sun, X. Ren, G. Su, V. Perot, J. Dy, and T. Pfister. 2022b. Learning to prompt for continual learning. In *Proceedings of the IEEE/CVF conference on computer vision and pattern recognition*, pp. 139–149.
355. Z. Wang, Y. Zhang, X. Xu, Z. Fu, H. Yang, and W. Du. 2023. Federated probability memory recall for federated continual learning. *Information Sciences*, 629: 551–565.
356. C. J. Watkins and P. Dayan. 1992. Q-learning. *Machine learning*, 8(3-4): 279–292.
357. Y. Wei, Z. Hu, Z. Wang, L. Shen, C. Yuan, and D. Tao. 2024. Free: Faster and better data-free meta-learning. In *Proceedings of the IEEE/CVF Conference on Computer Vision and Pattern Recognition*, pp. 23273–23282.
358. W. Wen, H. H. Yang, W. Xia, and T. Q. Quek. 2022. Towards fast and energy-efficient hierarchical federated edge learning: A joint design for helper scheduling and resource allocation. In *ICC 2022-IEEE International Conference on Communications*, pp. 5378–5383. IEEE.
359. C. Wu, A. Kreidieh, K. Parvate, E. Vinitsky, and A. M. Bayen. 2017. Flow: Architecture and benchmarking for reinforcement learning in traffic control. *arXiv preprint* arXiv:1710.05465, 10.
360. C. Wu, L. Herranz, X. Liu, J. van de Weijer, B. Raducanu, et al. 2018. Memory replay gans: Learning to generate new categories without forgetting. *Advances in Neural Information Processing Systems*, 31: 5962–5972.
361. Z. Wu, S. Sun, Y. Wang, M. Liu, X. Jiang, R. Li, and B. Gao. 2023. Survey of knowledge distillation in federated edge learning. *arXiv preprint* arXiv:2301.05849.

362. H. Xiao, K. Rasul, and R. Vollgraf. 2017. Fashion-mnist: a novel image dataset for benchmarking machine learning algorithms. *arXiv preprint* arXiv:1708.07747.
363. J. Xie, Z. Shen, C. Zhang, B. Wang, and H. Qian. 2020. Efficient projection-free online methods with stochastic recursive gradient. In *AAAI*, pp. 6446–6453.
364. T. Xie, N. Jiang, H. Wang, C. Xiong, and Y. Bai. 2021. Policy finetuning: Bridging sample-efficient offline and online reinforcement learning. *Advances in neural information processing systems*, 34.
365. J. Xu and Z. Zhu. 2018. Reinforced continual learning. *arXiv preprint* arXiv:1805.12369.
366. T. Xu, Z. Wang, and Y. Liang. 2020a. Improving sample complexity bounds for (natural) actor-critic algorithms. In *Proceedings of the 34th International Conference on Neural Information Processing Systems*.
367. T. Xu, Z. Wang, and Y. Liang. 2020b. Non-asymptotic convergence analysis of two time-scale (natural) actor-critic algorithms. *arXiv preprint* arXiv:2005.03557.
368. F. Yan, S. Sundaram, S. Vishwanathan, and Y. Qi. 2012. Distributed autonomous online learning: Regrets and intrinsic privacy-preserving properties. *IEEE Transactions on Knowledge and Data Engineering*, 25(11): 2483–2493.
369. B. Yang, X. Cao, X. Li, C. Yuen, and L. Qian. 2020. Lessons learned from accident of autonomous vehicle testing: An edge learning-aided offloading framework. *IEEE Wireless Communications Letters*, 9(8): 1182–1186.
370. B. Yang, O. Fagbohungbe, X. Cao, C. Yuen, L. Qian, D. Niyato, and Y. Zhang. 2021. A joint energy and latency framework for transfer learning over 5g industrial edge networks. *IEEE Transactions on Industrial Informatics*, 18(1): 531–541.
371. B. Yang, L. He, N. Ling, Z. Yan, G. Xing, X. Shuai, X. Ren, and X. Jiang. 2023. Edgefm: Leveraging foundation model for open-set learning on the edge. *arXiv preprint* arXiv:2311.10986.
372. L. Yang, A. S. Rakin, and D. Fan. 2022. Rep-net: Efficient on-device learning via feature reprogramming. In *Proceedings of the IEEE/CVF Conference on Computer Vision and Pattern Recognition*, pp. 12277–12286.
373. T.-J. Yang, Y.-H. Chen, and V. Sze. 2017. Designing energy-efficient convolutional neural networks using energy-aware pruning. In *Proc. of the IEEE Conf. on Comput. Vision and Pattern Recognit.*, pp. 5687–5695.
374. Z. Yang, Y. Chen, M. Hong, and Z. Wang. 2019. Provably global convergence of actor-critic: A case for linear quadratic regulator with ergodic cost. *Advances in neural information processing systems*, 32.
375. H. Yao, Y. Zhou, M. Mahdavi, Z. J. Li, R. Socher, and C. Xiong. 2020. Online structured meta-learning. *Advances in Neural Information Processing Systems*, 33: 6779–6790.
376. J. Ye, P. Wu, J. Z. Wang, and J. Li. 2017. Fast discrete distribution clustering using wasserstein barycenter with sparse support. *IEEE Trans. on Signal Process.*, 65(9): 2317–2332.
377. D. Yin, M. Farajtabar, A. Li, N. Levine, and A. Mott. 2020. Optimization and generalization of regularization-based continual learning: a loss approximation viewpoint. *arXiv preprint* arXiv:2006.10974.
378. M. Yip, S. Salcudean, K. Goldberg, K. Althoefer, A. Menciassi, J. D. Opfermann, A. Krieger, K. Swaminathan, C. J. Walsh, H. Huang, et al. 2023. Artificial intelligence meets medical robotics. *Science*, 381(6654): 141–146.
379. R. Yonetani, T. Takahashi, A. Hashimoto, and Y. Ushiku. 2019. Decentralized learning of generative adversarial networks from non-iid data. *arXiv preprint* arXiv:1905.09684.
380. J. Yoon, E. Yang, J. Lee, and S. J. Hwang. 2017. Lifelong learning with dynamically expandable networks. *arXiv preprint* arXiv:1708.01547.
381. J. Yoon, S. Kim, E. Yang, and S. J. Hwang. 2020. Scalable and order-robust continual learning with additive parameter decomposition. In *Eighth International Conference on Learning Representations, ICLR 2020*. ICLR.

382. J. Yoon, W. Jeong, G. Lee, E. Yang, and S. J. Hwang. 2021. Federated continual learning with weighted inter-client transfer. In *International Conference on Machine Learning*, pp. 12073–12086. PMLR.
383. C. Yu, J. Liu, S. Nemati, and G. Yin. 2021. Reinforcement learning in healthcare: A survey. *ACM Computing Surveys (CSUR)*, 55(1): 1–36.
384. F. Yu, Y. Zhang, S. Song, A. Seff, and J. Xiao. 2015. LSUN: construction of a large-scale image dataset using deep learning with humans in the loop. *CoRR*, abs/1506.03365.
385. T. Yu, X. Geng, C. Finn, and S. Levine. 2020. Variable-shot adaptation for online meta-learning. *arXiv preprint* arXiv:2012.07769.
386. Y. Yu, X. Si, C. Hu, and J. Zhang. 2019. A review of recurrent neural networks: Lstm cells and network architectures. *Neural computation*, 31(7): 1235–1270.
387. S. Yue, J. Ren, J. Xin, S. Lin, and J. Zhang. 2021. Inexact-admm based federated meta-learning for fast and continual edge learning. In *Proceedings of ACM MobiHoc 2021*, pp. 91–100.
388. E. Yurtsever, J. Lambert, A. Carballo, and K. Takeda. 2020. A survey of autonomous driving: Common practices and emerging technologies. *IEEE access*, 8: 58443–58469.
389. A. Zador, S. Escola, B. Richards, B. Ölveczky, Y. Bengio, K. Boahen, M. Botvinick, D. Chklovskii, A. Churchland, C. Clopath, et al. 2023. Catalyzing next-generation artificial intelligence through neuroai. *Nature communications*, 14(1): 1597.
390. G. Zeng, Y. Chen, B. Cui, and S. Yu. 2019. Continual learning of context-dependent processing in neural networks. *Nature Machine Intelligence*, 1(8): 364–372.
391. F. Zenke, B. Poole, and S. Ganguli. 2017. Continual learning through synaptic intelligence. In *International Conference on Machine Learning*, pp. 3987–3995. PMLR.
392. C. Zeno, I. Golan, E. Hoffer, and D. Soudry. 2018. Task agnostic continual learning using online variational bayes. *arXiv preprint* arXiv:1803.10123.
393. B. Zhang, R. Rajan, L. Pineda, N. Lambert, A. Biedenkapp, K. Chua, F. Hutter, and R. Calandra. 2021a. On the importance of hyperparameter optimization for model-based reinforcement learning. In *International Conference on Artificial Intelligence and Statistics*, pp. 4015–4023. PMLR.
394. C. Zhang, S. Bengio, M. Hardt, B. Recht, and O. Vinyals. 2021b. Understanding deep learning (still) requires rethinking generalization. *Communications of the ACM*, 64(3): 107–115.
395. G. Zhang, L. Wang, G. Kang, L. Chen, and Y. Wei. 2023. Slca: Slow learner with classifier alignment for continual learning on a pre-trained model. In *Proceedings of the IEEE/CVF International Conference on Computer Vision*, pp. 19148–19158.
396. Y. Zhang, R. J. Ravier, V. Tarokh, and M. M. Zavlanos. 2019. Distributed online convex optimization with improved dynamic regret. *arXiv preprint* arXiv:1911.05127.
397. Y. Zhang, Y. Qin, Y. Zhang, X. Zhou, S. Jian, Y. Tan, and K. Li. 2024. Oncenas: Discovering efficient on-device inference neural networks for edge devices. *Information Sciences*, 669: 120567.
398. Z. Zhang, Z. Pan, and M. J. Kochenderfer. 2017. Weighted double q-learning. In *IJCAI*, pp. 3455–3461.
399. Z. Zhang, S. Lin, M. Dedeoglu, K. Ding, and J. Zhang. 2020. Data-driven distributionally robust optimization for edge intelligence. In *IEEE INFOCOM 2020-IEEE Conf. on Comput. Commun.*, pp. 2619–2628. IEEE.
400. Z. Zhang, B. Guo, W. Sun, Y. Liu, and Z. Yu. 2022. Cross-fcl: Toward a cross-edge federated continual learning framework in mobile edge computing systems. *IEEE Transactions on Mobile Computing*.
401. W. X. Zhao, K. Zhou, J. Li, T. Tang, X. Wang, Y. Hou, Y. Min, B. Zhang, J. Zhang, Z. Dong, et al. 2023. A survey of large language models. *arXiv preprint* arXiv:2303.18223.

402. Y. Zhao, C. Yu, P. Zhao, H. Tang, S. Qiu, and J. Liu. 2019. Decentralized online learning: Take benefits from others' data without sharing your own to track global trend. *arXiv preprint* arXiv:1901.10593.
403. P. Zhou, X. Yuan, H. Xu, S. Yan, and J. Feng. 2019a. Efficient meta learning via minibatch proximal update. *Advances in Neural Information Processing Systems*, 32.
404. Y. Zhou, X. Ma, D. Wu, and X. Li. 2022a. Communication-efficient and attack-resistant federated edge learning with dataset distillation. *IEEE Transactions on Cloud Computing*, 11(3): 2517–2528.
405. Y. Zhou, J. Xiao, Y. Zhou, and G. Loianno. 2022b. Multi-robot collaborative perception with graph neural networks. *IEEE Robotics and Automation Letters*, 7(2): 2289–2296.
406. Z. Zhou, X. Chen, E. Li, L. Zeng, K. Luo, and J. Zhang. 2019b. Edge intelligence: Paving the last mile of artificial intelligence with edge computing. *Proc. of the IEEE*, 107(8): 1738–1762.
407. R. Zhu and M. Rigotti. 2020. Self-correcting q-learning. *arXiv preprint* arXiv:2012.01100.
408. M. Zinkevich. 2003. Online convex programming and generalized infinitesimal gradient ascent. In *Proceedings of the 20th international conference on machine learning (icml-03)*, pp. 928–936.
409. G. Zizzo, A. Rawat, N. Holohan, and S. Tirupathi. 2022. Federated continual learning with differentially private data sharing. In *Workshop on Federated Learning: Recent Advances and New Challenges (in Conjunction with NeurIPS 2022)*.
410. S. Zou, T. Xu, and Y. Liang. 2019. Finite-sample analysis for SARSA with linear function approximation. *Advances in neural information processing systems*, 32.

MIX
Papier aus verantwortungsvollen Quellen
Paper from responsible sources
FSC® C105338

If you have any concerns about our products,
you can contact us on
ProductSafety@springernature.com

In case Publisher is established outside the EU,
the EU authorized representative is:
Springer Nature Customer Service Center GmbH
Europaplatz 3, 69115 Heidelberg, Germany

Printed by Libri Plureos GmbH
in Hamburg, Germany